Decision making in natural resource management

Decision making in natural resource management: a structured, adaptive approach

Michael J. Conroy
Warnell School of Forestry and Natural Resources, University of Georgia, Georgia

James T. Peterson
US Geological Survey, Oregon Cooperative Fish and Wildlife Research Unit, Oregon State University, Oregon

WILEY-BLACKWELL

A John Wiley & Sons, Ltd., Publication

Library of Congress Cataloging-in-Publication Data

Conroy, Michael J. (Michael James), 1952-
  Decision making in natural resource management: a structured, adaptive approach / Michael J. Conroy, James T. Peterson.
    p. cm.
  Includes bibliographical references and index.
  ISBN 978-0-470-67175-7 (cloth : alk. paper) – ISBN 978-0-470-67174-0 (pbk. : alk. paper)
  1. Natural resources–Decision making. 2. Natural resources–Management. I. Peterson, James T. II. Title.
  HC85.C675 2013
  333.7–dc23
                            2012035084

A catalogue record for this book is available from the British Library.

Cover image: Beartooth Mountains in Montana. © Colton Stiffler/Shutterstock.com
Cover design by Design Deluxe

Set in 10.5/12 pt ClassicalGaramondBT by Toppan Best-set Premedia Limited
Printed and bound in Malaysia by Vivar Printing Sdn Bhd

1 2013

Contents

List of boxes

Preface

This book is intended for use by natural resources managers and scientists, and students in the fields of natural resource management, ecology, and conservation biology, who are or will be confronted with complex and difficult decision-making problems. This audience will find that you will be called upon to assist with solving problems because you have a technical expertise in a certain area. Perhaps you are a specialist in fish nutrition and physiology, or statistical modeling, or in spatial analysis; or, you may specialize in the human-dimensions side of the equation, dealing with people's attitudes, values, and behavior. Often you will be asked to provide input on just one narrow aspect of a problem, and you might assume that your client (e.g., the natural resources agency that pays your contract) knows how to take your information, apply it in the context of solving a bigger problem, and that all will be well. You would often be mistaken.

In our experience, agencies, NGOs, and other organizations dealing with conservation problems often seek technical solutions to problem solving, when in fact their difficulties lie at a deeper level. What these organizations typically lack is an understanding of how the components of their decision-making problem relate to one another, and to the overarching goals and mission of the organization. That is, typically their approach to decision making lacks *structure*. Besides being an inefficient use of resources (something we have little to spare in these days of economic belt tightening), this sort of ad hoc approach to decision making can play into the criticism emanating from some camps that conservation and natural resource management are not based on rigorous, repeatable methods and thus, need not be taken as seriously as "real" sciences. In fact, natural resource management draws from numerous scientific fields (ecology, biology, physics, and geography to name a few), as well as the quantitative (statistics, mathematics, computer sciences) and social sciences (economics, policy, human dimensions). However, when we see actual decision-making processes in action, they can appear fragmented and poorly focused, often using the (sometimes copious) information that is available from the sciences in an informal way. Our hope is that the methods describe in this book will help biologists and managers better focus the rich sources of knowledge we have from these fields to solving pressing conservation problems.

Acknowledgements

Many people have helped make this book possible, and we thank them. The authors thank their spouses, Liz and Rebecca, for putting up with us during this project. We thank our graduate students and colleagues at Georgia and Oregon State for their feedback and insights that help make this a better book. Between the two of us we have (either jointly or independently) now conducted over twenty workshops applying principles of Structured Decision Making to solving a wide range of natural resource problems. Each workshop has increased our understanding of how SDM works, and given us insights into why it occasionally does not work; this book is in large part the product of that experience.

We are especially grateful to the following colleagues who volunteered their time to provide us detailed reviews of each of the chapters: Paige Barlow, John Carroll, Sarah Converse, Jason Dunham, Andrea Goijman, Tom Kwak, Clint Moore, Rebecca Moore, Krishna Pacifici, Colin Shea, and Seth Wenger. Their comments were extremely helpful to us, both in catching errors as well as for insights on how to deliver our message with greater accuracy and clarity. Any remaining errors, which we hope are few and unimportant, belong to the authors. The use of trade, product, industry, or firm names or products is for informative purposes only and does not constitute an endorsement by the US Government or the US Geological Survey. The Oregon Cooperative Fish and Wildlife Research Unit is jointly sponsored by the US Geological Survey, the US Fish and Wildlife Service, the Oregon Department of Fish and Wildlife, the Oregon State University, and the Wildlife Management Institute.

Guide to using this book

This book is divided into three major parts: Introduction, Tools, and Applications, and we recommend some depth of reading for all users of all three parts. For Part I – Introduction, we recommend that all readers examine Chapters 1 and 2; however, those already familiar with the basics of SDM might quickly skim these sections, since presumably the major concepts will be familiar. We highly recommend that all readers who seek to actually develop decision models carefully read Chapter 3 on developing objectives, and those who plan to work with stakeholder groups should definitely read Chapter 4. We also recommend that administrators and policy makers read these sections, if for no other reason than to become familiar with the terminology of SDM, as well as to have a more realistic expectation of what can, and cannot be achieved.

Part II of the book gets into the nuts and bolts of how to assemble decision models and to use information from field studies and monitoring to inform decision making. These chapters should be read in depth and we recommend that everyone read the introductory sections of both chapters, scan the topic sentences for the remainders, and refer back in detail to specific sections as needed. For example, one not need have a detailed knowledge of linear modeling, to appreciate the fact that linear models can both capture essential hypothetical relationships as well as form testable predictions that can be used in decision making. Likewise, one need not know the details of dynamic programming to understand the basic principles of optimization, and appreciating that casting decisions in a dynamic framework greatly complicates this process. On the other hand, if one is actually constructing and applying linear models, or using dynamic decision models, a deeper understanding and a more comprehensive reading is essential.

Part III covers applications of these approaches, and should be read by all. In particular, our coverage of case studies that "worked" (Chapter 9) and those that were less than fully successful (Chapter 10) should provide important insights to those seeking to apply these methods.

We also have provided a glossary, several technical appendices, and an Electronic Companion, and we encourage readers to use all three of these resources. The glossary provides a comprehensive list of terms we have used, together with brief definitions for each; we think readers will find this a useful guide to navigating a sometimes confusing terrain. The appendices provide a level of technical detail that is important to have available, but was inappropriate to include in the body of the book, and should be referred to for elaboration on these topics. Finally, the Electronic Companion provides worked examples with computer code for all of the Box examples, except those with trivial solutions, some additional useful code and explanation, as well as links to other resources available on the Internet including example exercises (problems) for coursework.

Companion website

As noted above, we have provided a companion website for the book, which can be accessed via www.wiley.com/go/conroy/naturalresourcemanagement. Additional resources on the companion provide details for the Box examples, including data input and program output. In most cases (except commonly available commercial software like Microsoft Excel ®), the programs are freely available via the Internet. We have provided additional modeling software and examples that, while not directly referenced in the book, may be useful to readers. We also have provided links to both freely available as well as commercial software; readers should always obtain the most current versions of these applications. Finally, we have provided links to several workshops and courses we have conducted in this area, which should be of interest, especially to advanced undergraduates and graduate students seeking to use these approaches in their research.

PART I. INTRODUCTION TO DECISION MAKING

1

Introduction: Why a Structured Approach in Natural Resources?

In this chapter, we provide a general motivation for a structured approach to decision making in natural resource management. We discuss the role of decision making in natural resource management, common problems made when framing natural resource decisions, and the advantages and limitations of a structured approach to decision making. We will also define terms such as **objective, management, decision, model,** and **adaptive management**, each of which will be a key element in the development of a structured decision approach.

The first and obvious question is: why do we need a structured approach to decision making in natural resource management? We have thought a lot about this question, and realize that while the answer may not be obvious, it really comes down to some basic premises. For us, natural resource management is a developing field, and many aspects of it are not "mature." In many respects we think that conservation and natural resource management suffer from the perception that many have that it is an ad hoc and not particularly scientific field. In our view, we have a choice: we can either use ad hoc and arguably non-scientific means to arrive at decisions; or we can use methods that are more formal and repeatable. In our view, the latter will better serve the field in the long run.

We also want to emphasize that when we refer to "management" we are speaking very broadly. That is, "management" includes virtually every type of decision we could make about a natural resource system, which would include traditional game management tools (e.g., harvest and habitat management), but also reserve design, legal protection and enforcement, translocation, captive propagation, and any other action intended to effect a conservation objective. This means that we consider conservation and management as one and the

Decision Making in Natural Resource Management: A Structured, Adaptive Approach,
First Edition. Michael J. Conroy and James T. Peterson.
© 2013 John Wiley & Sons, Ltd. Published 2013 by John Wiley & Sons, Ltd.

same and believe that artificial distinctions only serve to confuse students and practitioners.

The role of decision making in natural resource management

Virtually all problems in natural resource management involve decisions: choices that must be made among alternative actions to achieve an objective. We will define "decisions" and "objectives" more formally in the coming chapters, but can illustrate each with some simple examples. Examples of decisions include:

- Location on the landscape for a new biological reserve.
- Allowable season lengths and bag limits for a harvested population.
- Whether to capture a remnant population in danger of extinction and conduct captive breeding.
- Whether to use lethal control for an exotic invasive limiting an endemic population, and if so, which type of control.
- Whether and how to mitigate the impact of wind turbines on bird mortality.

Note that in each case, there is a choice of an action, and that some choices preclude others. So for example, if we choose location *A* for our reserve, given finite resources and other limitations, we have likely precluded locations *B–D*. Similarly, if we close the hunting season we cannot at the same time allow liberal bag limits. If we capture the remnant population we have (at least immediately) foregone natural reproduction, and so on.

Also, each of the above decisions is presumably connected to one or more objectives. We will develop objectives more fully in Chapter 3, but broadly stated, the objectives associated with the above decisions might be, respectively:

- Provide the greatest biodiversity benefit for the available funds and personnel.
- Provide maximum sustainable harvest opportunity.
- Avoid species extinction and foster species recovery.
- Restore an endemic population.
- Minimize bird mortality while fostering "green" energy.

So, at a very basic level, decision making is about connecting decisions to objectives, and **structured decision making** (**SDM**; Hammond et al. 1999, Clemen and Reilly 2001) is just a formalized way of accomplishing that connection. For some of us this connection (and way of thinking) is so obvious that it hardly needs stating, and certainly doesn't require a book-length coverage. However, we have in our careers in academia and government, and working with natural resource management agencies, NGOs, and business, encountered numerous examples in which we believed that problems in the management of resources were exacerbated, and in some cases directly caused, by poor framing of the decision problem.

We also want to emphasize the important role of science in decision making. Science should inform decision making, but we must always recognize that science is a process and not an end. Thus, we can use science to inform decision making, but we must always be seeking to improve our scientific understanding as we make decisions. We sometimes use the analogy of a 3-legged stool of management, research, and monitoring to make this point (Conroy and Peterson 2009).

Common mistakes in framing decisions

Poorly stated objectives

It is apparent to us that, in many cases, the objectives of management are poorly stated, if they are stated at all. This can lead to decisions that lead nowhere – that is, they are not connected to any apparent objectives. This in turn means that the decisions do not address the management problem, waste resources, and potentially create unnecessary conflict among the stakeholders. The reverse also can occur when objectives are stated, but management decisions are apparently arrived at by an independent process. As a result, the objectives cannot be achieved because they are not connected to management actions. Again, the management problem is not addressed, resources are wasted, and unnecessary conflict created; additionally, **stakeholders** (parties who have an interest in the outcome of decision making, and who may or may not be **decision makers**) may feel disenfranchised, since apparently their input in forming objectives has been ignored.

Prescriptive decisions

A related situation arises in cases where "decisions" are formulated in a rule-based, prescriptive manner that presumes that certain sets of conditions (perhaps attributes measured via monitoring) necessarily trigger particular actions. Such formulaic approaches (common in many species recovery plans) may be useful tools in a decision-making process, but do not constitute decision making (except in the trivial sense of having decided to follow the formula).

Confusion of values and science

When attempts are made to define objectives, a very common problem that we see is the *confusion of values* (or objectives) with *science* (or data/ information). That is, conflating what we know (or think we know) about a problem, with what we are trying to achieve. Most natural resource professionals come from a background in the biological or earth sciences, and are more comfortable discussing "facts" and data than they are discussing values. As we will see, "facts" come into play when we try to connect candidate decisions to the objectives we are trying to achieve. Objectives, on the other hand, reflect our values (or the values of those with a stake in the decision whose proxies we hold). If we do not get the values (objectives) right, the "facts" will be useless for arriving at a decision. More insidiously, disagreements about "facts" or "science" are frequently a smokescreen or proxy for disagreement about values. One needs to look no further than the cases of the Northern Spotted Owl (*Strix occidentalis caurina*) or anthropogenic climate change. In each case, scientific belief (and supporting "facts") coincides remarkably with the values of the respective stakeholder communities, with for example timber industry advocates tending to be skeptical of the obligate nature of ancient forests for owls, and many political or social conservatives questioning the science of climate change (Lange 1993, McCright and Dunlap 2011, Martin et al. 2011, Russill 2011).

Poor use of information

Another very common disconnect we see is the *poor use of information from monitoring programs*. While some general-purpose monitoring can perhaps be justified (e.g., the Long Term Ecological Research Network [LTER; http://www.lternet.edu/] programs that provide baseline monitoring in relatively undisturbed areas), omnibus monitoring programs that are not connected to and do not support decision making are often unproductive (see also Nichols and Williams 2006). Rather, we agree with Nichols and Williams (2006) that changing the focus and design of monitoring programs as part of an overarching program of conservation-oriented science or management.

This is not to say that monitoring (of any kind) is an absolute requirement of decision making. In some cases, there are few data to support quantitative statements about a decision's impact, and little prospect that sufficient data will be acquired in the near term to allow unequivocal statements about management; many problems involving imperiled species and their habitats fall into this category. Nonetheless, it is incumbent on managers to make decisions given whatever data or other knowledge is available. Putting off a decision until more information is available is, of course, itself a decision, with potentially disastrous consequences ("paralysis by analysis" is another variant). The reality is that we can always learn more about a system; the trick is to use what we know *now* to make a good decision, while always striving to do better with future decisions.

What is structured decision making (SDM)?

SDM consists of three basic components. The first is explicit, quantifiable objectives, such as maximizing bear population size or minimizing human–bear conflicts. The second is explicit management alternatives (actions) (e.g., harvest regulations or habitat management) that can be taken to meet the objectives. The third component is models that are used to predict the effect of management actions on resource objectives (e.g., models predicting population size after various harvest regulations). Because knowledge about large-scale ecological processes and responses of resources to management are always imperfect, **uncertainty** is incorporated in SDM through alternative models representing hypotheses of ecological dynamics and statistical distributions representing error in model parameters and environmental variability.

Why should we use a structured approach to decision making?

Some decision problems have an obvious solution and need no further analysis. In such cases, two or more decision makers with the same objective would probably arrive at the same decision, perhaps without even consciously making a choice. Such decision problems probably do not require a structured approach.

However, we suggest that these types of problems are not typical of natural resource management. In our experience, natural resource decision problems are typically complex, and multiple decision makers can easily disagree on the best decision. Furthermore, the process by which natural resource decision makers arrive at decisions tends to be difficult to explain, which in turn makes it difficult to communicate. For example, a supervisor, who has much knowledge and experience to draw on, trying to explain decisions to a new employee, who has only a rudimentary understanding of issues. Inevitably, this results in miscommunication due to the ad hoc way decisions are typically made in natural resource management, which in turn makes them both difficult to convey as well as difficult to replicate. An SDM process can avoid these problems and foster better communication and knowledge transfer. For another example, before the advent of **adaptive harvest management (AHM)** for setting waterfowl harvest regulations, regulations were effectively decided by a small number of agency staff. While these staff received technical and other input, there was no clear, repeatable process by which decisions were reached, and thus decisions could appear arbitrary to outside observers.

A structured approach, on the other hand, clarifies the decision-making problem by decomposing it into components that are easier to understand and convey. A structured approach also provides *transparency* and *legacy* to the decision-making process, so that the process does not have to be reinvented every time there is institutional change or turnover. Finally, a structured approach should provide a clear linkage between research and monitoring components and decision making, and thus avoid waste and redundancy.

Examples of how SDM and **adaptive resource management (ARM,** defined below) can be, or are, currently applied to natural resource management include management of sustainable harvest from fish (Peterson and Evans 2003, Irwin et al. 2011) and wildlife (Anderson 1975, Williams 1996, Smith et al. 1998, Johnson and Williams 1999, Moller et al. 2009) populations, endangered species management (Moore and Conroy 2006, Conroy et al. 2008, McDonald-Madden et al. 2010, Keith et al. 2011), sustainable agriculture and forestry (Butler and Koontz 2005, Schmiegelow et al. 2006), river basin and watershed management (Clark 2002, Prato 2003, Leschine et al. 2003), water supply management (Pearson et al. 2010), management of air and water quality (Eberhard et al 2009, Engle et al. 2011), design of ecological reserves (McDonald-Madden et al. 2011,

McGeoch et al 2011), control of invasive species (Foxcroft and McGeoch 2011) and climate change (Wintle et al. 2010, Conroy et al. 2011, Nichols et al. 2011). This list is selective and not exhaustive, and non-inclusion of a resource area by no means suggests that SDM or ARM would not be useful in many other areas. Conversely, not every SDM application has been successful or even well executed. We will consider some of the reasons why these approaches can and might fail.

Limitations of the structured approach to decision making

Above, we have discussed a number of advantages of a structured approach to decision making and how a structured approach can ameliorate common problems in framing decisions. To summarize, these include:

- transparency and improved communication;
- a clearer connection of decisions to stated objectives;
- institutional memory in the decision making process;
- better use of resources (e.g., in monitoring programs).

However, a structured approach can be viewed as having disadvantages to the way business might be conducted currently. First, a structured approach requires a long-term institutional commitment to carry through, and there is always the risk that a future administration will undo the process. Also, a structured approach can, at least in the short term, be threatening to the institutional way of doing business that lacks transparency and operates under hidden assumptions. Of course, these are not really arguments against taking a structured approach so much as they are obstacles that must be overcome (or navigated around) to make SDM work.

Finally, readers should not get the idea that we are promoting structured decision making as a foolproof way of making "good" decisions. A distinction must be made between being "wrong" in the sense of obtaining a less-than-desirable outcome following a sound decision-making process and being "wrong" by following a flawed decision process that occasionally leads to good outcomes by accident. By following a "good process" we do not assure ourselves of good outcomes, because of uncertainty (Chapter 7). We hopefully will experience more good than bad outcomes, but the bad outcomes we do experience are understandable in the context of our decision process. Furthermore, as we will see, they provide us with opportunities to learn and improve future decision making. Following a "bad process" *will* occasionally result in desirable outcomes, but these will not be understandable in the context of the decision process, and provide no potential for learning or improvement of decision making through time.

No one can be assured of a good result from any specific decision, but we *can* assure you that if you follow a sound decision process you will a) do better in the longer run than if you do not, and b) be in a position to defend your decision even when the results are poor. The distinction between *process* and *outcome* is emphasized (albeit in somewhat tongue-in-cheek fashion) by Russo and Shoemaker (2001). These authors describe good and bad outcomes following a good process as, respectively "a deserved success" and "a bad break". By contrast, these same outcomes following a bad process are respectively characterized as "dumb luck" and "poetic justice."

Adaptive resource management

Adaptive resource management (ARM; Walters 2002, Walters 1986, Williams et al. 2002, Williams et al. 2009) extends SDM to the case where outcomes following decisions are uncertain, which we argue is common in natural resource management. This uncertainty is incorporated via the use of alternative models representing hypotheses of ecological dynamics and statistical distributions representing error in model parameters. Each model (hypothesis) is assigned a level of plausibility or probability. The optimal decision then is selected based on the current system state (e.g., bear population size) and a prediction of the expected future state following a management decision, taking into account various sources of uncertainty.

When management decisions reoccur over space or time (e.g., annual harvest regulations), model probabilities are updated by comparing model-specific predictions to observed (actual) future conditions. The adjusted model probabilities can then be used to predict future conditions and choose the optimal decision for the following time step. This adaptive feedback explicitly provides for learning through time and, ideally, the resolution of competing hypotheses with monitoring data.

Under ARM, monitoring data serve two purposes. First, they provide an estimate of the current system state and a means of monitoring the responses of the system to management. This aspect of monitoring is shared with SDM when decisions are recurrent and state dependent. Under ARM, monitoring provides the additional role of learning about system dynamics, which in turn improves future decision making. Because of its great potential for integrating monitoring programs into decision making, ARM has now been formally adopted by the U.S. Department of the Interior (USDI) for managing Federal resources (Williams et al. 2009).

There is some confusion in the literature about what "adaptive management" means. Some of the confusion arises from differences in the relative emphasis placed on "learning" (that is reducing **structural uncertainty**; see Chapters 7 and 8) versus seeking an optimal resource outcome (Williams 2011) and the degree to which practitioners of ARM assert that experimental "probing" is required (e.g., Walters 1986, Walters et al. 1992). We deal with these issues to some degree in Chapter 8 and Appendix E but largely take the view that these are differences without a distinction. We see no conflict between "learning" and "gaining", particularly when it is made clear (Chapters 7 and 8, and Appendix E) that system uncertainty detracts from the latter, and thus "learning" and "gaining" are more properly viewed as synergistically related than in competition with each other. More serious, we believe, are usages of "adaptive management" that detract from it as a meaningful concept. For example, we have heard ARM referred to as "trial and error", "seat of the pants", "conflict resolution", or "building stakeholder collaboration." Certainly, these can be aspects of an ARM process but do not themselves constitute such a process.

In our view, three features absolutely must be present for the process to be deemed ARM:

1. Decisions must be **recurrent**. We cannot envision a role for ARM for one-time decisions, simply because there is no opportunity for learning to influence future decision making.
2. Decisions must be based on predictions that incorporate structural uncertainty (Chapter 7). Often this will be represented by two or more alternative models or hypotheses about system functionality.
3. There must be a **monitoring** program in place to provide the data that will be fed back into adaptive updating, without which there, by definition, can be no updating. Programs that do not contain these essential elements, in our view, are not, and should not be called, "adaptive management." We note that these essential elements *are* part of the USDI adaptive management protocol, which we hold as a model for other agencies and groups (Williams et al. 2009).

Summary

In this chapter, we have presented a broad overview of SDM and ARM, explained why we think a structured approach may be beneficial to a wider range of natural resource decision problems, and provided a wide array of examples that are currently or potentially amenable to SDM and ARM.

In the next chapter, we describe the key elements of SDM, including development of a problem statement, elucidation of objectives, specification of decision alternatives, and establishment of boundaries (temporal, spatial) for the decision problem. We then discuss some general principles for evaluating and selecting among alternative decisions. Finally, we will introduce the use of predictive modeling in decision making and discuss the issue of uncertainty. All of these topics will be developed in greater detail in later chapters.

References

Anderson, D.R. (1975) Optimal exploitation strategies for an animal population in a Markovian environment. *Ecology* 56, 1281–1297.
Butler, K.F. and T.M. Koontz, (2005) Theory into practice: Implementing ecosystem management objectives in the USDA Forest Service. *Environmental Management* 35, 138–150.
Clark, M.J. (2002) Dealing with uncertainty: adaptive approaches to sustainable river management. *Aquatic Conservation-Marine and Freshwater Ecosystems.* 12, 347–363.
Clemen, R.T. and T. Reilly, (2001) *Making Hard Decisions.* South-Western, Mason, Ohio.

Conroy, M.J., R.J. Barker, P.J. Dillingham, D. Fletcher, A.M. Gormley, and I. Westbrooke, (2008) Application of decision theory to conservation management: recovery of Hector's dolphins. *Wildlife Research* **35**, 93–102.

Conroy, M.J. and J.T. Peterson, (2009) Integrating management, research, and monitoring: leveling the 3-legged stool. *Proceedings of Gamebird 2006*, Athens, Georgia.

Conroy, M.J., M.C. Runge, J.D. Nichols, K.W. Stodola, and R.J. Cooper, (2011) Conservation in the face of climate change: The roles of alternative models, monitoring, and adaptation in confronting and reducing uncertainty. *Biological Conservation* **144**, 1204–1213.

Eberhard, R., C.J. Robinson, J. Waterhouse, J. Parslow, B. Hart, R. Grayson, and B. Taylor, (2009) Adaptive management for water quality planning – from theory to practice. *Marine and Freshwater Research* **60**, 1189–1195.

Engle, N.L., O.R. Johns, M.C. Lemos, and D.R. Nelson, (2011) Integrated and Adaptive Management of Water Resources: Tensions, Legacies, and the Next Best Thing. *Ecology and Society* **16**, [online].

Foxcroft, L.C. and M. McGeoch, (2011) Implementing invasive species management in an adaptive management framework. *KOEDOE* **53(2)**, [online].

Hammond, J.S., R.L. Keeney, and H. Raiffa, (1999) *Smart Choices: A Practical Guide to Making Better Decisions*. Harvard Business School Press, Boston, Massachusetts.

Irwin, B.J., M.J. Wilberg, M.L. Jones, and J.R. Bence, (2011) Applying Structured Decision Making to Recreational Fisheries Management. *Fisheries* **36**, 113–122.

Johnson, F. and K. Williams, (1999) Protocol and practice in the adaptive management of waterfowl harvests. *Conservation Ecology* **3(1)**, 8. [online] URL: http://www.consecol.org/vol3/iss1/art8/.

Keith, D.A., T.G. Martin, E. McDonald-Madden, and C. Walters, (2011) Uncertainty and adaptive management for biodiversity conservation *Biological Conservation* **144**, 1175–1178.

Lange, J.I. (1993) The logic of competing information campaigns: conflict over old growth and the spotted owl. *Communication Monographs* **60**, 239–257.

Leschine, T.M., B.E. Ferriss, K.P. Bell, K.K. Bartz, S. MacWilliams, M. Pico, and A.K. Bennett, (2003) Challenges and strategies for better use of scientific information in the management of coastal estuaries. *Estuaries* **26**, 1189–1204.

Martin, J., P.L. Fackler, J.D. Nichols, B.C. Lubow, M.J. Eaton, M.C. Runge, B.M. Stith, and C.A. Langtimm, A. Catherine, (2011) Structured decision making as a proactive approach to dealing with sea level rise in Florida. *Climate Change*. **107**, 185–202.

McCright, A.M. and R.E. Dunlap, (2011) Cool dudes: The denial of climate change among conservative white males in the United States. *Global Environmental Change – Human Policy Dimensions*. **21**, 1163–1172.

McGeoch, M.A., M. Dopolo, P. Novellie, H. Hendriks, S. Freitag, S. Ferreira, R. Grant, J. Kruger, H. Bezuidenhout, R.M. Randall, W. Vermeulen, T. Kraaij, I.A. Russell, M.H. Knight, S. Holness, and A. Oosthuizen, (2011) A strategic framework for biodiversity monitoring in South African National Parks, *KOEDOE* **53(2)**, [online].

McDonald-Madden, E., I. Chades, M.A. McCarthy, M. Linkie, and H.P. Possingham, (2011) Allocating conservation resources between areas where persistence of a species is uncertain. *Ecological Applications* **21**, 844–858.

McDonald-Madden, E., W.J.M. Probert, C.E. Hauser, M.C. Runge, H.P. Possingham, M.E. Jones, J.L. Moore, T.M. Rout, P.A. Vesk, and B.A. Wintle, (2010) Active adaptive conservation of threatened species in the face of uncertainty. *Ecological Applications*. **20**, 1476–1489.

Moller, H., J.C. Kitson, and T.M. Downs, (2009) Knowing by doing: learning for sustainable muttonbird harvesting. *New Zealand Journal of Ecology* **36**, 243–258.

Moore, C.T. and M.J. Conroy, (2006) Optimal regeneration planning for old-growth forest: addressing scientific uncertainty in endangered species recovery through adaptive management. *Forest Science* **52**, 155–172.

Nichols, J.D. and B.K. Williams, (2006) Monitoring for conservation. *Trends in Ecology and Evolution* **21**, 668–673.

Nichols, J.D., M.D. Koneff, P.J. Heglund, M.G. Knutson, M.E. Seamans, J.E. Lyons, J.M. Morton, M.T. Jones, G.S. Boomer, and B.K. Williams, (2011) Climate Change, Uncertainty, and Natural Resource Management. *Journal of Wildlife Management* **75**, 6–18.

Pearson, L.J., A. Coggan, W. Proctor, and T.F. Smith, (2010) A Sustainable Decision Support Framework for Urban Water Management. *Water Resources Management* **24**, 363–376.

Peterson, J.T. and J.W. Evans, (2003) Quantitative decision analysis for sport fisheries management. *Fisheries* **28**, 10–21.

Prato, T. (2003) Adaptive management of large rivers with special reference to the Missouri River. *Journal of the American Water Resources Association* **39**, 935–946.

Russill, C. (2011) Truth and opinion in climate change discourse: The Gore-Hansen disagreement. *Public Understanding of Science* **20**, 796–809.

Russo, J.E. and P.J.H. Shoemaker, (2001) *Winning Decisions: Getting it Right the First Time*. Currency Doubleday, New York, New York.

Schmiegelow, F.K.A., D.P. Stepnisky, C.A. Stambaugh, and M. Koivula, (2006) Reconciling salvage logging of boreal forests with a natural-disturbance management model. *Conservation Biology* **20**, 971–983.

Smith, C.L., J. Gilden, B.S. Steel and K. Mrakovcich, (1998) Sailing the shoals of adaptive management: The case of salmon in the Pacific Northwest. *Environmental Management* **22**, 671–681.

Walters, C.J. (2002) *Adaptive Management of Renewable Resources*. Blackburn Press, New Jersey.

Walters, C.J. (1986) *Adaptive Management of Renewable Resources*. MacMillan.

Walters, C.J., L. Gunderson, and C.S. Holling, (1992) Experimental Policies for Water Management in the Everglades. *Ecological Applications* **2**, 189–202.

Williams, B.K. (1996) Adaptive optimization and the harvest of biological populations. *Mathematical Biosciences* **136**, 1–20.

Williams, B.K., J.D. Nichols, and M.J. Conroy, (2002) *Analysis and Management of Animal Populations*. Elsevier Academic.

Williams, B.K, R.C. Szaro, and C.D. Shapiro, (2009) *Adaptive Management: The US Department of Interior Technical Guide*. [Online] URL: http://www.doi.gov/archive/initiatives/AdaptiveManagement/TechGuide.pdf.

Williams, B.K. (2011) Passive and active adaptive management: Approaches and an example. *Journal of Environmental Management* **92**, 1371–1378.

Wintle, B.A., M.C. Runge, and S.A. Bekessy, (2010) Allocating monitoring effort in the face of unknown unknowns. *Ecology Letters* **13**, 1325–1337.

2

Elements of Structured Decision Making

In this chapter, we develop the key elements of structured decision making, including clear development of a problem statement, elucidation of objectives, specification of decision alternatives, and establishment of boundaries (temporal, spatial) for the decision problem. We discuss optimal decision making and general principles for evaluating and selecting among alternative decisions. We introduce the use of predictive modeling in decision making, and discuss the issue of uncertainty. The basic ideas presented here are by no means unique to natural resource management but are in common with decision making in other fields (e.g., Hammond et al. 1999, Clemen and Reilly 2001, Russo and Shoemaker 2001). Each of these topics is covered in general, conceptual terms, to be covered in more detail in the ensuing chapters.

First steps: defining the decision problem

In our view, many decision problems in natural resources management suffer, and some fail outright, because of the failure to appropriately define the decision problem at the outset. A problem statement turns a vague task – "Respond to declining fishing success in Green Lake" – into an affirmative statement that ties actions to measurable outcomes over a specified timeframe – "Use changes in creel limits, size restriction, and habitat management to increase fishing catch rate in Green Lake by 25% over the next 5 years within budgetary constraints." A problem statement should propose an **action** (or set of choices) that we **predict** will lead to **outcomes** that fulfill **objectives**. Our analysis of a decision problem starts with a problem statement of this generic form, which we will then decompose into its constituent elements.

Decision Making in Natural Resource Management: A Structured, Adaptive Approach,
First Edition. Michael J. Conroy and James T. Peterson.
© 2013 John Wiley & Sons, Ltd. Published 2013 by John Wiley & Sons, Ltd.

Once we have developed our problem statement we can then proceed to delineate the steps to "solve" the problem. Although we can start with any of the components, it is often most natural to start by asking what the **objectives** are. As we will see in the next chapter, this is actually more complicated than it appears at first. Essentially, by objectives we mean the achievement of particular, measurable outcomes in relation to the decisions we have made. However, it will be important to distinguish between **fundamental objectives** – which we desire because they represent our fundamental or core values – and **means objectives** – which are desirable to the extent that they help us fulfill fundamental objectives. Finally, objective setting is complicated by the fact that we typically will have **multiple objectives** that may compete or conflict with one another. We return to objective setting in more detail in Chapter 3.

We must also, of course, establish the range of actions or **decisions** that we have at our disposal. Actually, although this step seems obvious, we have encountered many situations in which resource managers claim to have one or more objectives they wish to achieve, but are unable to articulate actions by which those objectives could be achieved. We will elaborate on decision alternatives more fully in the next chapter. Briefly, they include obvious sorts of manipulative actions such as habitat and harvest management, but also other conservation actions that may include designation of reserves, legal protection of species and habitats, or public education. Finally, it must always be acknowledged that "no action" can be a choice – whether deliberate, or by default (e.g., hesitation due to a lack of information).

It is also essential to define the spatial, temporal and organizational bounds of a decision problem. For example, are we trying to solve a species conservation problem for a specific reserve; for a class of similar reserves; or for the entire range of the species? Are our objectives short term (e.g., achievement of a particular population target within the next 5 years) or longer term (achievement of a population target or other conservation goal over hundreds of years)? Of course, as we will see, temporal objectives can be linked, with fulfillment of short-term objectives viewed as a means to a longer-term one, rather than ends unto themselves. Finally, we must consider the resolution of the conservation problem, as in the spatial, temporal or organizational resolution at which we are making decisions, or at which results will be evaluated. For example, we could be concerned only with maximizing aggregate harvest over spatial units over some time frame; or we could instead be concerned about how harvest is allocated among spatial units, how it varies over time, or both. Finally, in setting out our decision problem, we should always be on the alert for "sticking points"- situations or factors that can (often will) cause the process to go off the rails. Common sticking points include, but are not limited to, poor or incomplete initial problem definition, failure to include key stakeholders or decision makers, hidden objectives, confounding values with science, and political interference. As we will see in Chapter 10, many of these are evident in hindsight, but with our "lessons learned" can perhaps be avoided in the future.

Up to this point we have acted as though once we delineate objectives and decision alternatives, we can simply proceed with solving the decision problem. That is, we have assumed that we have a good idea what will happen, and what will be achieved by the way of objectives, if we select a particular action. In reality, the relationship between decisions and outcomes is obscured by uncertainty, and the best we typically have is a set of beliefs or a **model** as to what we think we are most likely to achieve in terms of y, if we do x. We will spend a lot of time in later chapters on how we deal with this uncertainty, which we assert is unavoidable in resource management. Suffice it to say that uncertainty in and of itself is *not* an impediment to decision making. However, what *is* an impediment is when we confuse uncertainty – or belief – about how natural systems respond to actions, with outcomes that we value in our objective. As we will see, uncertainty – even when it is profound – can be dealt with, but *only when we have first agreed on the objective we are trying to achieve.*

Unfortunately, the history of resource management is rife with examples where **belief** about how natural resource systems function has been conflated with *values* that we are trying to achieve. It is no accident, for example, that belief that Northern Spotted Owls (*Strix occidentalis*) are (or are not) an ancient forest obligate has been strongly tied to values (resource utilization versus preservation) in Pacific Northwest forests (Lange 1993). For example, a stakeholder who claims that owls are not forest obligates may have evidence to support this contention, or may simply place a high value on forest utilization, and he perceives that any contrary belief may threaten his preferences. Clear separation of beliefs, which are resolved via objective data gathering and analysis, from **values,** which are negotiated among stakeholders, is, therefore, something that we not only encourage but insist upon be undertaken at the earliest stages of decision analysis.

General procedures for structured decision making

Once we have developed a concise statement of the decision problem, we are ready to initiate a structured decision-making approach. At this juncture we find it useful to decompose the problem statement into its constituent elements. All decisions have in common three elements: 1) an objective, 2) a set of decision alternatives for achieving the objectives, and 3) a model of decision influence that represents belief in how various actions will lead to outcomes that tend to fulfill (or not) the objective. A simple flow or network diagram can be useful at this point, and will help to ensure that we are including all the appropriate elements of the decision problem, as well as keeping separate their treatment (e.g., avoiding confusing values with beliefs). We illustrate this first with a generic decision problem, in which we have a resource objective whose fulfillment depends at least in part on our ability to influence a resource "state" (Figure 2.1). By "state" we simply mean some measurable condition(s) of nature, such as population abundance, habitat conditions, diversity, or other attribute. To take a more specific problem, we could be contemplating a prescribed burning program to improve habitat conditions for wildlife species of interest. Failure to burn is assumed to result in a decline of habitat quality (value −1), a moderate burn is assumed to result in status quo habitat (value 0), and intense burn in improvement (value +1). However, burning is assumed to have costs of 1, 2, and

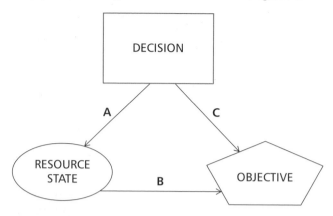

Figure 2.1 Schematic of a generic resource decision problem. Candidate decisions are thought to influence a resource state in some manner in order to achieve an objective. The decision also may influence the objective directly (e.g., through costs incurred). Arrows represent direction of causality/ influence: (A) influence of decision on resource state, (B,C) influence of combination of resource state and decision on objective value.

3 for no burning (e.g., baseline monitoring costs), moderate, and intense burning, respectively.

Once we have represented the relationship among decisions, measurable states, and objectives, we can begin to explore means of obtaining **optimal** solutions. By "optimal" we usually mean "select the decision that provides us the best objective value compared to all competing decisions." Here, of course, we are assuming that our objective adequately represents our values (e.g., the balance between a conservation goal and costs or other competing uses of resources). In later chapters we will explore more sophisticated means for optimizing decision problems, but for some problems the method for selecting the optimal decision is as simple as examining a list of values and selecting the decision associated with the highest or lowest value, depending on whether the goal is to minimize or maximize the objective value. In our burn example, we have a fairly simple relationship between the decisions, outcomes, and objective values, leading to the value for each candidate decision as below:

Decision	Resource value	Cost	Objective value (Resource value/Cost)
No burn	−1	1	−1
Moderate burn	0	2	0
Intense burn	1	3	0.333

(Figure 2.2). Assuming that the goal is to maximize resource value per unit costs, by simply examining the objective values it is clear that "intense burn" is the

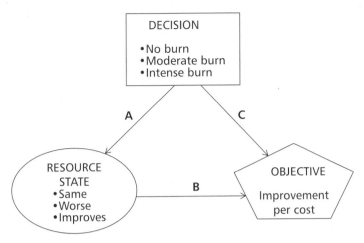

Figure 2.2 Schematic of a prescribed burning decision problem. Decision to burn is thought to improve habitat but also has costs proportional to intensity/ effort; both habitat improvement and cost are accounted for in the objective.

optimal decision. Notice that the identity of the optimal action is sensitive to belief about the relationship between the actions and outcome ("Resource value" column) and values placed on outcomes ("Objective value" column): other choices for these parts of the decision problem may point to another action as optimal.

Predictive modeling: linking decisions to objectives prospectively

Most resource problems are more complicated than this for several reasons, including the fact that the relationship between decisions and objective outcomes often involves tradeoffs and dynamics that cannot be described in a simple list or equation form. We will deal with more general approaches to optimizing decision problems in later chapters, but first let's return to an issue that we essentially ignored in the previous examples: the specific relationship between our proposed decisions and the outcomes we are trying to achieve. In the burn example, we assumed implicitly that there is a 1:1 relationship between the actions (no burn, moderate burn, intense burn) and outcomes (habitat degradation, status quo, and habitat improvement). Even if such assumptions are justified, we argue emphatically that they should be stated explicitly. We believe that the best way to be clear about key assumptions in the decision model is to formalize the assumed relationships in an explicit **model**.

To be clear, by "model" we simply mean "any conceptualization of the relationship between decisions, outcomes, and other factors". Models certainly can include formal mathematical models and computer programs, but also simple flow diagrams (e.g., Figures 2.1 and 2.2). By formalizing our assumption into a model (however simple), we can accomplish several things. First, key assumptions related to the decision-making process are now in the open, where they can be examined fully by us, by our colleagues, and by the stakeholders in the decision. In the process, we may reaffirm our assumptions, or we may decide to revise them, augment them with alternative ideas, or abandon them altogether. Secondly, a formal model of decision influence will be necessary for any quantitative decision analysis. Ordinarily this will require us to specify parameter values for key relationships, which in turn will often be derived from analysis of data from previous studies or monitoring programs (Part II).

Uncertainty and how it affects decision making

Up to this point we have acted as though there is a straightforward, absolutely consistent relationship between decisions that we make (or can make) and outcomes that we are endeavoring to achieve. Technically speaking, we have assumed that the problem is **deterministic** (not subject to uncertainty) as opposed to **stochastic** (influenced by uncertainty). In reality, virtually all except the most trivial decision problems, both in everyday life and in natural resource management, involve uncertainty. To take a simple example from everyday life, suppose we are faced with the decision whether to pack raingear for an upcoming trip. The value of the decision we make (pack or do not pack the raingear) will depend on an outcome (rain will, or will not occur) that obviously is in the future, and that cannot be predicted with certainty (Figure 2.3). Uncertainty affects both the decisions we make (or should make) and the value of decisions we do make. In the rain example, only if we have perfect foreknowledge of the rainfall outcome will we be able to select the "perfect" decision; in all other cases our decision will be based on an evaluation of the probability of the outcome, and sometimes we will be "wrong", in the sense that the **decision-outcome** (the combination of an action and a resulting outcome following decision making) will be less desirable than hoped for.

Type of uncertainty

The rainfall example illustrates one specific (but common) type of uncertainty, namely **environmental uncertainty**, due to the fact that many aspects of nature beyond the control of the decision maker vary in a random or stochastic manner and potentially influence outcomes. Here, we distinguish between **random** and **unpredictable**. Random events are predictable, but the predictions involve **probability** statements. Thus, we may have a weather forecast (prediction) for 60% chance of rain during our trip; this stochastic prediction may be useful to us in formulating a decision (Figure 2.4a). By contrast, a truly unpredictable event is one for which it is difficult to make an accurate probability statement, often because there is little data or experience on which to base such a statement. We return to this issue is Chapter 5, where we discuss **statistical models**, and Chapters 6 and 7, where we discuss other methods for quantifying uncertainty, such as **expert elicitation**.

Event

Decision	Rains	Does not rain
Pack raingear	Satisfied (We were prepared)	Unsatisfied (Unnecessary packing, inconvenience of carrying umbrella; possible loss of umbrella)
Do not pack raingear	Unsatisfied (We got wet)	Satisfied (We were not encumbered by raingear, which turned out to be unnecessary)

Figure 2.3   Decision making under uncertainty. Relative satisfaction (utility) of decision-outcome combinations. Decision making is under uncertainty because the outcome (rainfall or no rainfall) is random following either decision.

Another type of uncertainty, commonly encountered in natural resource management, is **partial controllability.** By this we mean that our intended action is itself not completely under our control, but may be influenced by other (often random) factors. In the prescribed fire example, we may **decide** (as in intend) to effect a moderate burn, but intervening weather or other events may result in the realization of a smaller, less intense or larger, more intense fire. Similarly in harvest management a decision may be made with the intent of achieving a particular harvest level, but other factors such as economic conditions or animal migration patterns may result in higher or lower rates of harvest than intended (Figure 2.4a). Partial controllability also occurs when decisions must be changed for reasons beyond the decision-maker's control, such as funding shortages, withdrawal of cooperators, or loss of equipment due to weather or fire.

A third and very important type of uncertainty is **statistical uncertainty,** also known as **partial observability.** This type of uncertainty occurs because of our typical inability to perfectly "see" nature and thereby evaluate either what decisions to make (when decisions are tied to the state of the system or **state dependent,** as they frequently are), or what outcomes have been achieved following decisions (and therefore, whether objectives have been met). For example, if we are responsible for restoring a stock of an endangered fish in a stream, we will

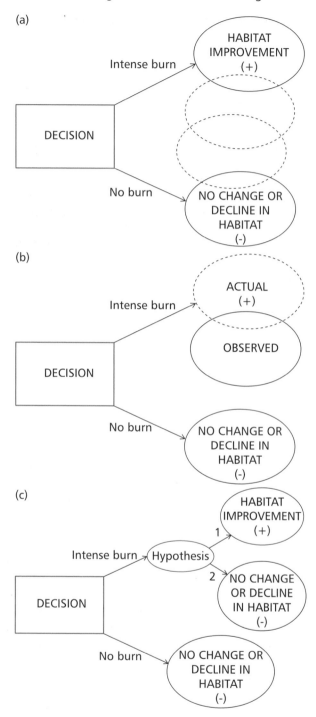

Figure 2.4 Types of uncertainty. (a) Environmental uncertainty and partial control. (b) Statistical uncertainty. (c) Structural uncertainty.

almost certainly be basing our decisions on population and habitat surveys that in turn are based on **statistical samples** and **estimates** rather than complete enumeration and perfect measurement of these attributes (Figure 2.4b). The estimates, therefore, are subject to statistical errors and possible biases that relate back to the specific data collection methods and statistical analyses used. We will return to the topic of statistical uncertainty in detail in Chapter 5.

Finally, a very important but often overlooked form of uncertainty is known as **structural uncertainty**. Structural uncertainty refers to uncertainty in the basic relationships in our predictive model of decision influence, and is often represented by alternative models (hypotheses). In the example of prescribed fire, one (perhaps preferred) hypothesis is that habitats (and presumably animal populations) will respond favorably to the physical conditions induced by the action. However, it may be plausible that fire will have none of its intended effects, perhaps for reasons not accounted for in our model (Williams et al. 2002). By entertaining the possibility of alternative, competing models (hypotheses) of decision influence we "hedge our bets" in decision making, rather than putting our full faith in a belief that could be wrong. We also set up the possibility of learning under **adaptive management**, which provides a framework for decision making under structural uncertainty, with feedback to reduce uncertainty and improve management through time (Walters 1986, Williams et al 2009).

Dealing with uncertainty in decision making

Above we have identified several types and sources of uncertainty that can be important to natural resource management. We will later see (Chapters 7 and 8) how these various types of uncertainty can both influence the decisions that are or can be made, as well as affect the value of the resource objective achieved in decision making. Here, we introduce the idea of how uncertainty can be accommodated in decision making, which we will return to more formally in later chapters.

Obviously, we can simply ignore uncertainty in our decision-making process, that is, essentially pretend that it does not exist. We strongly discourage managers from taking this approach, which we see as a form of denial. Unfortunately, as far as we can tell it is precisely how many professional managers in agencies, NGOs, and other organizations appear to treat natural resource management. At least, it is not common (yet) for uncertainty to be either explicitly recognized in the decision process, or formally incorporated into decision making. Rather, managers often choose decisions based on what is perceived as the least risky decision as a means of bet hedging against uncertain and undesirable outcomes. Part of our goal in writing this book it to attempt to change those attitudes, however incrementally.

Secondly, we can try to reduce or even eliminate various sources of uncertainty. In some cases, this will be feasible. For example, in the prescribed fire situation, we can wait for conditions that are optimal (and forecasted to remain so) for the burn, and forego burning at all other times. Essentially, this is incorporating an additional state of nature (the weather) into our decision model, and will reduce (but never eliminate) uncertainty. Similarly, we can reduce (but not eliminate) partial observability by conducting our monitoring surveys according to rigorous statistical designs, and analyzing the data using appropriate statistical models.

More commonly, we simply must acknowledge uncertainty as a factor we must deal with in decision making. There are a number of ways this can be done, and we will return to this topic in detail in Chapters 6 and 7. Briefly, our usual approach will be to represent uncertainty by **statistical distributions** and consider that the value of the decision-outcome will take on a value determined by these distributions. Because we cannot know the value in advance, we will operate under the assumptions that (1) the decision-outcome *could* take on any value specified by the distribution(s), but (2) some values will be more likely than others, based on our probability model. We then usually average the objective values over the statistical distributions for each decision, and finally, select the decision that appears to produce the best or optimal average or **expected value**. We can illustrate this with the prescribed fire example and a simple probability model that describes how, instead of certainty of a result following the burn, we are 90% sure of the outcome but retain a 5% probability of outcome for the other two possibilities. So for example, if we decide on a moderate burn, we are 90% confident in "status quo" but allow a 5% change of either degradation or improvement. We then calculate the expected values (resource value/ cost) for each decision as

$$\text{Expected value}(Decision) = \frac{\text{Value}(Same)}{\text{Cost}(Decision)} \times \text{Prob}(Same) + \frac{\text{Value}(Worse)}{\text{Cost}(Decision)}$$
$$\times \text{Prob}(Worse) + \frac{\text{Value}(Improves)}{\text{Cost}(Decision)} \times \text{Prob}(Improves)$$

$E(\text{No burn}) = (-1)0.9 + (0)0.05 + (1)0.05 = -0.85$

$E(\text{Moderate burn}) = (-1/2)0.05 + (0/2)0.9 + (1/2)0.05 = 0.0$

$E(\text{Intense burn}) = (-1/3)0.05 + (0/3)0.05 + (1/3)0.9 = 0.2833$

Note that in this example, the optimal decision (intense burn) does not change as a result of uncertainty, but its value is lower (0.28 vs. 0.33). We will return to this issue in more detail in Chapter 7.

There are alternatives to basing decisions on expected value optimization, and we will consider some of these later (Chapter 8). For example, rather than seeking to maximize or minimize the objective value on average, we may decide that it is more important to avoid extremely "bad" (in terms of objective value) outcomes. Nevertheless, the approaches have in common that they explicitly recognize uncertainty; represent it via probability distributions; and use the distributions together with an objective value measure to establish optimality rules for decisions.

Summary

In this chapter, we have given a broad overview of the decision problem, and have identified several elements that will receive attention in subsequent chapters. In each of the subsequent chapters, we will build on these ideas, first exploring in more detail how to develop achievable objectives, and how to solve decision-making problems in which objectives involve competing values, constraints, and linkage among decisions. In Part II, we will see how data from studies and monitoring programs can be used to build predictive models for use in decision making, and we will apply these ideas to the development of stochastic models of decisions, followed by detailed coverage of dealing with uncertainty in decision analysis and methods for optimization appropriate to a wide class of decision problems. Finally, we will return in detail to the topic of decision making under uncertainty and learning via adaptive management.

References

Clemen, R.T. and T. Reilly. (2001) *Making Hard Decisions*. South-Western, Mason, Ohio.

Hammond, J.S., R.L. Keeney and H. Raiffa. (1999) *Smart Choices: A Practical Guide to Making Better Decisions*. Harvard University Press, Cambridge, Massachusetts.

Lange, J.I. (1993) The logic of competing information campaigns: conflict over old growth and the spotted owl. *Communication Monographs* **60**, 239–257.

Russo and Shoemaker. (2001) *Winning Decisions: Getting it Right the First Time*. Currency Doubleday.

Walters, C.J. (1986) *Adaptive Management of Renewable Resources*. MacMillan.

Williams, B.K., J.D. Nichols, and M.J. Conroy. (2002) *Analysis and Management of Animal Populations*. Elsevier Academic.

Williams, B.K., R.C. Szaro, and C.D. Shapiro. (2009) Adaptive Management: The US Department of Interior Technical Guide. [Online] URL: http://www.doi.gov/archive/initiatives/AdaptiveManagement/TechGuide.pdf

3

Identifying and Quantifying Objectives in Natural Resource Management

In this chapter, we explore in detail how to develop quantifiable objectives in natural resource management. We discuss the importance of creativity in the decision-making process and the necessity of distinguishing fundamental and means objectives. We discuss how to overcome roadblocks that impede the identification of objectives and creative management decision alternatives. Finally, we address the issues of quantifying objectives and focus on approaches to deal with multiple and competing objectives by couching objectives as constraints and weighting objectives based on decision-maker values.

Identifying objectives

As introduced in Chapter 2, by **objectives** we mean specific, quantifiable outcomes that reflect the values of decision makers and stakeholders and relate directly to the management decisions. Proper definition of explicit objectives is of huge importance in decision making and so requires careful attention. In fact, failure to explicitly identify objectives can and will lead to gridlock and conflict among decision makers and stakeholders. Objectives are those things that matter most to the decision makers and stakeholders, and as such, they will be used to evaluate tradeoffs and identify the best decision and later to determine whether a management action was successful. Thus, the structured decision-making process can only proceed when decision makers agree on the objective or set of objectives. In short, it is not an understatement to say that everything, and we mean everything, is largely determined by the objectives.

Right now you might be thinking, "this should be easy, I already know my objectives." Unfortunately, it isn't easy. In our experience, we find that identifying

Decision Making in Natural Resource Management: A Structured, Adaptive Approach,
First Edition. Michael J. Conroy and James T. Peterson.
© 2013 John Wiley & Sons, Ltd. Published 2013 by John Wiley & Sons, Ltd.

objectives is the most difficult step in the structured decision-making process. Much of the difficulty generally arises from four basic problems. First, decision makers and stakeholders often confuse fundamental and means objectives. **Fundamental objectives** are those things that a decision maker truly values and wants to achieve, whereas **means objectives** are a means to *achieving* fundamental objectives. Failure to properly distinguish these two types of objectives can result in the implementation of ineffectual management actions. Secondly, stated objectives are often vague and imprecise. Clarity is crucial to help decision makers and stakeholders communicate their values and objectives. The failure to clearly articulate objectives can lead to potential conflict among stakeholders and worse, can result in the identification of ineffectual management strategies. Thirdly, decision makers and stakeholders in specific interest groups may reject the stated objectives of those other interest groups based on perceived conflicts with their own objectives. These perceptions are generally based on decision makers' and stakeholders' ideas or "mental models" of system dynamics that they preemptively use to assess tradeoffs. Fourthly, decision makers often restrict objectives to those things that they perceive as achievable or feasible based on available data or information. Such thinking blocks creativity in the structured decision-making process and can hinder decision makers from finding unique, perhaps win-win solutions to natural resource management problems. Below, we address each of the problems and outline approaches to overcoming them. Note that to successfully use these approaches in multi-stakeholder situations requires experience and skills, which we discuss in Chapter 4.

Identifying fundamental and means objectives

It is crucial to distinguish between **fundamental objectives** – which we desire because they represent our fundamental or core values – and **means objectives** – which are desirable to the extent that they help us fulfill fundamental objectives. We can sort out fundamental from means objectives by asking the questions *"why is that important"* and *"how can we accomplish that"* for each objective. If our answer to "why is that important" is along the lines of "to achieve another (perhaps more fundamental) objective", then it is probably a means objective. If the answer to "why is that important" is simply "because that is what we desire" or for many regulatory agencies, "because that is our legal mandate", then it is probably a fundamental objective. Similarly, if the answer to "how can we accomplish that" for an objective is "by fulfilling some other objective", then these other objectives are certainly means objectives (but they may also be fundamental in their own right). Another way of looking at the difference between fundamental and means objectives is to ask the question "would you be satisfied if y is achieved but x is not?"

An important thing to keep in mind is that means objectives *sometimes* help realize the fundamental objectives and they reflect the beliefs and "mental models" of system dynamics of the decision makers. In other words, they are often **hypotheses** about system dynamics. The key is to recognize the differences between fundamental and means objectives, realize that there may be more than one way to achieve a fundamental objective, and not get boxed in a corner by accepting a limited view of how the system works. For example, fishery biologists may believe that the streamflow regime affects ecosystem processes such that a natural flow regime will produce a natural (or pristine) stream fish community. However, a fundamental objective of "maintaining a natural flow regime" means that the decision makers value natural flow regimes for their own sake. What would happen to the fishery biologist if the flow regime was natural, but all the fish were dead? Remember the question: *would you be satisfied if natural flow is achieved but natural stream fish community is not?* More than likely, the fishery biologists would not be satisfied with the outcome. This means that the true fundamental objective is a "natural stream fish community" and there are several alternative means to achieving a natural stream fish community, such as maintain or increase water quality and improving instream habitats. In Chapter 7, we show how to incorporate alternative views of system dynamics (i.e., hypotheses) into decision making, but at this point in the process, you (and stakeholders) only need to be aware that this can be accomplished as part of the structured decision-making process. The important thing here is not to limit the objective identification process and to think creatively by considering multiple ways of achieving objectives – *even if they seem improbable at the time.*

We illustrate the distinction between fundamental and means objectives with an example of conservation of native biodiversity in New Zealand. Let's assume that a group of biologists held a workshop to determine the best ways to conserve native biodiversity. After a brainstorming session, they identified 6 objectives: maximize native biodiversity, maximize native bird diversity, maximize native plant diversity, reduce (or minimize) the number of non-native species that prey on birds, reduce numbers of possums, and reduce the number of exotic plants. To separate fundamental objectives from means objectives, the biologists ask the question *"Why is that important?"*. For example, maximizing native bird diversity was important because it would also help to achieve maximize native biodiversity. Similarly, maximizing native plant diversity was important because it would help maximize native bird diversity and native biodiversity. To separate means objectives from fundamental objectives, the biologists ask the question *"How can we accomplish that?"*. For example, native bird diversity can be increased or maximized by reducing predator numbers and plant diversity can be maximized by reducing the number of possums and exotic plants.

When separating fundamental and means objectives, it is useful to display their relations in an **objectives network** (Figure 3.1). An objectives network is useful for displaying the relations among objectives and for distinguishing fundamental and means objectives. Here, the fundamental objectives are displayed at the top of the network and means objectives at the bottom with each of the fundamental objectives fed by one or more means objectives. Objectives networks help to clarify that some objectives are qualitatively different from others. Specifically, it clarifies that we should be measuring "success" in terms of the fundamental objectives. In the case of New Zealand conservation (Figure 3.2), for example, it would *not* be a success if we managed to control predators and exotic plants but could show no measurable improvement in native bird or plant diversity. Thus, the fundamental objectives for the New Zealand conservation

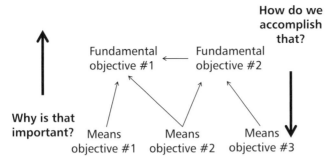

Figure 3.1   Relationship between fundamental and means objectives in an objectives network. Arrows indicated direction of influence; note that means objectives may influence more than 1 fundamental objective, and that a fundamental objective may also be a means objective.

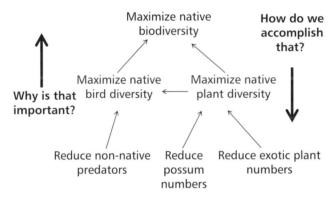

Figure 3.2   Objectives network for New Zealand conservation example.

example are: maximize native biodiversity, maximize native bird diversity, and maximize native plant diversity. Objectives networks also help to clarify the hypothesized relationship among objectives, including implicit assumptions about causality, and naturally lead to the development of an influence diagram (i.e., a graphical form of a decision model), something we will explore in depth in Chapter 6.

In the following sections, we place a special emphasis on fundamental objectives for three reasons. Fundamental objectives are those things that matter most to decision makers and will be the basis for judging the expected success of proposed management and identifying the course of action that is expected to best achieve their objectives. Changes in the fundamental objectives following implementation of a management action will be used to determine the success of the management action. Finally, and most importantly, fundamental objectives must be identified before proceeding with the structured decision-making process.

Clarifying objectives

Clarifying objectives may not be important when a single decision maker is involved because presumably, you (the decision maker) know what you are thinking. However, it is very important when decisions involve multiple decision makers and stakeholders. When we write or talk, the language we use is often vague and filled with phrases or words that can be interpreted multiple ways. For example, consider the fundamental objective of the New Zealand conservation example "maximize native biodiversity." Biodiversity can mean multiple things, such as the number of native species or the relative distribution of individuals among species also known as evenness. Similarly, which native species are included in the definition, all species including both micro- and macrofauna and flora, or only the species the public cares about? This language imprecision is known as **linguistic uncertainty** (Chapter 7) and it can profoundly affect communication among decision makers and stakeholders.

When working with stakeholders, we find that they tend to agree on general or vague objectives, such as "do good things for the environment." However, we soon find that the definition of "good things" differs substantially among stakeholders. Therein lies one of the problems with vague objectives – stakeholders may believe that they have the same objectives when, in fact, they do not. This can, and very likely will, lead to problems later in the process after the stakeholders realize that they are using the same words or phrases to represent different objectives. Remember that the whole process depends on properly identifying objectives, so this misunderstanding can cause a reset in the entire process. Conversely, stakeholders may use different phrases or terms to define the same underlying objective, which can result in conflict until the objectives are clearly defined. One easy way to ensure that objectives are clearly defined is by asking the question, *"How do you define that?"*. For example, the New Zealand biologists would be asked, how do you define biodiversity? Based on the objectives, it appears that diversity is restricted to native birds and plants (Figure 3.2), but the biologists would need to define the attributes (characteristics) of each fundamental objective, such as the identity of the bird species. The importance of clearly defining objectives will become very apparent when we discuss quantifying objectives later in this chapter.

An essential approach to clarifying fundamental objectives is by organizing them into hierarchies. A **fundamental objectives hierarchy** displays the relation among objectives and their **attributes** into a hierarchy with upper levels representing the fundamental objectives stated in general terms and the lower levels of the hierarchy describe the attributes or elements of the upper-level objectives. In other words, the lower-level elements explicitly define the attributes of the fundamental objectives. For example, the upper fundamental objective of the New Zealand conservation example is "maximize native biodiversity" and the lower-level objectives are "maximize native bird diversity" and "maximize native plant diversity" (Figure 3.3). These lower-level objectives define what is meant by maximize native biodiversity and they are further defined by a third level of the hierarchy. For instance, maximizing native bird diversity can be expanded to include the factors that describe native bird diversity, such as the

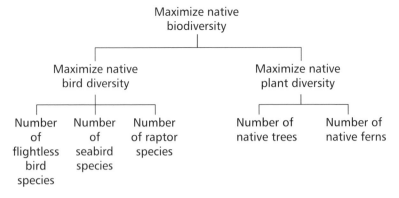

Figure 3.3 Fundamental objectives hierarchy for New Zealand conservation example.

number of flightless, sea, and raptor bird species. These three lower-level fundamental objectives define what is meant by their upper-level objective: native bird diversity. Notice that the lower-level fundamental objectives in this example are relatively unambiguous quantifiable factors. This is intentional because these fundamental objectives will be used to quantify changes in the fundamental objectives in response to management actions (more later). Developing a fundamental objectives hierarchy is essential when dealing with multiple fundamental objectives, and as we shall see, it can be used to assess and quantify the relative importance of multiple objectives (multi-criteria objective valuation, below).

Separating objectives from science

When working with multiple stakeholders, one of the most common problems during objectives identification is that stakeholders have difficulty separating objectives from their "mental models" of system dynamics. As a result, there is often disagreement among stakeholders when they perceive potential conflicts among objectives. For example, consider a stream water management problem where two stakeholders, a fish biologist and a farmer, are identifying objectives. The fish biologists does not believe that maximizing farm income is a reasonable objective because he believes that achieving this objective means that the farmer will use more water to irrigate, which (he believes) will negatively affect fish populations. Alternatively, the farmer cannot agree that maximizing fish populations is a reasonable objective because she thinks that achieving that objective means there will be less water for irrigation and (she believes) greater economic losses to her family. Notice that each stakeholder is using a mental model of cause and effect to pre-emptively assess the potential effects of what they *believe* will be the management actions and consequences before potential actions are even identified or models built. We refer to this as "pre-emptively rejecting" objectives and it is a very real and potentially, major source of conflict in multi-stakeholder natural resource problems.

Dealing with the pre-emptive rejection of objectives is relatively easy to explain, but is much more difficult in practice. The idea here is to get the decision makers and stakeholders to disregard their notions of cause and effect (for now) and focus on identifying and defining what they want to achieve. Remember that the focus should be on fundamental objectives not means objectives. When stakeholders appear to disagree on objectives, we ask the question: *"if you can achieve objective x, is it alright if they can achieve objective y?"*. Returning to the water-management problem we could ask the fish biologist, if you have large fish populations, is it ok if the farmer has a large income for her family? For stakeholders representing natural resource agencies, we could ask the question this way: *does your agency have any policy against farmers with a large income?* Stakeholders will usually be indifferent to the fundamental objectives of other stakeholders provided that their fundamental objectives are achieved, unless they have personal (e.g., moral, ethical) or policy (for agency stakeholders) objections. If these objections exist, structured decision making may not be the appropriate approach.

Earlier, we alluded to the fact that dealing with the pre-emptive rejection of objectives is much more difficult in practice. Decision makers and stakeholders may not trust you or one another, particularly during the early stages of the process, so getting them to disregard their ideas of system dynamics can be difficult. We have the greatest success when we explain to decision makers and stakeholders that primarily our goal is (if possible) to find win-win situations through the process, so it is in their best interest to keep an open mind and think hard about their objectives. We also find that it helps to provide them with examples to illustrate important points about the process that can provide them with the big picture. For example returning to the water-management problem, we might explain that achieving stakeholder objectives may involve management options that do not involve water-allocation tradeoffs, but nonetheless can help them achieve their objectives. For instance, the farmer could be provided with water-saving irrigation equipment to produce the same amount of crops (or more!) using less water. Alternatively, the fishery biologists could be provided with the means to improve stream habitat and increase fish populations for a given amount of streamflow. The idea is to convince the stakeholders not to allow pre-emptive rejection to limit creative thought because it is through creative thought that we can find novel solutions to management problems. We caution, however, there is no single approach or formula that will work in all situations, so these procedures will not always work. Keep in mind that working with stakeholders requires the right skills, experience, and occasionally trial and error, so you need to keep an open mind and be prepared to ask for help. In Chapter 4, we go into greater detail on working with multiple decision makers and stakeholders.

Barriers to creative decision making

Creativity is the key to successful structured decision making whether the decision involves a single person (you) or multiple decision makers. Unfortunately, decision makers (you) can and often do restrict their thinking due to a variety of self-imposed barriers. Here, we focus on those barriers that potentially inhibit decision makers from identifying objectives. In group environments such as workshops, participants (i.e., decision makers and stakeholders) may be hesitant to articulate their values and objectives because they fear criticism from others (Adams, 1980) or alternatively, because of perceived power discrepancies. For example, employees may fear suggesting objectives that may not be acceptable to supervisors. One means of overcoming these barriers is to allow anonymity during portions of the objectives identification process. Anonymity can be accomplished through low- and high-tech options. Our favorite low-tech approach is to have participants write their values and objectives on unsigned slips of paper. The lists are then collected and compiled by a third party, usually a facilitator (Chapter 4) and the participants modify the combined list of objectives together. A similar high-tech version used by increasing numbers of professional facilitators is to provide participants with keyboards that are connected to a central database. The participants then enter objectives into the database anonymously. Regardless of the format, the idea here is to encourage decision makers and stakeholders to open up and think creatively without having to worry about criticism.

Another significant barrier to creative thought when identifying objectives is the tendency of decision makers to focus on those things that have always been perceived as important. This is sometimes referred to as "status quo bias" and is primarily due to the fact that humans tend to focus on the familiar rather than think creatively (e.g., Samuelson and Zeckhauser 1988). Similarly, decision makers often make the mistake of focusing excessively on the specifics of the problem at hand when identifying values and objectives. This often results in myopia and can prevent decision makers (you) from identifying creative solutions to problems. For example, consider a management problem where climate change is affecting the distribution of thermally sensitive gamefish species, such as trout. The State fish management biologists are unhappy because the temperatures have reduced the size and distribution of fishable trout populations, which has made the local anglers dissatisfied. If the biologists myopically focus on the problem, they might come up with a list of objectives that only address the trout fishery issue, such as maximize catchable trout abundance and minimize stream temperatures. However, if the fish managers thought more broadly about their personal or agency objectives, they might expand their list of objectives to include maximizing the satisfaction of all uses of the area (e.g., hiking, boating, fishing) or license (e.g., fishing, boating) sales. Notice that these objectives do not necessarily involve trout or even fishing and this expanded list could lead to creative solutions unrelated to the stated problem. To aid in the process, decision makers and stakeholders should create a list of objectives that includes personal as well as professional (organization, agency) objectives. While formulating the list, decision makers should think about: what would make them happy; what things they're trying to avoid; what impacts are their objectives likely to have on others they care about; what are their long- and short-term goals; and what are the constraints or potential roadblocks that are keeping them from their objectives. It also is useful if decision makers allow some time for their objectives and ideas to incubate. It is amazing how often a good night's sleep can help decision makers think more clearly about objectives. In group settings, it can be very effective to get participants to disperse and think individually about their objectives for some time period (e.g., overnight), being sure to create an objectives list. This allows individuals to think freely and clearly about their objectives without distractions and undue influence from others. The participants can then reconvene and compile an objectives list using their individual lists.

Types of fundamental objectives

Most, if not all, natural resource decisions involve multiple fundamental objectives that are often related or may have no apparent connection to the problem at hand. Dealing with multiple fundamental objectives can be difficult, so it is important to recognize the types of fundamental objectives to prevent problems in the future. For instance, it is fairly common in natural resource management to encounter situations where objectives are either in competition or in direct conflict with one another. **Competing objectives** are fundamental objectives that require satisfaction, but there are insufficient resources to fulfill all. Examples of competing objectives include: managing two endangered species but there are insufficient funds to fully implement the actions for both species and managing the allocation of water resources for both ecological and human needs. **Conflicting objectives** are superficially similar to the competing objectives, but differ in that fulfillment of one (or more) fundamental objective is in direct conflict with the other. Examples of conflicting objectives include: managing two endangered species with one species as the prey for the other and managing habitat on a refuge for two species that are each negatively affected by the preferred habitat of the other species. Although competing and conflicting objectives appear to be similar, they have very different consequences for decision making. Because competing objectives involve competition for limited resources rather than direct conflict, decision alternatives can be identified that may not involve tradeoffs. For example, if funds are the competing resource, a decision alternative that involves "obtaining more funds" or "using existing funds better" could result in the fulfillment of all competing objectives without having to balance the achievement of one objective versus the partial or non-achievement of another. In contrast, conflicting decisions involve direct conflict because achievement of one objective necessarily negatively affects another. Thus, conflicting objectives will always involve tradeoffs. Later in this chapter we discuss how to deal with conflicting objectives.

Decisions are often functionally linked (that is, achievement of one objective depends on achievement of another), or otherwise dependent in time or space and are defined as **linked decisions**. For example, a species restoration plan calls for reintroductions to a depopulated area. However, for the reintroduction to succeed habitat must first be restored to a suitable state (Figure 3.4a). A variation on this occurs in sustainable harvest management, in which harvest opportunity available in later years is dependent on the harvest in earlier years. Achievement

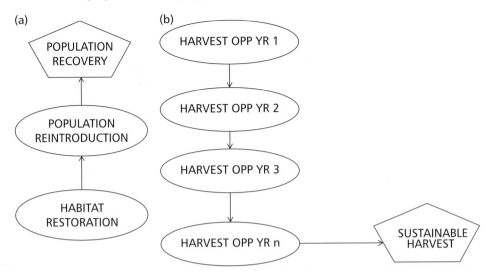

Figure 3.4 Linked objectives (a) Population recovery fundamental depends on 2 means objectives being fulfilled sequentially (first habitat restoration then reintroduction). (b) Sustainable harvest fundamental objective depends on linkage between harvest/ population outcomes in earlier years, which provide (or remove) harvest opportunity in subsequent years.

of the fundamental objective of maximum sustainable harvest is by definition dependent on the accumulation of harvest in each year, with suboptimal results in any year leading to an overall, suboptimal result (Figure 3.4b). Solving problems with many linked objectives and decisions poses some technical issues that we will deal with when we cover optimization in Chapter 8.

In addition to helping to clarify the distinction between means and fundamental objectives, creating and graphing an objectives network can reveal (hopefully, early in the decision process) problems with the way the decision problem has been defined. Two kinds of problems can be revealed at this stage, and both are critical to resolve before moving on. The first is the existence of **hidden objectives** – (fundamental) objectives that are clearly important to the decision maker/ decision stakeholder, but are not "on the table" (Figure 3.5a). Hidden objectives often can be revealed by asking the question "If all the fundamental objectives are satisfied, has the problem been solved?" If the answer is "no" then there are almost certainly hidden objectives that need to be revealed. Unfortunately, hidden objectives are often revealed toward the end of the process, so you should be ready for the unexpected.

A second issue that arises is that of **stranded objectives:** fundamental objectives that have no connection to means objectives or decisions, and thereby no means of fulfillment. Analysis at this stage sometimes will reveal that the scope of the problem has been too narrowly defined or the proper decision maker is not included in the process. For example, it may occur that the means for fulfilling a fundamental objective lie beyond the control or purview of the team in charge

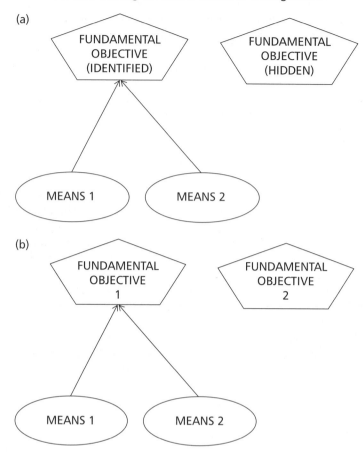

Figure 3.5   Hidden and stranded objectives. (a) Hidden objective – fundamental objective not included in initial analysis of problem and (b) stranded objective – fundamental objective has no linking means objective or decision.

of the decision. In such cases the decision makers and stakeholders must either narrow the scope of the problem so that all objectives lie within the responsibility of the group, or expand the existing group to include individuals who can contribute to the fulfillment of the additional objectives (Figure 3.5b).

Identifying decision alternatives

The next step in laying out the decision problem is to explicitly identify the decision alternatives. Here too, creative thought is *the* most valuable tool to use when identifying alternatives. Several books have been devoted to creative decision making, and among these we find Adams (1980) Keeney (1992), and Hammond et al. (1999) particularly useful. Decision makers should be encouraged to think creatively and all of the approaches we discussed earlier for identifying objectives are applicable here. For instance, group participation when identifying decision alternatives can lead to group-think and inhibit creativity. Thus, participants should be encouraged to think about decision alternatives individually; record their ideas on a list; allow their thoughts to incubate for a short period. The group can then meet and combine candidate alternatives into a single list. In addition, decision makers should be encouraged to challenge constraints and identify potential decision alternatives that initially appear to be unfeasible. The most straightforward way to do this is to assume that a constraint does not exist and then think about potential decision alternatives in its absence. For example, the Army Corps of Engineers (ACOE) wanted to develop hydropower on a large Southeastern reservoir that would increase water temperatures in the upper portion of a reservoir. Fishery managers attempted to stop the project because it would eliminate an important striped bass fishery at the only location in the reservoir where cold oxygenated water was in sufficient amounts to allow the fishery. The ACOE assumed that the constraint for the location of fishery did not exist, and identified an alternative that included oxygen injection of the cold hypolimnion in another portion of the reservoir that allowed the fishery to exist. Thus, the ACOE and fishery managers could have the hydropower operation and the striped bass fishery on the same reservoir.

Once a list of decision alternatives is created, decision makers need to think about how they relate to the fundamental objectives. We do this by connecting the decision alternatives to the objectives using the means objectives network as a base and examining the connections. At this point, we are not necessarily focused on making quantitative statements about decision influences (that will come later), but rather we are again trying to assure that objectives are tied to management actions, and vice versa. So, if we start at a candidate decision alternative and do not find a path to a fundamental objective, we have incorrectly framed the problem: either the decision is not needed, or we have excluded an objective. Likewise, if we start with a fundamental objective and cannot find a reverse path to a decision, the problem is mis-specified, and we must either include an enabling decision or eliminate the objective.

We illustrate this using the New Zealand biodiversity conservation example. Because we needed to identify the quantitative attributes of our fundamental objective (maximize native diversity), we use the fundamental objectives hierarchy as a base of the network. Let's assume that during the workshop, the participants also identified the candidate decision alternatives: lethal (e.g., trapping, poisoning) or and non-lethal methods (e.g., fencing, repellents, sterilants) for controlling non-native animal species; and chemical treatment and burning for controlling non-native plant species. Assuming that we have properly specified and separated means objectives, each means objective requires 1 or more decisions that are presumed capable of completely or partially achieving the objective. For example, 2 of the means objectives (reducing non-native predator number and reducing possum numbers) are addressed by lethal (e.g., trapping, poisoning, or shooting) or non-lethal control measures (e.g., fencing, repellents, sterilants), separately or in combination (Figure 3.6a). To keep things from getting overly complex at this point, we consider each type of decision as "on/

(a)

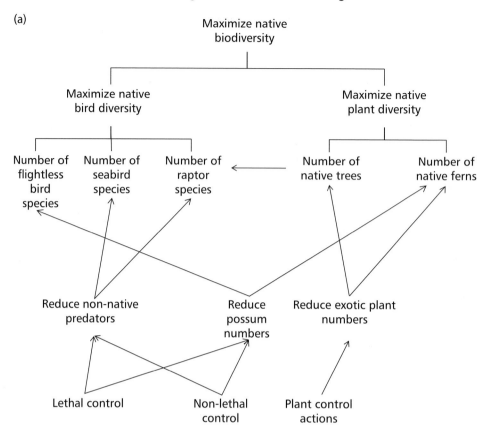

Figure 3.6 Example decision network for New Zealand conservation problem. (a) Specific management actions are linked to means objectives, which in turn help to fulfill fundamental objectives. Follow-up steps include specification of relationships between decisions and objectives and the calculation of utilities for each combination of decision and outcome. (b) Inclusion of cost as a fundamental objective in decision network.

off" or all or nothing: we either use lethal methods or we do not, non-lethal methods or not, plant control or not. Again, at this point we are not specifying a quantitative relationship between the decisions and the objectives, only that a relationship exists.

Before we finish linking alternatives with objectives, let us return to the objectives statement for the NZ conservation problem, because we have overlooked something. Notice that our fundamental objectives all relate to "maximizing native diversity." Taken literally, this means that we will select decisions that in fact produce maximum diversity. Leaving aside for the moment the question of whether (or how) diversity can (or even should) be "maximized", it follows that 1) we have unlimited resources to achieve this objective, and 2) we would be equally satisfied by objective values, irrespective of the costs or other tradeoffs required to achieve the stated result. Because neither of these

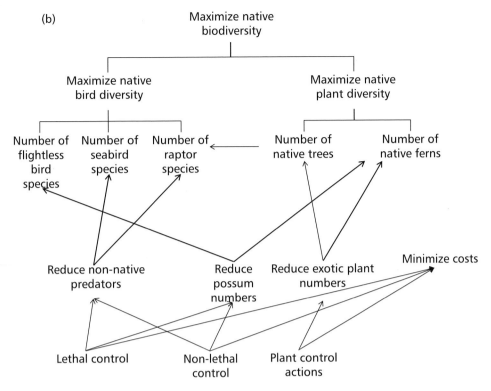

(b)

Figure 3.6 *(Continued)*

will ever be the case, we need to add a component to our fundamental objectives, which we will entitle "minimize cost" (Figure 3.6b). We make this point because in our experience conservationists are much more adept at envisaging grand conservation goals than they are at taking into account costs and other tradeoffs. A "bottom-line" perspective is needed if for no other reason than poor choices waste resources and will ultimately detract from future decision-making opportunities.

After connecting the candidate decision alternatives to the means objectives network, it should be apparent that the network resembles a rough graphical representation of how decision makers think the system works. This graphical representation is defined as a **decision network** and it will form the basis for modeling the effect of management decisions on alternatives. Most networks at this stage in the process are rough representations that will require modification and adding more components (usually, means objectives) to capture the dynamics of the system. We cover these steps in Chapter 6. Note too that additional alternatives can be identified during this process by examining the network and asking: *How?* In fact, we recommend that you continuously look for alternatives throughout the process, even during model building because you never know when a creative decision may emerge.

Quantifying objectives

When identifying objectives, we always need to be mindful that the fundamental objectives or their attributes need to be *quantifiable* states of nature or other elements that we can predict, observe, and measure. That is, to make intelligent decisions (e.g., evaluating tradeoffs among competing choices), we first need to determine the units that we will use to quantify the objectives. Eventually, we will create a model or set of models for predicting changes to our quantifiable objectives in response to alternative management actions (Chapter 6), but for now we are interested in quantifying our objectives.

Quantifying objectives is relatively straightforward when decisions involve a single fundamental objective. For example, consider the management of a threatened species with the fundamental objective as maximize species persistence. To translate this into a quantifiable objective, we need some more specificity as to what is meant by "persistence", which can be problematic. Since "persistence" largely depends on the risk tolerance of decision makers (Chapter 7), the definition will be largely subjective and vary from decision maker to decision maker. It is therefore necessary to recast the objective in terms of attributes that *can* be estimated, such as species occupancy. For example, a reasonable (and quantifiable) fundamental objective might be restated as

- "Maximize occupancy of suitable sites for the species."

The expected (estimated) number or proportion of occupied suitable sites can then be used to compare the relative merits of alternative management decisions. The number or proportion of occupied suitable sites also can be measured after a management action is implemented to determine success of the action. Single fundamental objectives, however, are very rare in natural resource decision making and most often managers need to satisfy multiple objectives. We will deal with the issue of multiple objectives next.

Dealing with multiple objectives

Fundamentally, there are three factors that come into play when we have multiple objectives:

- the different objectives will not necessarily be expressed (or even possible to express) in the same units (currency);
- some of the objectives may be in conflict with other objectives; and
- some objectives are more important and deserve greater consideration or weight when evaluating decisions.

The first issue means that we cannot simply add or otherwise summarize the objective components to find a meaningful composite objective. However, this creates a fundamental problem in that we cannot simultaneously maximize "apples" and "oranges": optimization requires the objective function to be a single response, regardless of the number of inputs (our decision controls). The second issue means that we nevertheless need metrics that can accommodate objectives with different units, precisely because there are tradeoffs: we usually cannot "have it all". The third issue means that not all objectives are created equal, so we often need a method for identifying and quantifying the relative importance of objectives. As we will see, these issues are closely related, as are several of the solutions.

There are numerous approaches to dealing with multi-objective decisions and no single approach will work for all decisions. Here, we describe approaches that we believe to be among the most useful for natural resource decision making, with the caveat that the choice of approach is ultimately up to the decision maker(s). We broadly categorize these approaches as integrated and constrained approaches, and they can be used in combination. The former involves the development of an **objective function** that integrates (or combines) all fundamental objective attributes into a single value, possibly after conversion of objectives using a **utility function**. The second approach treats some of the objectives (like cost) as limitations or **constraints** on the other objectives, and can be appropriate in cases where we have **competing objectives** that cannot all be fulfilled due to a cost or other limitation, or where we decide that fulfillment of one objective is desirable only when some other objective is maintained within an acceptable range. We describe each approach briefly but refer readers to other references (e.g., Clemen and Reilly 2001) for more in-depth coverage and examples, particularly for the development of **utility**, which we will return to in more detail below.

We can illustrate these 2 approaches with an example of a conservation problem where we have 2 species of birds with opposing habit requirements: the Wood Thrush (WOTH; *Hylocichla mustelina*), which is dependent on dense shrub understory and mid-story, and the Red-cockaded Woodpecker (RCWO; *Picoides borealis*), which is dependent on savannah-like conditions (grassy understory, little or no mid-story vegetation) (Powell et al. 2000, Moore and Conroy 2006). The hypothetical relationship between percent mid-story cover and suitability for each species is illustrated in Figure 3.7a. Assuming that a management control (such as prescribed fire) is capable of manipulating mid-story cover, obviously decisions that would be optimal for RCWO are detrimental to WOTH and vice versa. Furthermore, the 2 species suitability curves cross (i.e., are equal for the same value of mid-story) at very low suitability for either. We could decide to essentially split the difference between the 2 species' needs by summing (equivalently averaging) their requirements, in which case the objective would resemble Figure 3.7b. Although this approach would allow us to avoid a decision that would be disastrous for both (around 40% cover), we are left with 2 decision regions where the outcome is apparently the same: low cover values and high cover values. Weighting one species higher in the averaging eliminates this, but results in a situation where one resource value will universally dominate the other, RCWO in this case (Figure 3.7c). We could introduce a curvilinear form to the objective function, resulting in a clear choice of an optimum (Figure 3.7d); however, we would need to derive a function and its coefficients that represents our satisfaction in different combinations of the objective value. We will return

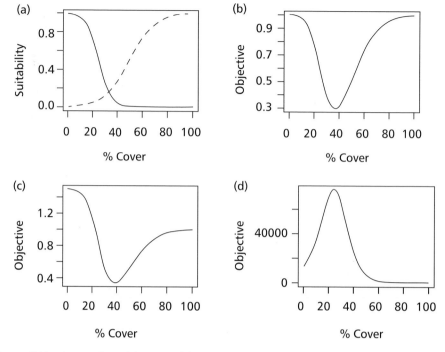

Figure 3.7 Competing objectives. (a) Response of 2 species, Red-cockaded Woodpecker (RCWO – solid line) and Wood Thrush (WOTH – dotted line) to increasing mid-story vegetation cover. (b) Common objective with equal species values; (c) Common objective with value 1.5 for RCW and 1.0 for WOTH; (d) quadratic function with interaction between RCW and WOTH.

to this issue in a bit, when we discuss in more detail **multi-attribute objectives** and **utility functions.**

The second approach, which treats one objective as a constraint, turns out to be a sensible one, because while one of the species (WOTH) is of conservation concern, the other (RCWO) is a federally endangered species; particularly given that these are federal (National Wildlife Refuge) lands, attainment of RCWO objectives would seem to take precedence. However, managers are also mandated to follow principles of "ecosystem management" and not to focus solely on management of one species to the detriment of others. Thus, in this instance, it might be more appropriate to view the objective as "maximize habitat suitability for RCWO, while keeping habitat suitability within acceptable bounds for WOTH." We can illustrate this approach in the case of the RCWO/WOTH tradeoff, first seeking to maximize RCWO suitability but maintaining WOTH above a threshold, and then reversing the procedure, where we are trying to maximize WOTH suitability, while maintaining RCWO suitability above a threshold. For the thresholds chosen, the optimal decision changes somewhat but not hugely (Figure 3.8). Of course the thresholds themselves have to be chosen and can have a major impact on the sorts of alternatives that are even

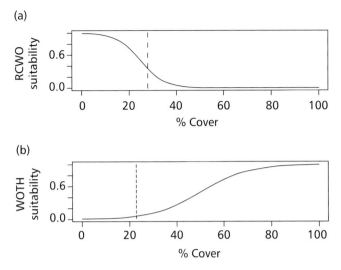

Figure 3.8 Maximizing an objective subject to a constraint. (a) RCWO suitability, subject to constraint that WOTH suitability ≥0.1 (all cover values to the right of dashed line); (b) WOTH suitability subject to RCWO suitability ≥0.6; (all cover values to the left of the dashed line).

considered for selection. In this example, however, given the strong legal mandates to conserve RCWO one would assume that managers would assign a higher priority to achieving that outcome than competing ones. We will revisit this issue further in Chapter 8, where we will see the importance of appropriate choice of decision constraints (and of not "over-constraining" decisions).

Finally, there is a third and perhaps more obvious way of dealing with problems such as the RCWO/WOTH tradeoff, and that is by recognizing up front that simultaneously achieving habitat suitability for both species on the same unit of land may be impossible, and that attempting to do so in a spatially uniform fashion would lead to a mediocre outcome everywhere. Thus, in a sufficiently large refuge, we might allocate some portions of the refuge to management targeted for RCWO, while leaving others suitable for WOTH; or if the refuge is too small, we might determine that certain refuges have priority for RCWO and other for WOTH. Finally, if such solutions do not work because we simply lack land or other resources over the refuge system, our decision problem may need to include other decision alternatives, such as the acquisition of additional lands.

Multi-attribute valuation

In multi-stakeholder decisions, quantifying multiple objectives that are measured on different units (e.g., user satisfaction and population size) can be particularly problematic, especially when the relative importance of each objective differs.

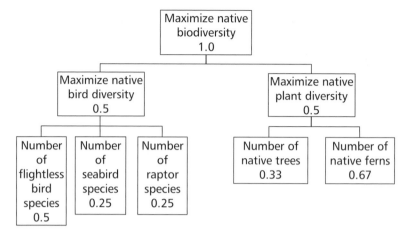

Figure 3.9 Fundamental objective hierarchy for New Zealand conservation decision with associated attribute weights.

One approach to dealing with this situation is known as multi-attribute valuation (also described as multi-criteria valuation). We begin with the fundamental objectives hierarchy and make a series of pairwise comparisons among attributes *within each hierarchical level*. The relative importance of each attribute is subjectively weighted by the decision makers and stakeholders with the most important attribute receiving a greater score. The sum of all weights within a level on a branch, however, should equal 1. To illustrate, consider the NZ biodiversity conservation example. Here, the biologists decided that the relative importance of maximizing native bird diversity and maximizing native plant diversity was equal and each received weights of 0.5 (Figure 3.9). However, the biologists valued the number of flightless bird species twice that of seabirds and raptors with the latter two being equal. Thus, the number of flightless bird species was weighted as 0.5 and the weights for seabirds and raptors were 0.25 each.

The combined utility of a decision is calculated using the expected outcome of the decision and weights for each attribute. The expected outcome could be on the real unit measure if all of the attributes were on equivalent units of measure (e.g., number of species) or alternatively, could be based on ranked outcomes. We illustrate equivalent units of measure using the NZ conservation decision. To calculate the combined utility of a decision, assume that the expected outcome of the decision was as follows:

> Native bird diversity
>> Number of flightless bird species, 5
>> Number of seabird species, 7
>> Number of raptor species, 3
> Native plant diversity
>> Number of tree species, 12
>> Number of fern species, 18

The utility (U) of each attribute are calculated using the weights:

$$U_{bird} = 5 \times 0.5 + 7 \times 0.25 + 3 \times 0.25 = 5$$
$$U_{plant} = 12 \times 0.33 + 18 \times 0.67 = 16.02$$

The combined utility then is the weighted sum of each diversity attribute as

$$U = 5 \times 0.5 + 16.02 \times 0.5 = 10.51$$

We illustrate the use of ranked outcome in Box 3.6.

Utility functions

A common approach to dealing with multi-objective decisions is to convert the various objective attributes into common units via a **utility function**. In some cases the choice of a utility function is obvious. For example, suppose our objective is to achieve a conservation threshold while minimizing the cost of a reserve design, and we are given the per-unit costs c_1, c_2 of 2 design inputs d_1, d_2. Fairly obviously, the utility function

$$U(d_1, d_2) = c_1 d_1 + c_2 d_2$$

will convert the controls d_1, d_2, which would typically be on different scales, to a common monetary scale. We would seek decisions that minimized this function while still meeting the conservation threshold (now viewed as a constraint). Again, however, we could reverse the problem, and view costs as a constraint on our ability to achieve a conservation outcome x while staying within a budgetary constraint, that is maximizing x (or some function $U(x)$) subject to the condition

$$c_1 d_1 + c_2 d_2 \leq C$$

where C is our total budget or other resource limit.

As suggested above, we also are typically interested in evaluating the utility of a decision in terms of some outputs $\underline{x}$ that occur (or are predicted) to occur following the decision. That is, we are interested in

$$U(\underline{x}) = U(f(\underline{d}))$$

where $f(\underline{d}) = \underline{x}$ maps the predicted relationship of the decisions to outcomes. Again, the $\underline{x}$ attributes may be assignable monetary or other common units. If we have a single outcome attribute $\underline{x}$ and a decision with costs C we might be able to convert the problem to single units using cost ratios, but these can be problematic and potentially obscure the objectives of the decision makers, so they should be used very cautiously.

In many cases, price, marginal value, and other economic metrics are suitable measures for determining decision utility in natural resources. However, simply converting **attributes** to dollars or other currency does not solve the problem, if for no other reason than different people (or even the same people at different times) have different **utility functions** for money (Lindley 1985). The authors suspect that Ted Turner would be less concerned about a $1000 loss or gain of income than any of the readers of this book.

Often there is no obvious means for putting attributes on a common scale, but there is an awareness of what would constitute a "best" or "worst" outcome. If so, we can use a method known as **proportional scoring** (Clemen and Reilly 2001). For each attribute, we can calculate an individual utility as

$$U(x_i) = \frac{[x_i - worst(x_i)]}{best(x_i) - worse(x_i)}$$

where x_i is the measurement on the original attribute scale and $(worst)x_i$ and $(best)x_i$ are the least and most desired values of the attribute over the anticipated range. Proportional scoring provides a linear relationship between the values of the attribute x and utility. However, in many cases the utility function will be nonlinear, indicating that the marginal value of the outcome varies depending on the outcomes' current value. The actual shape of the utility function will be determined by a number of factors, including the relative risk tolerance or aversion of the decision maker, (Lindley 1985), and must be determined for the decision maker at hand, via a process known as **values elicitation**, related to risk profiling (Chapter 7).

Regardless of the specific form of the utility function for each attribute, when the objective is multi-attribute we still have to decide how to combine these into an objective function. A simple (but not the only; see Clemen and Reilly 2001) approach is to calculate a weighted sum of utilities

$$U(\underline{x}) = \sum_{i=1}^{n} k_i U(x_i)$$

where k_i are weights that sum to one and represent the relative importance of each attribute, which may be taken as equal unless there are reasons for giving some objective attributes more weight than others (we will consider some, below). We apply the proportional scoring approach with additive and multiplicative utility functions in Box 3.1.

Determining weights by pricing out or indifference scores

In the previous example, we wound up giving the utility components equal weight because we had no basis for other weighting. A further analysis of the problem, however, will often reveal that there *is* an objective basis for giving unequal weights: one that is determined by the values of the decision maker. One device is to ask the question "how much would I be willing to pay (or otherwise tradeoff) in one utility value, in order to achieve a particular gain in

**Box 3.1   Utility function with proportional scoring:
NZ conservation example**

Here, we illustrate how objectives whose components are measured on different attribute scales can be converted to a common objective function by **proportional scoring**.

We consider the problem of species conservation in New Zealand, in which the 3 fundamental objectives, each measured on a different attribute scale, are summarized below.

Attribute	Attribute scale	Range (worst – best)
Diversity	Number of species	10 – 60
Environmental impact	% Negative impact	75% – 0
Cost	NZD × 1M	$10 000 – $0

For each attribute, we calculated an individual utility as

$$U(x_i) = \frac{[x_i - worst(x_i)]}{best(x_i) - worse(x_i)}$$

where x_1 is the measurement on the original attribute scale and $worst(x_i)$ and $best(x_i)$ are the least and most desired values of the attribute over the anticipated range. Note that here we are describing utility in terms of an outcome x and not directly in terms of a decision variable, d (e.g., fencing, lethal control); thus an additional model describing the relationship between d and x would be needed.

This results in utilities of 0, 0.5, and 1.0 for diversity values of 10, 35, and 60, respectively; for impacts of 75%, 37.5%, and 0, respectively; and for costs of $10 000, $5000, and $0 respectively with other values obtained proportionally. Note that these utilities are based on a linear relationship between objective outcome and utility under **proportional scoring**, and that more general, nonlinear relationships may result from the process of **values elicitation** (see Chapter 4).

We considered 2 overall utility functions, first a strictly additive function

$$U_1(\underline{x}) = k_1 U(x_1) + k_2 U(x_2) + k_3 U(x_3)$$

where k_i is the relative importance of each attribute, and $\sum_{i=1}^{3} k_i = 1$. In this example, we'll assume that conservation, cost, and environmental impact are equally important so $k_i = 1/3$.

(Continued)

This function results in the highest score (1) being assigned to the outcome $x_1 = 60$, $x_2 = 0\%$, $x_3 = \$0$ as expected (maximum diversity, minimal cost and impact), with the least value (0) to the outcome $x_1 = 10$, $x_2 = 75\%$, $x_3 = \$10\,000$ (minimum diversity, maximum cost and impact). However, other combinations produce apparently counterintuitive results, e.g., a value of 0.67 for the combination $x_1 = 10$, $x_2 = 0\%$, $x_3 = \$0$, which adequately reflects the economic and impact values, but effectively ignores the conservation value. While it is possible that such a result is a true reflection of decision-maker values, more likely further refinement is needed. For example, if failure to achieve (even partially) any one of the objective attributes leads is clearly undesirable, then a multiplicative function incorporating interactions among the 3 utility components is needed

$$U_2(\underline{x}) = U(x_1) \times U(x_2) \times U(x_3)$$

This now scales overall utility to a maximum of 1, still corresponding to $x_1 = 60$, $x_2 = 0\%$, $x_3 = \$0$; however, all outcomes that are "worst" on any attribute scale result in overall zero utility (e.g., $x_1 = 10$, $x_2 = 0\%$, $x_3 = \$0$ would have zero utility because of the unfavorable conservation outcome). Our approach to optimization then will be to seek the combination of decision controls (fencing, trapping, poisoning, and other available actions) that maximize the selected objective function $U_1(\underline{x})$ or $U_2(\underline{x})$.

another one?" Essentially what we are saying is that overall we are *indifferent* to the loss of the one value in exchange for the other one, and so they should be equally valued; we can then solve for the weights that result in this indifference condition being true. The indifference concept can be applied to more complex problems, in which indifference curves or contours are developed that describe combinations of attributes which are equally valued (e.g., Hammond et al. 1999:609). We apply the pricing out approach to revising the weights for the previous example in Box 3.2.

Marginal gain

In many decisions, managers would like to maximize management benefit on a per unit cost basis. For example, the fundamental objective of stream restoration activities in Northern California is to maximize the salmon production. Managers, however, have multiple decision alternatives with widely varying costs and sometimes less-costly management actions, such as gravel augmentation, can be applied over greater spatial extents compared to more intensive activities such as streambank reshaping. Thus, managers would like to identify the action or combination of actions that give them the biggest bang for the buck. To account for the relative cost, managers could estimate the marginal gain, which is the expected increase in the fundamental objective (outcome) minus the expected

---

Box 3.2   Utility function with weights based on pricing out: NZ conservation example

We return to the NZ conservation problem, focusing only on the tradeoff between biodiversity conservation and cost. From Box 3.1 we already have individual utility scores for each of these attributes; we put them together into an additive utility function with (for now) unspecified weights and proportional scoring on the component utilities.

$$U(\text{diversity, cost}) = k_1 U(\text{diversity}) + k_2 U(\text{cost})$$

Because the weights must sum to one this simplifies to

$$U(\text{diversity, cost}) = kU(\text{diversity}) + (1-k)U(\text{cost})$$

Suppose that we are willing (or indifferent) to spending an additional $5000 in order to achieve a gain of 10 in diversity. Take as a baseline diversity = 35 and cost = $5000; what we are saying is that this should have equal value to diversity = 45 with cost = $10000. In order to satisfy this condition we must have

$$kU(\text{diversity} = 35) + (1-k)U(\text{cost} = 5000)$$
$$= kU(\text{diversity} = 45) + (1-k)U(\text{cost} = 10000)$$
$$k(0.5) + (1-k)0.50 = k(0.7) + (1-k)0$$
$$k = 0.714$$

Therefore, a utility function

$$U(\text{diversity, cost}) = 0.714 \times U(\text{diversity}) + 0.286 \times U(\text{cost})$$

satisfies this condition and is appropriate under this indifference assumption.

We note that use of non-additive utilities and non-proportional scoring (i.e., nonlinear functions for the objective component utilities) will require more complex analysis (Clemen and Reilly 2001).

---

outcome if no management action is taken divided by the cost of the action. Marginal gain is estimated as:

$$U(i) = \frac{[x_i - x_{no\ action}]}{c_i}$$

where U is the marginal gain, x is the outcome, and c is the cost for action i and *no action*.

Swing weighting

Another approach to utility weighting, known as **swing weighting**, involves a "thought experiment", in which the decision maker envisions hypothetical alternatives in which attributes "swing" from worst to best case outcomes one at a time. The procedure starts by envisioning a "worst case" scenario in which the outcomes occur at the worst level for each objective attribute. This provides a benchmark against which all the other hypothetical alternatives can be compared. The procedure still involves the specification of subjective ranking and scoring values, but the juxtaposition of "worst" and "best" outcomes is useful in keeping these values tied to the decision maker's perspectives (rather than being set as equal or other arbitrary values). We apply the swing weight approach to the NZ conservation problem in Box 3.3.

Box 3.3   Utility function with weights based on swing weights: NZ conservation example

We illustrate swing weighting with the NZ conservation example. Again, the objectives are defined on 3 different attribute scales: diversity (10–60), impact (0–75%), and cost ($1000–$10 000). We start with the "worst possible case" in terms of outcomes on each of the 3 attribute scales: low diversity (10), high impact (75%), and high cost ($10 000). This receives a rank of 4 and a rating score of 0. We then "swing" each of the attributes from worst to best one at time, starting with diversity to create "imaginary" alternatives.

| | Consequence | | | | | |
	Diversity	Impact	Cost	Rank	Rate	Weight
Benchmark (worst)	10	75	10 000	4	0	
Diversity	60	75	10 000			
Impact	10	0	10 000			
Cost	10	75	1000			

Next, we rank the relative desirability of these outcomes from 1 (best) to 4 (worst, already set as the benchmark). This is a subjective exercise; below the weights indicate that we would be most satisfied with high diversity outcome even if accompanied by low costs and low impact, less so with a low impact outcome accompanied by low diversity and high costs; and even less so by a high impact outcome with low diversity and low cost.

	Consequence			Rank	Rate	Weight
	Diversity	Impact	Cost			
Benchmark (worst)	10	75	10 000	4	0	
Diversity	60	75	10 000	1		
Impact	10	0	10 000	2		
Cost	10	75	1000	3		

Next, we assign a rating score to each scenario, indicating its relative distance from best (100) or worst (0); again, we have already assigned the benchmark scenario a zero; similarly, the highest rank scenario is by definition 100. In the rating scores below the rates for the last 2 scenarios indicate that we are relatively more concerned about impact than we are about cost.

	Consequence			Rank	Rate	Weight
	Diversity	Impact	Cost			
Benchmark (worst)	10	75	10 000	4	0	
Diversity	60	75	10 000	1	100	
Impact	10	0	10 000	2	75	
Cost	10	75	1000	3	25	

Finally, the rates are normalized to produce weights that sum to 1 for each of the attributes. For example, the Diversity weight is formed by dividing the ratings score for diversity by the sum of scores across all 3 attributes.

$$100 / (100 + 75 + 25) = 0.5$$

	Consequence			Rank	Rate	Weight
	Diversity	Impact	Cost			
Benchmark (worst)	10	75	10 000	4	0	
Diversity	60	75	10 000	1	100	0.5
Impact	10	0	10 000	2	75	0.375
Cost	10	75	1000	3	25	0.125

The resulting weights are now used in the objective function with the individual attribute utilities defined as before, i.e.,

$$U(\text{diversity, impact, cost}) = 0.5 \times U(\text{diversity}) + 0.375$$
$$\times U(\text{impact}) + 0.125 \times U(\text{cost})$$

Other approaches

There are many other ways that objective functions can be constructed, and the best approach will depend on the application. Keep in mind too that these approaches are not mutually exclusive, and that they can and often should be used in combination. As we saw earlier, one objective attribute (x_1) may be of primary interest, while a second one (x_2) is viewed as a constraining factor or to varying degrees an ancillary factor. Our earlier approach was to maximize or minimize x_1 while keeping c within acceptable bounds by imposing constraints. This approach effectively prohibits decisions that result in x_2 being outside the acceptable bounds. An alternative, "softer" approach is to penalize the utility of x_1 by a function that becomes increasingly "harsh" as the x_2 approaches its boundaries. A simple device is to use a multiplier $U(x_2)$ that maps x_2 into the range 0–1. The utility for x_1 can either be expressed in the original units or rescaled (as in the previous example); for simplicity we will use the original units. Overall utility is then computed as

$$U(\underline{x}) = x_1 \times U(x_2)$$

which effectively gives full value to x_1 only when $U(x_2) = 1$ and otherwise discounts x_1 proportionally. In turn, $U(x_2)$ can be any function (e.g., a nonlinear one) that expresses the desirability of x_2 remaining in a given range. We use this approach for a problem involving harvest management of ducks in Box 3.4.

Box 3.4   Non-additive utility function: harvest of American black ducks incorporating a population value

Here, we consider an alternative approach to constructing a utility function, in which one of the objective attributes (x_1) is of primary interest and the other (x_2) is used to compute a utility function $U(x_2)$, with range 0–1. Utility for x_1 can be either be expressed in the original units or rescaled (as in the previous example); for simplicity we will use the original units.
   Overall utility is then computed as

$$U(\underline{x}) = x_1 \times U(x_2)$$

which effectively gives full value to x_1 only when $U(x_2) = 1$ and otherwise discounts x_1 proportionally.
   The example comes from the adaptive harvest management of American black ducks (Conroy et al. 2002, Conroy 2010). The primary objective was

to maximize the long-term harvest of black ducks (x_1 = cumulative harvest of ducks in thousands). However, it was also deemed desirable to maintain populations (x_2 = projected spring abundance of ducks in thousands) above specified population goals (NAWMP). The harvest and population goals were integrated into a common objective function

$$U(\underline{x}) = x_1 \times U(x_2)$$

where

$$U(x_2) = \frac{1 - \exp(-\gamma x_2 / x_2^*)}{1 - \exp(-\gamma)}$$

where γ is a coefficient determining the shape of the population utility and x_2^* is the population goal (currently set as 830 thousand). Utility curves for a range of values $-10 < \gamma < 10$ are plotted below, which managers may examine and select to represent their decision-making preferences.

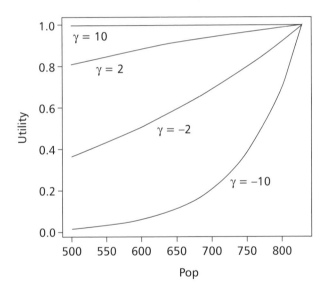

The resulting $U(\underline{x})$ fully values harvest when $x_2 \geq x_2^*$ with discounting of the harvest value for lower predicted population sizes depending on the value of γ. The discounted objective value is used in turn in program ASDP to calculate optimal state-specific harvest policies (season length and bag limits) taking into account the population goal.

Box 3.5   Cost ratio utility function with constraints: NZ conservation example

We return to the species conservation example (Box 3.3), in which the fundamental objective has 3 components, each measured on a different attribute scale, summarized below.

Attribute	Attribute scale	Range (worst – best)
Diversity	Number of species	10 – 60
Environmental impact	% Negative impact	75% – 0
Cost	NZD X 1M	$10 000 – $1000

We will now convert the Diversity and Cost objectives to a ratio

$$U(diversity, cost) = \frac{diversity}{cost}$$

which can be thought of as a measure of the marginal gain in diversity; we retained the Environmental Impact objective as a constraint. Thus, the form of our optimization will be to maximize $U(diversity, cost)$ subject to environmental impact $< x\%$, for example, 40%.

   The choice of which components to use as constraints and which to maximize or minimize should reflect the values and needs of the decision maker. In some cases the constraining variable will be obvious. For example, in the above problem if it had a fixed budget it would probably make more sense to view cost as a constraint and maximize a $U(diversity, impact)$ utility; the latter would, of course, require deciding on common units and weighting for the utility. The point is that we should first clearly define the relationships among **fundamental objectives**, and the metrics we will use to evaluate their achievement, before deciding on a particular mathematical form for the utilities and objective function.

Additional considerations

Based on our descriptions, it may appear that quantifying objectives is primarily a mathematical exercise. It is not. Quantification of objectives should reflect the values of the decision makers and as such, it can be a largely subjective process. Decision makers also often are concerned that the methods used (e.g., weights, ranks) to combine multiple objectives can affect what decision is deemed best. Later in the book we will cover methods for evaluating how alternative methods for incorporating multiple objectives affect decision making using sensitivity

Box 3.6   Multi-attribute valuation with ranked outcomes

We illustrate the use of ranks for multi-attribute valuation using data on the preference of largemouth bass fishermen in Georgia from Peterson and Evans (2003). The fundamental objectives hierarchy for fishery biologists with the state is shown below.

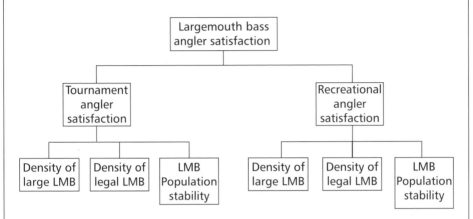

Here, the fundamental objective was largemouth bass (LMB) angler satisfaction, which consisted of 2 user groups, tournament and recreational anglers. The satisfaction of each user group was met by maximizing the number of large LMB, the density of legal LMB (i.e., larger than the minimum length limit), and the stability of the LMB population. To develop the weights, the state fishery managers conducted the following angler survey.

	Attribute	Tournament anglers	Recreational anglers
Number of respondents		54	43
Rank in order of importance to you the following qualities of a bass fishery (3 = most important, 1 = least).	Consistency in the fishery year after year (population stability)	2.55	2.59
	More bass above the length limit, but fewer very large bass (density of legal LMB)	1.81	1.66
	More large bass, but fewer bass overall (Density of large LMB)	1.64	1.74

Fifty-six percent of the respondents were tournament anglers and 44% were recreational anglers. These percentages were used to weight each user group. The group-specific weights were obtained by normalizing the average ranks

(Continued)

of each used group. For example, the weights for the tournament anglers were calculated as:

Attribute	Weight
Population stability	2.55/(2.55 + 1.81 + 1.64) = 0.425
Density of legal LMB	1.81/(2.55 + 1.81 + 1.64) = 0.302
Density of large LMB	1.64/(2.55 + 1.81 + 1.64) = 0.273

Completing the same calculations for recreational anglers we have the following attributes weights.

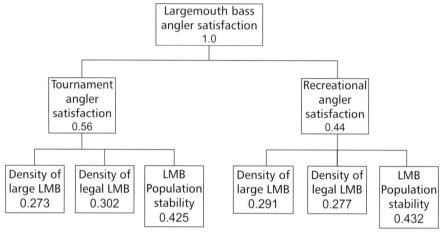

To calculate the combined utility, the estimated density of large and legal LMB and population stability were grouped into 3 categories low–high based on quantiles and outcomes were ranked from least desirable (1) to most desirable (3) as shown below.

Density of large LMB		Density of legal LMB		Population stability	
Outcome	Rank	Outcome	Rank	Outcome	Rank
Low	1	Low	1	Low	1
Medium	2	Medium	2	Medium	2
High	3	High	3	High	3

These ranks and the attribute weights were then used to calculate the utility. For example, the outcome low density of large LMB (rank = 1), medium density of legal LMB (rank = 2) and high population stability (rank = 3) was calculated as:

$$0.56 \times (0.273 \times 1 + 0.302 \times 2 + 0.425 \times 3)$$
$$+ 0.44 \times (0.291 \times 1 + 0.277 \times 2 + 0.432 \times 3) = 2.147$$

The maximum score is 3, so the relative satisfaction is 2.147/3 = 71.6% satisfaction.

analysis and indifference curves (Chapter 7). These should always be performed to ensure that the relative value of outcomes (e.g., utilities) is consistent with the decision-maker values. These analyses should alleviate decision-maker concerns when decisions involve multiple stakeholders. However, in certain rare circumstances, stakeholders may not agree on a method for combining multiple objectives into a single value and creating constraints. In these instances, stakeholders can be shown the expected outcomes of each fundamental objective predicted for each decision alternative and chose the alternative that is acceptable. It will become apparent in later in the book that this approach should be avoided, if possible, because it is wasteful and prevents analyses that will provide greater insight into the decision-making problem.

Decision, objectives, and predictive modeling

At this point in the decision process we have

- Identified what our objectives are and distinguished between fundamental and means objectives in an objective network.
- Identified candidate decision alternatives.
- Graphically represented the relations among candidate decision alternatives and means and fundamental objectives in a decision network.
- Conducted a preliminary evaluation and analysis to determine if we are missing critical objectives or decisions; if any alternatives or objectives are "stranded" (decisions not linked to objectives or vice versa);
- Quantified and weighted our fundamental objectives.
- If necessary, converted a multi-attribute objective into a single-objective function in common units, possibly with accompanying constraints.

We are nearly ready to embark on the next step of decision analysis: the development and evaluation of a formal decision model, the topic of Part II of this book. Before we do, in Chapter 4 we will take up the important issue of working with stakeholders in the development of the decision problem.

References

Adams, J.L., (1980) *Conceptual Blockbusting*, 2nd edn, Norton, New York, New York.

Clemen, R.T. and T. Reilly. (2001) *Making Hard Decisions*. South-Western, Mason, Ohio.

Conroy, M.J., M.W. Miller, and J.E. Hines. (2002) Review of population factors and synthetic population model for American black ducks. *Wildlife Monographs* 150.

Conroy, M.J. (2010) Technical support for adaptive harvest management for American black ducks. Final Report from University of Georgia to USGS. December, 2010. Athens, GA. http://coopunit.forestry.uga.edu/blackduck/final10.pdf.

Lindley, D.V. (1985) *Making Decisions*. John Wiley & Sons Ltd, Chichester.

Hammond, J.S., R.L. Keeney, and H. Raiffa. (1999) *Smart Choices*. Harvard Business School Press, Boston, Massachusetts.

Keeney, R.L. (1992) *Value-Focused Thinking: A Path to Creative Decision Making*. Harvard University Press, Cambridge, Massachusetts.

Moore, C.T. and M.J. Conroy. (2006) Optimal regeneration planning for old-growth forest: addressing scientific uncertainty in endangered species recovery through adaptive management. *Forest Science* **52**, 155–172.

Peterson, J.T. and J.W. Evans. (2003) Decision analysis for sport fisheries management. *Fisheries* **28(1)**, 10–20.

Powell, L.A., J.D. Lang, M.J. Conroy, and D.G. Krementz. (2000) Effects of forest management on density, survival, and population growth of wood thrushes. *Journal of Wildlife Management* **64**, 11–23.

Samuelson, W. and R. Zeckhauser (1988). Status Quo Bias in Decision Making. *Journal of Risk and Uncertainty* **1**, 7–59.

4

Working with Stakeholders in Natural Resource Management

Most natural resource issues involve a wide variety of decision makers and stakeholders and their input is critical for properly framing a problem and identifying objectives. Working with stakeholders, however, can be a difficult task – particularly when their objectives conflict. In this chapter, we examine the types of stakeholders and their roles in the natural resource decision-making process and present approaches for identifying important stakeholders. We also discuss the importance of developing a governance structure for stakeholder groups and highlight the advantages and disadvantages of common governance structures. Finally, we provide practical advice for organizing and facilitating stakeholder meetings, with emphasis on the role of the facilitator.

Stakeholders and natural resource decision making

Stakeholders are persons or organizations with a vested interest in the outcomes of management decisions. For natural resource management, these can generally be grouped into the following categories:

- Consumers: members of the general public that use natural resources either consumptively or non-consumptively. Examples of natural resource consumers include: anglers, hunters, bird watchers, hikers, and boaters, residents of rural landscapes, and citizens who are interested in natural resources (e.g., through nature programs on television) but do not directly participate in outdoor activities.
- Non-governmental organizations (NGOs): organizations that advocate for the conservation or wise use of natural resources that are not part of any

Decision Making in Natural Resource Management: A Structured, Adaptive Approach,
First Edition. Michael J. Conroy and James T. Peterson.
© 2013 John Wiley & Sons, Ltd. Published 2013 by John Wiley & Sons, Ltd.

governmental entity and are generally not-for-profit. Examples of environmental NGOs include: International Union for the Conservation of Nature and Natural Resources, The Nature Conservancy, and Ducks/Trout Unlimited.

- Natural resource management agencies: local, state/provincial, or federal agencies charged with managing the natural resources within their legally mandated jurisdiction. Examples of natural resource management agencies include: the US Fish and Wildlife Service, New Zealand Department of Conservation, Fisheries and Oceans Canada, state/ provincial fish and wildlife agencies, and city parks and recreation departments.
- Political: elected or appointed representatives of local, state/provincial, and federal governments that can include: members of congress or parliament and their staff, county commissioners, mayors, and city managers.
- Economic: individual business, business organizations, and landowners whose members potentially experience monetary loss or gain from a natural resource decision. Examples of economic stakeholders include: National Marine Manufacturers Association, Energy Networks Association, Canadian Lobster Council, chambers of commerce, and farmers.
- Scientists, biologists, agronomists, technical consultants and others who have a technical knowledge or interest in the resource issue but are not directly affected by the results of decision making.

We offer these groupings to use as a general framework when thinking about the stakeholders to include in the decision-making process. Participants in stakeholder meetings are likely to be a member of more than one stakeholder group and hence, are likely to have multiple **objectives** (Chapter 3) that correspond to their stakeholder group memberships. For example, a member of a fishing-tackle manufacturers association with economic objectives can also be an angler with an entirely different set of objectives, such as increasing access to the fishery. The benefit of including a stakeholder with multiple group membership is that they may facilitate communication and understanding among different stakeholders. The potential downside is that the participant may portray themselves as representing one stakeholder group when many of their objectives are unrelated to that group, which can lead to conflict. Thus, it is important to keep in mind that while stakeholders are represented by one or more people, a person does not necessarily equal a stakeholder.

Why involve stakeholders?

Natural resource decisions usually involve trust resources. That is, resources that are publicly owned, so the members of the general public have vested interest in the outcomes of decisions. Natural resources also have multiple uses that can lead to competition among user groups and of course, potential conflict among groups. By implementing a stakeholder-driven process, we can help those with vested interests understand how decisions are arrived at and participate in the decision-making process. Stakeholders can also be a vital part of the process of identifying **fundamental objectives** (Chapter 3), a crucial step in SDM. Although it is possible to anticipate and identify many objectives without including stake-holders, involving them will typically assure a more comprehensive analysis of objectives, and also will help us anticipate and, if possible, to resolve or at least minimize potential conflicts among competing user groups. Finally, and perhaps most importantly, we build public support and ownership of the decision when the public is explicitly involved in the decision-making process.

Stakeholder analysis

All stakeholders are not necessary equal. For example, all decision makers are stakeholders in that they have a vested interest in the outcome of a management decision. However, not all stakeholders are decision makers. Decision makers often have the legal authority or responsibility (mandate) to carry out a manage-ment action, providing the resources such as funds, personnel and equipment, or have roles as landowners. Thus, decision makers generally have greater responsibility and accountability than other stakeholders. We make the distinc-tion here because failure to identify and involve all decision makers at the begin-ning of decision-making process can (and likely will) cause substantial problems throughout the process and may force stakeholders to begin the process over. So the question is, how do we know if the right stakeholders are involved? We do that by conducting a stakeholder analysis.

Stakeholder analysis is used to identify and assess the importance of people, groups of people, or organizations/ institutions that may significantly influence or be influenced by the management decision. There are many types of stake-holder analyses that have been categorized as normative, descriptive, and instru-mental, and the use of each depends largely on the objectives of and the resources available to the analyst (Reed et al. 2009). In this chapter, we focus on norma-tive approaches because they are among the most widely used in natural resource management applications. Normative approaches emphasize stakeholder partici-pation and empowerment as a means to increase the legitimacy and acceptance of the process and ultimately, the decisions. We also focus on the use of stake-holder analyses matrices because they are among the most flexible and cost-effective methods that explicitly represents the power dynamics of stakeholders and provides a means to prioritize stakeholders (Freeman 2010).

Stakeholder analysis matrices are useful tools for identifying and evaluating the importance of stakeholders. Potential stakeholders are placed in a matrix based on two factors: the ability of the decision to affect the stakeholder and the ability of the stakeholder to affect the decision (Figure 4.1). Potential stake-holders that are strongly affected by a decision or have a strong effect on the decision are essential to involve in the final stakeholder group, whereas it is not important to involve those that have little or no influence on the decision or are not affected by the decision. To create a stakeholder analysis matrix, the first step is to develop a list of potential stakeholders. At this stage, you should include as many potential stakeholders as possible and not limit your list to those that you perceive to be important. A useful method for identifying potential stake-holders is to ask the following questions:

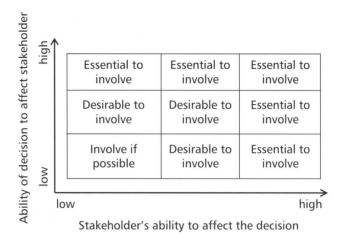

Figure 4.1   A graphical representation of stakeholder rankings based on their ability to affect the decision and the effect of the decision on the stakeholder.

- What are the different interest groups potentially affected by the decision?
- Which interest groups are usually involved in similar decisions and which ones are usually excluded?
- Who has the knowledge of how the system works (e.g., scientists, biologists)?
- What entities (e.g., management agencies) or people (e.g., farmers) have the legal authority and resources to implement management actions or make recommendations?

The last question is particularly important because the answer to that question is the identity of the decision makers, which we distinguish from other stakeholders later in the chapter. Another useful method for identifying potential stakeholders is to communicate directly with pre-identified stakeholders to identify other previously unknown stakeholders (Harrison and Qureshi 2000). Once the list is created, the candidate stakeholders are qualitatively ranked based on the two attributes:

1) The ability of the decision to affect the stakeholder.
2) The stakeholder's ability to affect the decision.

Potential stakeholder	Ability of the decision to affect the stakeholder	Stakeholder's ability to affect the decision
Stakeholder A	high	low
Stakeholder B	high	medium
Stakeholder C	none	high

It is usually best to keep the qualitative rankings simple with few categories, such as low, medium, and high. Remember that this is a screening process designed to provide you with a means to evaluate the relative importance of stakeholders and the power dynamics of the decision. Once the matrix is created, the rankings are examined to distinguish the stakeholders that are essential to involve in the decision-making process from those that are desirable. To illustrate, we provide an example of a stakeholder analysis as applied to water and endangered-species management in the southeastern US in Box 4.1.

Box 4.1   Stakeholder analysis example: water management and mussel conservation in the southeastern US

To illustrate the use of a stakeholder analysis matrix, consider the problem of water management and endangered species conservation in the Flint River Basin in the southeastern US. The upper portion of the basin is experiencing substantial urban development and associated increases in water-resource development, and the lower portion of the basin is primarily intensively irrigated and highly productive agricultural lands. The basin also contains five endangered freshwater mussel species. To conserve and recover the endangered species, the US Fish and Wildlife Service (USFWS) and Georgia Department of Natural Resources (GADNR) needed to develop a conservation plan for the basin that could affect multiple stakeholders throughout the basin. The first step in the decision-making process was to create a list of potential stakeholders – those entities or groups of people that were potentially affected by a management actions intended to recover the endangered mussel species. Nine potential stakeholders were initially identified by the decision makers. However, discussions with two potential stakeholders, Flint River Water Council and the Georgia Farm Bureau, revealed two additional stakeholders that were previously unknown: Spring Creek Watershed Partnership and South Georgia Businesses for Economic Development.

Potential stakeholder	Ability of decision to affect stakeholder	Stakeholder's ability to affect the decision
USFWS	high	high
GADNR	high	high
US Department of Agriculture	medium	high
Georgia Water Planning & Policy Center	high	high
Local chambers of commerce	medium	low
Flint River Water Council	high	medium
The Nature Conservancy	low	low
Georgia River Network	low	low
Georgia Farm Bureau	medium	medium
Spring Creek Watershed Partnership	high	medium
South Georgia Businesses for Economic Development	medium	low

(Continued)

Six potential stakeholders were ranked "high" in at least one of the two key attributes and were deemed essential to involve in the decision-making process. Two potential stakeholders were ranked "low" on both attributes and a third was ranked "low" and "medium", so their involvement was not necessary. This did not preclude the involvement of these groups; it merely indicated that these two entities were much lower priority than the essential stakeholders. The remaining potential stakeholder, Georgia Farm Bureau, received a medium ranking for both attributes, suggesting that their involvement also is desirable. Thus, seven stakeholders were included as stakeholders throughout the decision-making process.

Stakeholder governance

Up to this point, we discussed the importance of stakeholders and provided a means for identifying the key stakeholders. That is, those stakeholders who must be involved in the decision-making process. These stakeholders will likely include groups or individuals with conflicting objectives or points of view. Thus, an important consideration when working with multiple stakeholders is the governance system. Governance defines the way stakeholders interact in the process. It establishes ground rules by which everyone agrees to abide. For example, should decisions only be made with the agreement of all stakeholders (i.e., consensus) or will a simple majority vote suffice. When developing a governance system, the focus should be on process, such as how will decisions be made, who represents a stakeholder interest group, how the leaders and representatives are chosen, rather than on who is the leader.

Before we investigate stakeholder governance, it is important to identify and define the types of decision making. Here, we define four types of decision making: autocratic, consultative, democratic, and consensus and discuss the advantages and disadvantages of each.

Autocratic decision making is when a single decision maker (i.e., person or entity) maintains absolute control of the decision. This means that the decision maker is completely responsible for the outcome of the decision whether good or bad. This style of decision making is probably most familiar to you because you make personal decisions using autocratic decision making. For example, deciding what to take to school or work for lunch. You have absolute control over these decisions and suffer the consequences of poor decisions, such as hunger. The advantages of autocratic decision making are that decision making is typically fast and that there is a single person or entity responsible for the outcome. There are several disadvantages to autocratic decision making in natural resource management. These can include less-than-desired effort from the people that execute the decision (i.e., the managers), a lack of transparency from the decision-making process, and a high potential for conflict among user-groups. Governance also is unnecessary for autocratic decision making, because it is not a participatory process.

Consultative decision making involves a single decision maker, but decisions are made after input is received from others. Similar to autocratic decisions, one person or entity still makes the decision, but others are solicited for ideas and suggestions. For example, the administrator of a wildlife agency that is in charge of setting harvest regulations may consult with other biologists and the general public before setting hunting season length. Consultative decision making is the most common form of decision making in natural resource management. The advantages of consultative decision making are that the input helps the decision maker understand the issues at hand and stakeholder concerns and stakeholders (i.e., biologists, scientists, the public) generally appreciate having their opinions heard and acknowledged. The disadvantages of consultative decisions making include the potential for leaving stakeholders with the assumption or impression that they have a say in the final decision. This can, and often does, lead to conflict. Consultative decision making is a participatory process in so far as the stakeholders are provided with the opportunity for input. However, a formal governance structure is not necessary and may, in fact, be counterproductive as it may reinforce the misimpression that stakeholders have a final say in the decision.

Democratic decision making involves majority rule where the choice of a preferred decision is arrived at by a vote of stakeholders. For example, stakeholders may vote on which management objectives are more important than others. There are several advantages of democratic decision making. First, it is generally perceived as the fairest form of decision making as everyone has equal input. However, there also have to be winners and losers in the process, which can lead to conflict among stakeholders. Democratic decision making, however, is often necessary when there is an impasse or gridlock in the stakeholder group. Democratic decision making is, of course, a participatory process that requires a governance structure to establish rules for how collective decisions are made.

Consensus decision making occurs when decisions are reached by collective agreement among all stakeholders. For example, all stakeholders agree on the set of management objectives. In some ways, consensus decision making can be thought of as an extreme form of democratic decision making that requires 100% agreement. It is by far the preferred method for decision making as all stakeholders are in agreement and all have ownership and responsibility in the decision and outcome. Although it is generally considered the preferred method, consensus has several disadvantages. First, reaching consensus takes time and a great deal of discussion and facilitation. Consensus also requires the willingness of stakeholders to share their objectives and ideas/beliefs openly. Finally, consensus can easily lead to gridlock as decisions can be forestalled by a single stakeholder. Consensus decision making is a participatory process that is necessary when establishing a governance structure. Consensus also is generally necessary for identifying and structuring objectives, but might not be feasible when ranking the importance of objectives, particularly when objectives conflict.

Governance systems should be familiar to you. Most nations, corporations, and interest groups have well-defined governance systems. For example, legislatures have developed institutions and rules for conducting political discourse and negotiation. Consensus is an important factor when developing a stake-holder governance system. All parties should agree with the rules of engagement before the potential for conflict arises, so the first order of business when stake-holders meet is to identify and define a governance structure. The stakeholders should address such questions as:

- Should the stakeholders be led by a board of directors or a chairperson?
- Who makes up the board?
- How are the board members and chairperson selected?
- Are there other important roles, such as a secretary?

The roles and responsibilities of officers, board members, and common stake-holders then need to be defined along with guidelines for conducting meetings. Considerations for conducting meetings include determining:

- Determining how meetings are conducted.
- Establishing procedures for closing old business and introducing new topics.
- Establishing time limits or ground rules for discussions.

The latter of these is particularly important to minimize filibusters and allow time for input from multiple stakeholders. A process should be established for resolving conflicts and making decisions. For example:

- Should disputes be resolved by the entire stakeholder group or an arbiter?
- Should all decisions be based on consensus, majority vote, or super majority vote?

Remember that each of these has tradeoffs as we discussed earlier. Finally, procedures need to be established for changing the governance structure if the existing structure is found to be unsuitable. For example, stakeholders may be dissatisfied with using consensus to make decisions and may want to change the threshold to a super majority vote. Developing these rules at the start is important because unforeseen problems often arise during the first steps of a multi-stakeholder driven decision-making process. Stakeholders need the ability to adjust the governing structure to meet those needs.

Developing governance can be challenging, particularly when the decision involves multiple user groups. Quite often multiple people that represent a single interest group will attend stakeholder meetings. For example, a meeting about a decision to restrict motorized vehicle access from public lands will likely include many members of the general public that enjoy recreating with all-terrain vehicles (ATV). All of these attendees are members of the same interest group (ATV enthusiasts). Does each person get to vote on a decision? If so, then all an interest group needs to do is recruit enough members to attend meetings and they potentially can impose their will on competing user groups. Thus, it can be necessary to identify stakeholder groups based around interest in a particular area and establish guidelines for selecting the representative of each stakeholder group. This is not an easy process, particularly if the management decision is complex and involves many stakeholders. However, it is not impossible and it requires patience and hard work. To illustrate, we describes the governance structure of a diverse group of stakeholders interested in the outcomes of alternative management actions for a hydropower dam in the southeast US in Box 4.2. This was a group of stakeholders that did not agree on their objectives, nor did they trust one another prior to initiating the decision-making process. Nonetheless, they were able to communicate more effective once they agreed upon the governance structure and more importantly, they made decisions.

Box 4.2   R.L. Harris Dam stakeholder governance

The R.L. Harris Dam is located on the Tallapoosa River in east-central Alabama. The dam was constructed in 1983 primarily as a hydropower facility, with alternative potential benefits that included flood control, recreational opportunities, and economic growth associated with the reservoir that was created by the dam. Natural resource management issues in the study reach below Harris Dam were primarily concerned with the effects of regulated flows on the persistence of native aquatic biota and the health of the ecosystem. However, there were several other concerns of local people and organizations that included: excessive downstream bank erosion, downstream flows for boating, upstream reservoir levels, and the local economy.

Since the construction of Harris Dam, there was an active and ongoing debate about how the dam should be managed. This cumulated in the organization of a Harris Dam stakeholder workshop in 2003 (Kennedy et al. 2007). The purpose of the workshop was to discuss Harris Dam management issues and to identify and organize stakeholders. During the workshop, 11 stakeholders were identified and are listed below.

Alabama Department of Conservation and Natural Resources	Lake Wedowee Properties Owners' Association
Alabama Rivers Alliance	Middle Tallapoosa River Conservation Association
Alabama Power Company	Randolph County Commission
US Fish and Wildlife Service	Randolph County Industrial Relations
Conservation Unlimited	Upper Tallapoosa Watershed Committee
Emerald Triangle Commission	

These stakeholders had different perspectives and objectives and had clashed on previous occasions. Thus, the first order of business was to establish a governance structure, which consisted of a board of directors with a single member from each stakeholder group and a charter that explicitly laid out the

(Continued)

ground rules for the interaction among stakeholders and the decision-making process. The charter is reproduced below.

Stakeholder Charter

CHAPTER I: PURPOSES AND PRINCIPLES

<u>Article 1</u>: The Purposes of the R.L. Harris Stakeholders Board are:
1. To seek a balance between river restoration, hydropower generation, and reservoir needs through the implementation of an Adaptive Management Model for information-based consensus decision making;
2. To improve and refine this model through long-term, continual examination of decision model results and impacts.

<u>Article 2</u>: The Board Members, in pursuit of the Purposes of the Board, shall act in accordance to the following Principles.
1. Membership on the Board shall not preclude the ability of Members to act independently and to exercise their rights (individually or on behalf of their represented group).
2. Established and newly created laws, regulations, and other legal agreements will be respected and incorporated into discussions and decisions.
3. The decision process will be long-term and continuing; Members of the Board will make commitments accordingly.
4. Board Members will seek to communicate openly and honestly about the needs of their interest groups. If needs are not addressed, they will not be served.
5. Members of the Board will strive for candid discussion of difficult issues in face-to-face situations. Confrontational public approaches will be recognized as generally unproductive to the process.
6. Board Members will make every effort to be flexible and open to new ideas and to the input of fellow Members. No extreme positions that would result in dramatic win/lose proposals for Board Members will be introduced into Board discussions.

CHAPTER II: MEMBERSHIP

<u>Article 3</u>: The Members of the R.L. Harris Stakeholder Board shall be appointed, elected, or clearly identified spokespersons for their respective interest groups. There shall be one spokesperson per interest group.

<u>Article 4</u>
1. Membership in the R.L. Harris Stakeholder Board will be open to all groups who have a direct interest in the decision process, i.e. specific recreational, economic, or ecological interests in the water resources impacted by the R.L. Harris Dam.
2. The process will not regress due to the entry of new Members. New stakeholders will familiarize themselves with the process-to-date and contribute to the discussion from their point of entry.

Article 5: A Member of the R.L. Harris Stakeholder Board who has missed two (2) consecutive meetings, and who has not provided an adequate Alternate, will be asked to resign.

CHAPTER III: RULES OF ENGAGEMENT

Article 6
1. Members of the Board will make at least a five- to seven-year commitment to stay engaged in the decision process.
2. Board Members will commit some level of time, talent, or treasure (resources) to the effort.

Article 7
1. A Technical Advisory Group (or Groups) (TAGs) will be established. The TAGs will consist of model builders, biologists, economists, and other technical experts as seen fit by the Board.
2. The TAGS will not act as decision-making bodies, but will solely serve an advisory role.

Article 8
1. A facilitator will be employed to guide the early stages of model development.
2. A project manager will be sought to coordinate Board activities.
3. A Board Chairperson may be elected in the future if the Board deems it necessary.

Article 9
1. Regular agendas and times for Board meetings will be planned and posted well in advance to enable maximum participation.
2. Before being incorporated into decision-making, scientific findings will be distributed well in advance of Board meetings to enable adequate technical preparation by Board Members.

CHAPTER IV: DECISION-MAKING

Article 10
1. The Board will seek consensus in all decisions, but when a vote is required, a two-thirds (2/3) majority will constitute a decision on the model and basic objectives.
2. A quorum will consist of more than half (1/2) the Membership of the Board. With quorum, meetings may take place, but decisions will always require two-thirds (2/3) majority of the Membership.

Article 11
1. Members of the Board may bring Alternates or Technical Advisors as non-voting participants.
2. Alternates may vote if the Board Member is not present, provided the Alternate has attended Board sessions regularly and/or is well informed on Board issues.

(Continued)

Article 12: Post-decision minority positions will be captured for later review.

Article 13
1. Proxy voting will be acceptable only on issues pre-determined by the Board. An alternate vote will not be considered a proxy vote.
2. Teleconferencing will be acceptable only under special circumstances, as determined by the Board.

Article 14: Public input will be part of ongoing meetings and operations, but any such input will be strictly non-voting.

Working with stakeholders

As we alluded in the previous sections, most stakeholder interaction and communication occurs at stakeholder meetings. During these meetings, stakeholders will – among other things – establish a governance structure and identify their objectives. To help these meetings run smoothly requires a **facilitator** – particularly when the decisions involve competing stakeholder groups and when the potential for conflict is high. Ideally, a facilitator is a neutral party to the decision. That is, the ideal facilitator is someone who does not stand to gain or lose anything from the outcome of a particular decision. For example, financial loss or gain (other than a salary or compensation) or loss or gain in professional standing. A stakeholder or decision maker should never be a facilitator. The facilitator is someone who sees to it that the stakeholder group is able to work effectively and collaboratively. To accomplish this, the facilitator must promote group participation, trust, mutual understanding, and shared responsibility among the stakeholders. The facilitator is not the leader of the stakeholder group but is the 'first among equals' that acts as a guide in the decision-making process.

The facilitator should attempt to ensure that everyone is included or at least feels included in discussions and has an opportunity to participate in stakeholder meetings. The facilitator works closely with stakeholders to set up meetings, with duties that often include: arranging meeting times and places, notifying participants, developing an agenda (with stakeholder input) and preparing and distributing materials for the meeting, such as background readings or reports. The facilitator also has important roles in running the meeting. Chief among these is attempting to maintain a civil environment and – equally important – keeping to the agenda. Nothing is as frustrating to stakeholders as running over in time and drifting off to topics that are not on the agenda. Thus, a good facilitator is essential to the decision-making process. In what follows, we describe the skills of a good facilitator and explain why these skills are important.

Characteristics of good facilitators

A good facilitator should be able to work with all the stakeholders, whose backgrounds can range from hyper-technical administrators to plain-speaking members of the general public and vice versa, hyper-technical members of the general public to plain-speaking administrators. The facilitator must be able to bridge the communication gap between disparate groups of people with different life experiences and skills. To bridge this gap, a good facilitator should be able to listen to a stakeholder and summarize their ideas so that they are understandable to every member of the stakeholder group. This often is no easy task and it requires that the facilitator has broad, but not necessarily in-depth knowledge of a variety of disciplines and viewpoints. For example in many situations, a facilitator may need to understand a stakeholder explaining a technical aspect of a decision, such as how a dam operates or how legal mandates relate to the decision, and translate the technical language into something that can be understood by every member of the group. The facilitator also should have strong organizational skills because he or she will be responsible for organizing and conducting stakeholder meetings. Because each stakeholder group and decision situation is unique, a good facilitator also needs to be flexible, with the ability to adapt plans to fit changing conditions and potential opportunities. For example, progress during initial stakeholder meetings can be very slow as stakeholders get to know one another and their viewpoints. It is during these initial meetings that opportunities for potential breakthroughs, such as agreement on principles or objectives, can occur. A good facilitator must be able to recognize and take advantage of these opportunities, even if it means diverging from the agenda. A good facilitator also should be sensitive to the mood of stakeholders. When creating and maintaining an atmosphere of trust and respect, the facilitator needs to be aware of how stakeholders are responding to the topics and the opinions and response of other stakeholders. Most people will not express their frustration or anger but will withdraw from the discussion. The facilitator needs to be sensitive to the mood of stakeholders. And finally, a good facilitator should have a good sense of humor and even be able to laugh at themselves. Meetings with stakeholders can, at times, be tense or tedious and humor can be useful for breaking the tension and tedium.

A good facilitator should not be afraid to take on bullies. Many stakeholder groups will contain one or more members who believe that their ideas, opinions, and objectives are *the* most important and will try to pressure other stakeholders into adopting their perspective. In these instances, a good facilitator should be able to delineate the bounds for civil discourse and have consequences for breaking those rules. In other words, the facilitator should make sure: 1) that the stakeholders know and understand the rules and 2) that they (the facilitator) follow through and apply these rules consistently and fairly. For example, we have observed several good facilitators that use rules, such as 3 strikes and you're out, as a means to constrain aggressive personalities. An important point that needs to be emphasized is that a good facilitator applies the rules consistently and fairly. If the stakeholders perceive that the facilitator is fair, then they will be more willing to comply with the ground rules. This also can create an atmosphere where peer pressure further constrains antagonistic stakeholders. Of course, no one hopes to facilitate a stakeholder group with aggressive personalities. Nonetheless, good facilitators should be prepared and have a plan to deal with a variety of stakeholders personalities. On the other end of the spectrum, a good facilitator should be able draw out quiet stakeholders and help them communicate their ideas and objectives. This is perhaps more difficult than dealing with aggressive personalities and requires the facilitator to read each stakeholder and adjust the pace of the discussion when they perceive interest by shy stakeholders and perhaps, directing questions to those stakeholders. Delineating the bounds for civil discourse and following through also can embolden shy or quiet stakeholders and get them to participate in discussions. Finally, a facilitator with knowledge of natural history, biology and natural resource management is a bonus. Although not necessary, facilitators with these skills and background can be invaluable because they can often help stakeholder groups avoid pursuing issues that are not biologically realistic or feasible and suggest alternatives that have been used successfully to address similar decision problems.

We assume that you are reading this book because you would like to develop the skills to be a structured decision-making practitioner. As a practitioner, you should be able to fulfill the role as a facilitator. However, there are also decision-making problems for which, in our opinion, it may be a better idea to obtain the services of a professional facilitator. We will start by assuming that you, or someone in your team, serves this role, and end with some suggestions if you decide that professional facilitation is more appropriate.

First, there are a few questions that need to be answered to determine if you are the right person for the role of facilitator for the particular decision at hand. The chief among these is "are you a neutral party to the decision?" Or more importantly, "do or will the stakeholders view you as a neutral third party?" The answer to the latter question is very important because you may consider yourself a neutral third party, but your effectiveness will be severely hindered if one or more of the stakeholders perceive a bias. The next important question to answer is: "do you have the appropriate skills and experience to be a facilitator?" Here, you need to be honest with yourself. Go through the list of skills described in the previous two paragraphs and see if you have what it takes to be effective. In most instances, these skills need to be developed and honed through experience. Learning while doing is an option, but it can be a painful learning experience and can stall or derail the decision-making process. Facilitating is not an easy task and it can be frustrating at times. One option for gaining experience that we find effective is to identify a potential mentor and offer to work with them on real-world decisions.

Another important consideration in determining if you should be the facilitator is whether you have another role in the decision-making process. Frequently, a facilitator can assume an additional neutral role (i.e., a dual role), such as an analyst, that may affect their effectiveness as a facilitator. For example, a facilitator may need to chastise a particularly forceful stakeholder. This, in turn, could affect the relationship between the facilitator and stakeholder that could compromise the willingness of the stakeholder to cooperate with the facilitator turned analyst or willingness of the stakeholder to believe the results of an analysis conducted by the analyst. These are all important things to consider when evaluating your fitness as a potential facilitator.

If, after addressing the above questions, you come away with doubts about your appropriateness as a facilitator, then it is probably time to consider the services of a professional. If you do decide to retain a professional, you should keep several things in mind. First, "facilitation" is not the same as "mediation". That is, we are trying to develop a decision-making process that incorporates stakeholder values that may be in conflict, but we are not necessarily trying to resolve those conflicts. Embarking on SDM as if it is a negotiating process will inevitably lead to decisions that "split the difference" and forces compromises that in fact may not even be necessary, once all the facts are assembled. Secondly, the facilitator should be familiar with at least the basic principles of SDM, and be "on board" that his or her role is to develop a *process* and not mediate a solution. Finally, it is always beneficial if the facilitator has a basic knowledge of resource management, as long as they are able to remain objective and dispassionate. We suggest that anyone considering retaining a professional facilitator for SDM carefully review candidate's credentials with these criteria in mind.

Getting at stakeholder values

Early in this chapter we alluded to the role stakeholders can play in helping to identify **fundamental objectives**. First, as we already suggested, stakeholders will be more positively engaged and cooperative if they have the sense that their objectives count in the process. Thus, from a purely tactical point of view, serious inclusion of stakeholders in developing objectives seems like a good idea – as long as they don't get the notion that they are being "taken for a ride." More positively, we strongly believe that it is *not* a decision-maker's role to dictate objectives to stakeholders, but rather to listen closely, and to help them organize their values into clear fundamental and **means objectives**, ideally leading to a clear separation of these via an objectives network and metrics for the fundamental objective **attributes** (Chapter 3). This can be an interesting and even fun (if sometimes frustrating) part of the stakeholder process, but we have found it to be extremely helpful. Many of the tools for ranking objectives and comparing the consequences of decision alternatives that we discussed in Chapter 3 can be useful in this process, and a part of a general process known as values elicitation (Keeney 1992, Wright and Goodwin 1999). However, we agree with Wright and Goodwin (1999) that these tools will be of limited assistance to stakeholders (and decision makers), particularly for the sorts of complex, multi-attribute problems that we typically deal with, without an ability to visualize the consequences of different choices. This visualization, in turn, will depend on tying the decisions to the objectives by means of a **model**. We will explore in detail the construction and use of models for decision making in Chapter 6, where we also discuss **expert elicitation**; and in Chapter 7 we will cover the additional and critical topic of taking **uncertainty** into account in decision making. Readers will want to have a thorough understanding of these chapters, and of Chapter 3, before diving into their first stakeholder meeting.

Stakeholder meetings

As we discussed earlier, the first step to planning a stakeholder meeting is identifying and recruiting the key stakeholders. However, it is equally important to identify the representatives of the stakeholders to invite. For example, one typical stakeholder for natural resource management decisions is the state/ provincial fish and wildlife management agency. Whom do you invite: the chief administrator of the agency or the biologist most familiar with the system of interest? Each person will provide different perspectives on the decision or problem. In general, but not always, it is best to involve stakeholder representatives that are most familiar with the actual management (i.e., background, history, current activities) and other decision making involving the system of interest. For example, biologists and managers in agency field offices are generally most familiar with the day-to-day management and issues. Upper-level administrators are not normally the best representatives to include in technical workshops, but they can be invaluable for identifying the best person to send from their agency. In many situations, technical advisors also will be needed at the meeting to provide information on the dynamics of the system being managed. For most natural resource management problems, they can include but are not limited to: biologists, ecologists, foresters, physical scientists and engineers, statisticians and modelers, and economists. For example, hydrologists or geomorphologists may be necessary to explain to the stakeholders how riparian areas function to stabilize stream banks. These key technical advisors are often useful when attempting to explain complex physical or biological process and resolve or prevent misunderstandings. They can also prevent delays in the decision-making process that are caused by a lack of information about a technical issue. Finally, meeting organizers should try to include a recorder at the stakeholder workshop. The job of a recorder is to take notes from the stakeholder meeting and assist the facilitator with routine tasks, such as passing out meeting materials, assistance with audio visual equipment, and helping keep track of time. The job of recorder is ideal for a facilitator in training as it provides them with valuable real-world experience.

Before we go any further, let's discuss the role and characteristics of good technical advisors. The technical advisors are only effective if they are viewed by stakeholders as neutral third parties that provide advice or clarification on technical issues. Technical advisors often take an active role in conducting analyses and developing models that will be used in the decision-making process. Thus, their neutrality is crucial to developing trust with the stakeholders and believability of the products that they create. For example, let's assume that a technical advisor created a model that predicted that the management action favored by a particular stakeholder had a negative effect on a sensitive or economically important species. To accept this less than desirable outcome, the stakeholder must have sufficient faith in the neutrality and expertise of the advisor. Technical advisors may of course also be stakeholders, to the extent that they may be directly or indirectly affected by decision outcomes, or even to the extent that they may gain financially depending on the need for future advisory input. However, their roles of **expert** or **stakeholder** must be clearly delineated, to avoid confusion between **science** and **values**. If necessary, formal guidelines should be communicated to make clear these distinctions, which would for example permit a technical advisor to recuse him or herself on a particular matter in which they have a personal stake. Otherwise, the modeling results could potentially lead to conflict. Note that later in this book we will learn how to resolve potential conflicts stemming from scientific disagreement though multi-model inference and adaptive management (Part II). Suffice it to say though, that in all cases the stakeholders must believe that the technical advisors are fair and are playing by the rules with respect to their disciplines. Facilitators and stakeholders should choose the technical advisors wisely. A general rule of thumb is to keep the number of advisers small and limited to those experts in subjects that are relevant to the problem or decision at hand. In addition, facilitators should try to limit substantial duplication of technical expertise. Remember that the most effective stakeholder meetings have relatively few participants. Finally, technical advisors should possess good people skills and should be able to interact well with fellow professionals and non-professionals alike. In other words, facilitators should avoid inviting technical advisors with strong personalities. Otherwise, they will waste valuable meeting time reminding the advisors of their proper role in the meeting. In fact, choosing the wrong technical advisors can set back the entire process, so choose wisely.

When planning a stakeholder meeting, it is very important to consider the objective or purpose of the meeting, which also can determine the type of meeting that is feasible. Although there are multiple reasons for having a stakeholder meeting, they generally fall into two broad categories: informational/organizational and technical. Informational meetings are generally used to introduce larger stakeholder groups to the management problem of interest and provide them with an overview of the issues and the approach that will be taken to identify the best management option. Informational meetings also are most useful when disseminating of the results of decision-making process. Organizational meetings are used to organize stakeholders and develop a governance structure. Informational/organizational meetings are the only type of meetings that can accommodate a relatively large number of stakeholders and hence, are generally the only type that can be feasibly open to the general public. Thus, these meetings can be invaluable in that they foster buy-in by all the stakeholders and ensure transparency.

In contrast, technical meetings are used to conduct most of the decision-making process that is detailed in this book, from structuring and quantifying objectives (Chapter 3) to evaluating and identifying the best decision alternative (Chapters 6–8). To be effective, technical meetings should have relatively few participants, definitely fewer than 20 individuals and we've found that 10 or fewer is optimal. Access at technical meetings should be limited to the representatives of the key stakeholder groups (e.g., the stakeholder governing board), the technical advisors, and of course the facilitator. We define this as the stakeholder working group. One important element of technical meetings is maintaining consistency in the stakeholder working group. Each member should attend meetings and be part of the process from start to finish to: 1) maintain consistency and 2) foster an atmosphere of trust, understanding, and teamwork. New members to the stakeholder working group also can disrupt the process by making the stakeholder group cover issues that have already been resolved and by unbalancing the group dynamic. Therefore, we strongly suggest that any substitution of stakeholder representatives or technical experts in the working group be made as early as possible in the process. Facilitators also should make it clear to members of the stakeholder working group that they should be committed to completing the process.

Another important feature of a successful meeting is the location and duration of the meeting. We focus here on technical meetings, that is– the meetings where all of the hard work is accomplished. The location should be comfortable, convenient, and free from distractions. Quite often, the most productive meetings take place away from typical daily distractions, such as email, phone calls, etc. Similarly, scheduling a few, multiple-day meetings tend to be more productive than conducting several, single-day meetings. This is generally because it takes a day or so for participants to become reacquainted with the problem *and one another*. However, we caution that holding very long meetings can be counterproductive. We find that participants tend to burn out after day 3 or 4, but that too depends on the particular problem and the stakeholder working group. In addition, organizing a social event during the meeting, such as a dinner, can help facilitate a team environment, thereby increasing productivity.

When planning a meeting, the facilitator should develop a detailed agenda that includes a review of progress to date and new items for discussion and include explicit timelines (Box 4.3). These serve as benchmarks for participants and as a device that helps keep the discussion on track. Remember, a good facilitator will treat the time of the meeting participants with respect and will use it wisely. The agenda should be prepared well in advance and in collaboration with the stakeholder working group. Background material, such as reports and publications, also should be made available to the stakeholder working group well in advance of the meeting. However, the facilitator should attempt to keep the background material to only those items that are necessary. Dumping large amounts of background material on members of the stakeholder working group will only ensure that very few will have reviewed the material before the workshop. The agenda should include time for a brief synopsis of background material to ensure that all members of the stakeholder working group are up to speed.

The first workshop

Involvement of stakeholders in the decision-making process begins with the very first workshop. This workshop can be crucial to successfully finding a solution to the natural resource management problem. The first workshop should be informational and should begin with introductions of the attendees. The participants should then be provided with the relevant background to the problem from the perspective of the key stakeholders, the management history (e.g., previous management actions), and background on what is known about the ecological, physical, and socioeconomic factors that are likely to affect the

---

Box 4.3   Example agenda for first stakeholder meeting

Below is the agenda from a 3-day stakeholder meeting on conservation of aridland aquatic species in the southwest US. The first day of the workshop included more than 60 participants from a variety of stakeholders and covered essential background on the ecology and management of aridland wetlands and on the structured decision-making process. Attendees for days 2–3 were limited to the stakeholder representatives and focused on the first steps in the decision-making process. This agenda is typical of many natural resources first stakeholder meetings.

AGENDA

Day 1
- Introduction to the issues
- Desert fishes and wetlands
- Synthesis of known wetlands with sensitive fish and amphibian species
- Physical/hydrological attributes of arid wetlands
- Determining the effects of livestock grazing on aridland amphibians
- Effects of full and partial cattle exclosures on amphibians
- Using livestock exclosures to restore wetland habitats
- Prescriptive livestock grazing as a management tool
- Introduction to structured decision-making and adaptive management

Day 2
- Preview of the day's agenda
- Identify decision situation, management alternatives, and conservation objectives
- Develop conceptual models of system dynamics

Day 3
- Review progress to date
- Refine and combine ecological response models/ideas with management alternatives and objectives
- Identify existing sources of data to parameterize model
  - Identify known uncertainties
  - Identify and task technical advisors
- Next steps and timelines

---

decision. The workshop participants also should be provided with a description of the decision-making process to let the participants know why the process is useful for the decision at hand and inform the participants where they are headed. The steps just described should take no more than a single day. The next steps in the workshop depend largely on the type of decision and the stakeholders involved. For stakeholder groups that are likely contentious, the next step would be to establish governance and identify the stakeholder representatives. For smaller and less-contentious decisions, the next step would include identifying and structuring objectives, which we covered in Chapter 3.

We are now ready to embark on the development and evaluation of a formal decision model, the topic of Part II of this book. First, in Chapter 5 we will cover some basic concepts in probability and statistics that are necessary background for the developments to follow. Then in Chapter 6 we will develop and analyze models of the stochastic influence of decisions and other factors on objective outcomes. In Chapter 7 we will deal at length with the issue of uncertainty, and specifically how different types of uncertainty affect decision making and need to be accounted for in the decision process. In Chapter 8 we will describe in detail the different approaches used to obtain optimal decisions for decision models that have been specified using the methods in this chapter and in Chapters 5–8. Finally, in Chapter 9 we will apply the concepts developed to several case studies in natural resource management.

References

Freeman, E.R. (2010) *Strategic Management: A Stakeholder Approach*. Cambridge University Press, Cambridge.

Harrison, S.R. and M.E. Qureshi. (2000) Choice of stakeholder groups and members in multi-criteria decision models. *Natural Resources Forum* **24**, 11–19.

Keeney, R.L. (1992) *Value Focused Thinking*. Harvard University Press, Cambridge, Mass.

Kennedy, K.D., E.R. Irwin, M.C. Freeman, and J.T. Peterson. (2007) Development of a Decision Support Tool and Procedures for Evaluating Dam Operation in the Southeastern United States Final Report submitted to US Geological Survey Reston, Virginia and US Fish and Wildlife Service, Atlanta, Georgia.

Reed, M.S., A. G., N. Dandy, H. Posthumus, K. Hubacek, J. Morris, C. Prell, C. H. Quinn, and L.C. Stringer. (2009) Who's in and why? A typology of stakeholder analysis methods for natural resource management, *Journal of Environmental Management* **90**, 1933–1949.

Wright, G. and P. Goodwin. (1999) Rethinking value elicitation for personal consequential decisions. *Journal of Multi-criteria Decision Analysis* **8**, 3–10.

Additional reading

Kaner, S., L. Lind, C. Toldi, S. Fisk, and D. Berger. (2007) *Facilitator's Guide to Participatory Decision-Making*, 2nd edn. Jossey-Bass, San Francisco, California.

PART II. TOOLS FOR DECISION MAKING AND ANALYSIS

5

Statistics and Decision Making

In this chapter, we focus on the use of statistics in decision making. First, we provide an overview of terms and concepts from statistics, such as expectation, independence, probability distributions, and linear models that contribute to the toolbox needed to make decisions in the face of uncertainty. We then transition to describing the use of statistics to quantify uncertainty and to use data to inform views of uncertainty. Examples include using models and data to describe biological/ecological relationships in management and conservation and to develop and parameterize probability models. We conclude with an introduction to more advanced modeling/statistical techniques, such as hierarchical modeling and Bayesian inference that provide more flexibility, creating a robust decision-making framework. Our goal is to merge two seemingly disparate topics (statistics and decision making) in a useful and practical manner providing researchers, managers and conservationists with the necessary tools to work in their area of expertise.

In Chapters 2 and 3 we developed a basic the framework for decision making whereby we

1) formally state our objectives,
2) present decision alternatives available to use in an attempt to reach our objectives,
3) create a model or set of models to encapsulate our belief in how various actions will lead to outcomes to fulfill our objectives.

Obviously, step number 3 requires models that depict our understanding of system dynamics and system responses, while incorporating different sources of

Decision Making in Natural Resource Management: A Structured, Adaptive Approach,
First Edition. Michael J. Conroy and James T. Peterson.
© 2013 John Wiley & Sons, Ltd. Published 2013 by John Wiley & Sons, Ltd.

uncertainty. If we think back to our burn example in Chapter 2, we assumed a 1:1 relationship between the decisions or actions (no burn, moderate burn, and intense burn) we wish to make and outcomes (habitat degradation, status quo, and habitat improvement). We briefly noted that this is not realistic (i.e., it is completely deterministic) and instead there will be different sources of uncertainty influencing both the set of actions we make and the value of the decisions we make (i.e., stochastic). Here we have two main objectives: 1) provide a set of foundational tools that allow us to formally quantify the different sources of uncertainty and develop models that characterize the necessary relationships needed in decision analysis and 2) provide approaches to estimate those important relationships using observed or historic data. Data plays an important role in decision making and we will focus on three different approaches to using data: 1) using data to develop probability models to describe uncertainty; 2) using data to parameterize and refine assessments regarding parameters of probability models; and 3) using data to understand, model, and describe relationships between variables. These concepts and tools will reappear in other chapters as they provide the foundation for handling uncertainty in a formal decision analysis.

Basic statistical ideas and terminology

Probability

Before we delve into the world of probability models, we need to first acquaint ourselves with the notion of **probability**. We can think of probability as the measure of one's belief in the occurrence of a **chance** or **random event**, that is, **outcomes** that we are uncertain about. Returning to another example from Chapter 2 in which we need to decide to pack or not to pack rain gear for an upcoming trip, we can think of whether or not it will rain as the chance event. We can assign a value as the possibility of rainfall for our upcoming trip, say 20% of rain and 80% of no rain. In this example, we used probability to quantify our uncertainty about whether or not it is going to rain by assigning a specific numeric value to each of the two events (rain and not rain) that have not occurred. Although this example is simple, it clearly illustrates several important rules of probability, which are more formally stated in Appendix A:

1) probabilities are numerical quantities, defined on a set of outcomes or events;
2) probabilities are nonnegative (e.g., ≥ 0) and must lie between 0 and 1 (here we expressed them as percentages, $20\% = 0.2$);
3) probabilities must add up over mutually exclusive events (only one event can happen not both); and
4) probabilities must add to 1 over all possible mutually exclusive outcomes (e.g., probability of rain + probability of no rain must equal 1).

These four properties of probability ensure a coherent framework, and thus provide a robust set of tools to quantify uncertainty. Note that these four rules state only the conditions an assignment of probability must satisfy; it does not tell us how to assign specific probabilities to an event (we will address this to some degree later). It is therefore important at this point to advance our definition of probability to clearly differentiate among two different ways to view probability and thus different ways to assign probabilities to events; each is important and will be used later on.

In a classical **frequentist** sense, we can think of probability as the relative frequency of an event occurring in a long series of trials. It is impossible, for example, to predict with certainty the occurrence of heads on a single toss of a balanced coin, but we would be willing to state with a fair measure of confidence that the fraction of heads in a long series of trials would be close to 0.5. If we observed that a particular event (say, getting a head on one toss of a coin) has occurred 90 times out of 100 then we would be more willing to state the probability of getting a head on the next toss is 0.9. Thus, we are using the relative frequency of an event occurring over a long series of trials as a way to encapsulate our uncertainty about predicting the outcome of that event occurring on a single future unobserved trial.

An alternative way to view probability is to think of probability as a subjective (often associated with **Bayesian** statistics, which we will discuss later) measure of uncertainty. Suppose the Chicago Bears (NFL Football team) are playing the Buffalo Bills in the Super Bowl, what is the probability the Chicago Bears win? How would we view this probability using the frequentist interpretation? Although it is possible to do so, it is not completely intuitive. Instead we would want to assign a **subjective probability** to the event. We imagine that when we posed this question almost all of you came up with a probability in your head and that these probabilities differed among almost all of you. Each of you had some idea about how good each team is and then assigned a probability based on that assessment. In this sense, we can use probability as a measure of uncertainty for any kind of unknown event regardless of whether or not it has happened in the past or if we have any previous information about its frequency of occurrence. In either case, frequentist or subjective, the guiding principle is that the state of knowledge about anything unknown is described by probability.

Before we continue to probability distributions we want to describe five important probability concepts that will be useful for manipulating and thinking about probabilities in a decision context. This presentation is informal (i.e., for readers unfamiliar with statistical concepts); a more formal development is provided in Appendix A.

Conditional probability is the probability of outcome A occurring given that outcome B has occurred, so, the event B is known. Mathematically, this is represented by $P(A|B)$ with the pipe "|" in the parentheses translated as the word "given" (Appendix A). For example, imagine we are interested in the probability of rain tomorrow. We note that has it rained the last two consecutive days and a tropical storm is expected. Now we have extra information related to whether or not it rains tomorrow and instead are interested in the conditional probability that it will rain tomorrow given this new information. The conditional probability should be higher than the **unconditional probability** of rain tomorrow, which is the probability of rain without having any extra information (Figure 5.1).

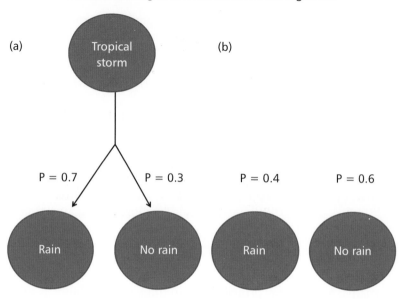

Figure 5.1 Conditional probability. (a) Probability of rainfall is conditional on whether a tropical storm has been sighted or not. (b) Unconditional probability of rainfall.

Independence signifies that the probability of an event is unaffected by the occurrence or non-occurrence of another event. Formally, independence can be defined in three ways:

a. $P(A|B) = P(A)$
b. $P(B|A) = P(B)$
c. $P(A \cap B) = P(A)P(B)$

where $\cap$ signifies "and" (i.e., the intersection of the events A and B). We can combine the first two concepts to define conditional probability and independence. Suppose we are interested in modeling the events of catching a fish or seeing a seal during a fishing trip, each of which is conditional on (i.e., depends on) whether the day is sunny or cloudy. If we have determined that the day is sunny, the events "Catch a fish" and "See a seal" are conditionally independent, in that knowledge of one event occurring now adds new information about the occurrence of the other event (Figure 5.2). Conditional independence is an important property that we will take advantage of in the construction of influence diagrams and belief networks, in which we can "modularize" components of a complex relationship into conditionally independent pieces (Chapter 6).

Two events A and $\bar{A}$ are **complementary**, if the fact that event A does not occur assures that $\bar{A}$ must occur and vice versa. Because the probabilities must add to 1 (rule 4 from above), $P(\bar{A}) = 1 - P(A)$. For example, suppose that the probability of catching a fish on a fishing trip $= p_1 = 0.6$ and the probability of seeing a seal on a fishing trip $= p_2 = 0.2$. Assuming that the events "catch a fish"

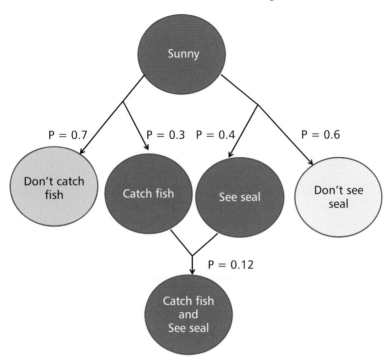

Figure 5.2 Conditional independence. The events "Catch a fish" and "See a seal" are both conditional on the event "Day is sunny". Given the even "Day is sunny" the events "Catch a fish" and "See a seal" are independent, and the joint probability of these events occurring is the product of their separate conditional probabilities.

and "see a seal" are independent, the probability of catching a fish *and* seeing a seal is $p_1 \times p_2 = 0.12$. The probability of not catching a fish, but seeing a seal is $(1 - p_1) \times p_2 = 0.08$. The probability of catching a fish, but not seeing a seal is $p_1 \times (1 - p_2) = 0.48$ and the probability of not catching a fish and not seeing a seal is $(1 - p_1) \times (1 - p_2) = 0.32$.

The **law of total probability** specifies that the overall probability of an event of interest occurring is the weighted sum of the conditional probabilities of the event, with the weights determined by the probabilities of the event that the event of interest is conditioned on. We can illustrate this with the fishing example, where we are interested in whether we catch a fish, with the probability higher on sunny days than cloudy days (Figure 5.3). In this example, over a 100-day season we could expect 60 sunny and 40 cloudy days. On the sunny days, we would expect $60(0.3) = 18$ days where we caught fish, on the cloudy days $40(0.2) = 8$ days, for 26 days total or total probability of 0.26.

The ideas of conditional and total probability come together in a very important theorem in probability known as **Bayes' theorem** (or **Bayes' rule**), which forms a general relationship between conditional and unconditional probabilities, and has many useful applications (Appendix A, Chapter 7). We illustrate

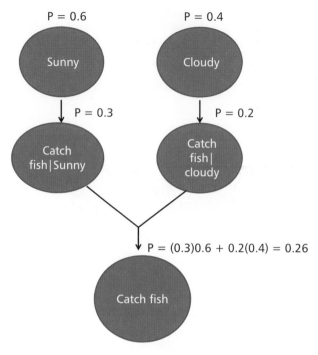

Figure 5.3 Total probability. The event "Catch a fish" is dependent on the complementary events "Day is sunny" and "Day is cloudy". The total probability of the event "Catch a fish" is then the weighted sum of the conditional probabilities, with the weights being the probability that the day is sunny or cloud.

the basic idea behind Bayes' theorem with the fishing example (Figure 5.4). Bayes' theorem specifies that

$$P(Catchfish \,|\, Sunny)P(Sunny) = P(Sunny \,|\, catchfish)P(catchfish)$$

Another very useful role for Bayes' theorem is in helping us to update knowledge (e.g., about a parameter value) from information (e.g., sample data). So if we let θ stand for the value of a parameter for example, and x stand for sample data, we can rewrite Bayes' theorem as:

$$P(\theta \,|\, x) = \frac{P(x \,|\, \theta)P(\theta)}{P(x)}$$

Viewed this way we have the following components:

- $P(\theta)$ expresses uncertainty about the parameter in the absence of data (before collecting), and is known as the **prior.**
- $P(x \,|\, \theta)$ expresses the probability or **likelihood** of having obtained the data result, given a particular value of the parameter.

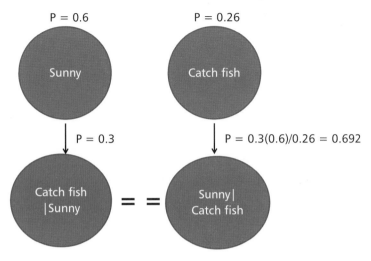

Figure 5.4 Bayes' theorem. By Bayes' theorem, P[Catch a fish| Sunny] P[Sunny] equals P[Sunny|Catch a fish]P[Catch a fish]. Given P[Sunny] = 0.6, P[Catch a fish| Sunny] = 0.3, and P[Catch a fish] = 0.26 from the law of total probability (Figure 5.3), P[Sunny|Catch a fish] = 0.692 for the equality to be true.

- $P(\theta|x)$ expresses uncertainty about the parameter in the presence of data (after collecting) and is also known as the posterior.
- $P(x)$ is the total probability of observing event x.

To illustrate Bayes' theorem, consider a situation where you went bird watching at a local park and failed to see your favorite species. Two possible things could have happened that led to your failure to see the species (the event), it was present in the park but you didn't see it or it was truly absent. Given the above notation, you would like to know the probability that the species occurs in the park, given it was not seen *P(present|not seen)*. We estimate this using Bayes' theorem as:

$$P(present \mid not\ seen) = \frac{P(not\ seen \mid present)P(present)}{P(not\ seen \mid present)P(present)+(1-P(present))}$$

This requires a prior probability that the species is present in the park, *P(present)*, and the probability of not seeing the species given it was present, *P(not seen | present)*. The total probability of the event then is the probability of the species being present and not seen, *P(not seen|present)P(present)*, plus the probability that the species was not present and was not seen, *P(not seen|not present)* (1 − *P(present)*). Notice that the probability that the species is not present is one minus the probability of presence because they are complementary. The probability of not seeing a species if it is not present is one (unless misidentified) so the total probability simplifies to: *P(not seen|present)P(present)* + (1 − *P(present)*).

Often Bayes' theorem is simplified to the following $P(\theta|x) \propto P(x|\theta)P(\theta)$ or *Posterior $\propto$ Likelihood x Prior* where the symbol $\propto$ signifies "is proportional to". We will talk more about Bayes' theorem later when we talk about probability distributions and estimation and we will continue to see that this relationship plays a critical role in how information is updated.

Probability distributions

Now that we have seen some basic rules and properties of probabilities we are ready to introduce **probability distributions**. A probability distribution is a model that describes the relationship between the set of all possible outcomes of a random event and those probabilities associated with each outcome. Probability distributions can either be classified as **discrete** or **continuous**. Discrete distributions model outcomes that occur as discrete integer values or in discrete classes and are called **probability mass functions** (i.e., **pmf**) while continuous distributions model outcomes that have continuous outcomes or values and are called **probability density functions (pdf)**. Table 5.1 has the most common probability distributions including common uses. Appendix B and the web supplement have more detail on probability distributions and their statistical properties, and Appendix G includes code in the free software package R (http://www.r-project.org/) which allows users to calculate and plot a variety of common distributions. We illustrate the use of R performing calculations and plotting for 3 commonly used distributions in Boxes 5.1–5.3.

A conventional notation for pmfs or pdfs is $f(y|\theta)$, which indicates that the probability of an observed value y depends on the parameter θ or parameters, such as mean and variance, used to specify the distribution of the **random variable** Y. Distributions have two restrictions. First, the pmf or pdf must have a probability that takes on nonnegative values over the range of the random variable, so

$$f(y|\theta) \geq 0$$

and second, the area under the distribution must equal 1 when summed (or integrated) over the range of the random variable (although of course the random variable itself can sum to any value); that is

$$\sum_y f(y|\theta) = 1 \text{ for discrete distributions}$$

and

$$\int_{-\infty}^{\infty} f(y|\theta)dy = 1 \text{ for continuous distributions.}$$

Remember this follows from Rule 4 from above, which states that probabilities must sum to 1 across all mutually exclusive outcomes. We can talk about the

Table 5.1 Features of some important statistical distributions.

Type	Distribution	Applications	Support	Parameters
Continuous	Normal	Continuous attributes with no theoretical upper or lower limit; approximating form for other distributions	$-\infty < x < \infty$	μ – location σ – scale
	Uniform	Attribute takes on any value in range with equal probability; random number generation	$a < x < b$	a, b – lower and upper bound (0 and 1 for random number generation)
	Exponential	Time till event	$0 \leq x \leq +\infty$	λ rate = 1/mean
	Gamma	Generalization of the Exponential distribution; prior distribution for Poisson	$0 \leq x \leq +\infty$	b scale c shape
	Beta	Probability outcomes Prior for binomial	$0 < x < 1$	α, β location, shape and scale
Discrete	Bernoulli	Success or failure in a single trial (e.g., survival, capture–recapture)	0, 1	p – probability of success in trial
	Binomial	Number of successes in n independent Bernoulli trials (e.g., survival)	$0, \ldots, n$	n – number of trials p – probability of success in 1 trial
	Poisson	Counts	$0, \ldots \ldots n_\infty$	λ – mean (density)
	Geometric	Number of trials before first success	$0, \ldots \ldots n_\infty$	p – probability of success in 1 trial

support of a distribution as the region where $f(x) > 0$. For example, the normal distribution (Table 5.1) has support from $-\infty$ to $+\infty$ meaning it can take any real value while the beta distribution has support between 0 and 1 restricting values to be between 0 and 1 inclusive. All other values have a probability of 0 and thus no support. The probability density function for a distribution describes the probability that the random variables take on particular values (for a discrete distribution) or are in the neighborhood of a value (for a continuous distribution; see below). For example, consider a situation where you moved to

Box 5.1   Modeling continuous outcomes: the normal distribution

Here we demonstrate modeling a continuous random variable with the normal distribution, using built-in functions in the R statistical package and Microsoft Excel®. The normal distribution is perhaps the most familiar statistical distribution. It is symmetric about the mean, with the familiar bell-shaped curve, and is used to model continuous, real values with theoretical range from negative to positive infinity. It is the limiting distribution of many test statistics and functions and is commonly used as an approximation, even when the data are thought to follow some other distribution, often after transformation to reduce skewness or discontinuities in the data. The normal density is determined by the 2 parameters μ and σ ($\sigma > 0$) as

$$f(x; \mu, \sigma^2) = \frac{1}{\sqrt{2\pi\sigma^2}} \exp\left(\frac{-(x-\mu)^2}{2\sigma^2}\right), \quad -\infty \le x \le \infty$$

For example, the normal density function over −50, 50 for $\mu = 5$ and $\sigma = 15$ is produced by

```
> #normal distribution
> #generate equally spaced values between -50 and 50
> x<-seq(-50,50,0.01)
> mu<-5
> sigma<-15
> density<-dnorm(x,mu,sigma)
> plot(x,density,type='l',ylab='f(x)')
```

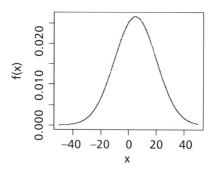

```
> distrib<-pnorm(x,mu,sigma)
> plot(x,distrib,type='l',ylab='F(x)')
```

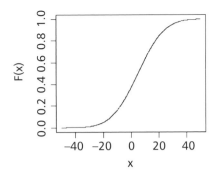

Specified probability quantiles are easily obtained from the qnorm() function, for example

```
>prob_levels<-c(0.001,.05,.25,.5,.75,.95,0.999)
> quants<-qnorm(prob_levels,mu,sigma)
> quants
[1] -41.353485 -19.672804  -5.117346   5.000000  15.117346
29.672804  51.353485
```

Equivalently, we could say that we are 90% confident that x is between −19.67 and 29.67, with 10% probability (5% in each tail) outside this range.

Notice that in the density dnorm() and distribution pnorm() functions, we passed the data as a list to the function, for scalar (1-dimensioned) values of the parameter. Generally speaking any of these function arguments can be lists, and it will make sense below (under the likelihood function) to reverse which ones are.

To complete the same functions in Excel, you have the following:

Density

```
=NORM.DIST(x,5,15,0), where x corresponds to the value of x
```

Cumulative distribution

```
=NORM.DIST(x,5,15,1), where x corresponds to the value of x
```

Quantiles

```
=NORM.INV(p,5,15), where p corresponds to the desired
quantile
```

Random number generation

The easiest way to generate random normal numbers is by using the built-in R function rnorm().

(*Continued*)

```
#Generate 100 random numbers for mu=5 and sigma =10
#method 1
> n<-100
> mu<-5
> sigma<-10
> x<-rnorm(n,mu,sigma)
```

Another (and just as valid) way is to first generate 100 random uniform(0,1) numbers,

```
>U<-runif(100)
```

and then treat these as probability values in qnorm(), which functions as the inverse distribution function to return values of *x* given *U*.

```
>x<-qnorm(U,mu,sigma)
```

Similarly, in Excel we have

```
=NORM.INV(rand(),5,10)
```

You should be able to test these 2 approaches to convince yourself that they produce equivalent results.

R code for this example is provided at the book website.

a new region of the country and during your first winter, there were 4 snowstorms that closed schools and businesses. You would like to know if this is as unusual a winter as your new friends and colleagues are telling you. The snowstorms were discrete events (i.e., snowed or it did not) so to estimate the probability of 4 snowstorms occurring in a given winter you would use a discrete distribution, such as a Poisson (Table 5.1). After combing through weather records for the past 50 years, you find that the average number of snow storms is 2.1, so for a discrete random variable (e.g., from a Poisson distribution with mean $\lambda = 2.1$) we have

$$f(4) = 0.099$$

indicating that the value of 4 is taken with probability 0.099. In other words, the probability of 4 snowstorms in any given winter is 9.9% or about 1 in 10 years, not all *that* unusual. With continuous distributions we cannot talk about the probability that a specific point value occurs. This is in fact equal to zero ($P(Y = y) = 0$) because there are an infinite number of possible outcomes and so the probability of any particular value must be infinitely small. For example, there are an infinite number of values between 1–2 for a normal distribution: 1.01, 1.001, 1.0001, and so on. Instead for continuous distributions, we are usually interested in the probability an event occurs within a particular interval,

---

Box 5.2   Modeling integer (count) outcomes: the Poisson distribution

The Poisson distribution is a very important discrete distribution that models outcomes that take on nonnegative integer values (0, 1, 2, . . ., n_∞). Examples include counts of animals, plants, and other objects where the process generating the counts is "random" in the sense that counts are not spatially clustered or dispersed except by chance.

Density, distribution, and quantiles

The Poisson distribution is specified by the single parameter λ that is equal to both the population mean and variance. Thus, sometimes the ratio of the sample mean to the variance is used as evidence (or lack thereof) of Poisson assumptions, with values of this ratio of about 1 taken as support for a Poisson count model. The probability mass function of the Poisson is given by

$$f(x; \lambda) = \frac{\lambda^x e^{-\lambda}}{x!}, x = 0, 1, 2, 3, \ldots\ldots$$

where e is the base of the natural logarithm, $\lambda > 0$, and $x!$ denotes the factorial function $x(x - 1)(x - 2)\ldots..1$. The distribution function is simply given by summation over the discrete values of x of the density

$$F(x) = \sum_{k=0}^{x} f(k; \lambda) = \sum_{k=0}^{x} \frac{\lambda^k e^{-\lambda}}{k!}, x = 0, 1, 2, 3, \ldots\ldots$$

The Poisson mass and distribution functions are easily implemented in R, for example for $\lambda = 5$:

```
#poisson distribution
#generate a sequence between 0 and 20
>x<-0:20
>lambda<-5
#mass function
>density<-dpois(x,lambda)
>plot(x,density,"s")
```

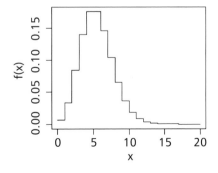

(*Continued*)

```
> #distribution function
> distrib<-ppois(x,lambda)
> plot(x,distrib,"s",ylab='F(x)')
```

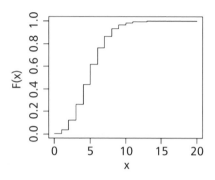

Likewise, standard quantiles are easily computed-

```
> #quantiles
> prob_levels<-c(0.001,.05,.25,.5,.75,.95,0.999)
> quants<-qpois(prob_levels,lambda)
> quants
[1]  0  2  3  5  6  9  13
```

In Excel we have,

Density

```
=POISSON.DIST(x,5,0),
```
where x corresponds to the value of x

Cumulative

```
=POISSON.DIST(x,5,1),
```
where x corresponds to the value of x

There are no functions for estimating the quantiles of a Poisson distribution in the standard version of Excel ®.

Random number generation

Given the discrete nature of the random variable in the Poisson distribution, there are several options for generating random variables, some of them quite simple, others not so simple but more flexible. We illustrate 2 methods of generating 100 random values from a Poisson(10) distribution.

Method 1 – built-in R function

First, we can of course rely on the standard, built-in R function rpois().

```
#Generate 100 random numbers for lambda=10
#method 1
>n<-100
>lambda<-10
>x<-rpois(n,lambda)
```

Method 2 – Uniform deviate, quantile (inverse distribution function)

The second method also relies on an R-function, the quantile or inverse distribution function, but performs the calculations by first computing 100 random(0,1) numbers and then transforming them with the inverse distribution (quantile) function.

```
#method 2
n<-100
lamdbda<-10
U<-runif(100)
x<-qpois(U,lambda)
```

Box 5.3   Modeling "success/failure" outcomes: the binomial distribution

The **Bernoulli distribution** is the natural distribution for modeling outcomes that can occur in 1 of 2 classes, such as success or failure, lived or died, heads or tails, male or female. The Bernoulli distribution has a single parameter p that describes the probability of a "success" (however it is defined). The **binomial distribution** defines the number of successes that occur in n independent Bernoulli trials, each with the same probability of success p. The binomial is thus based on summing Bernoulli distributions, and has 2 parameters, n and p. Because these 2 distributions are so closely related we will consider them together below.

Density, distribution, and quantiles

The Bernoulli random variable x takes on 2 possible values, either 1 (indicating success) or 0 (failure), and has a single parameter, p, denoting the probability of success. The probability density function is written as

$$f(x; p) = p^x (1-p)^{1-x}, x = 0, 1$$

(*Continued*)

which simplifies to $f(0;p) = 1 - p$ and $f(1;p) = p$. Note that we assume that there are only 2 possible outcomes, a success with probability p and a failure with probability $1 - p$, and that by definition the probability that it is either a success or failure adds to 1.

The mean of the Bernoulli distribution is $E(x) = \mu = p$ and the variance is $\mathrm{Var}(x) = p(1 - p)$.

The binomial distribution is closely related, with the binomial variable x defined as number of successes in n independent Bernoulli trials, each with probability p of success. The binomial thus has 2 parameters (n and p) though one of these (n) ordinarily is known and will not be estimated from data. The binomial probability mass function is

$$f(x; n, p) = \binom{n}{x} p^x (1 - p)^{n-x}, \, x = 0, n$$

The binomial distribution function is

$$F(x; n, p) = \sum_{k=0}^{x} \binom{n}{k} p^k (1 - p)^{n-k}, \, x = 0, n$$

The mean and variance are given by $E(x) = \mu = np$ and the variance is $\mathrm{Var}(x) = np(1 - p)$. The binomial density and distribution are easily implemented in R by the dbinom() and pbinom() functions (there is no separate Bernoulli function in R, with the Bernoulli simply being a binomial with a single trial ($n = 1$).

For example, for a Bernoulli with $p = 0.4$

```
> #Bernoulli
> p<-0.4
> x<-0:1
> density<-dbinom(x,1,p)
> distrib<-pbinom(x,1,p)
> plot(x,density,"h",ylab="f(x)",ylim=c(0,1))
>plot(x,distrib,"h",ylab="F(x)",ylim=c(0,1))
```

This produces plots for the mass and distribution of:

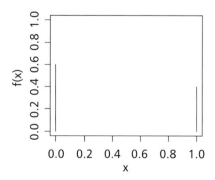

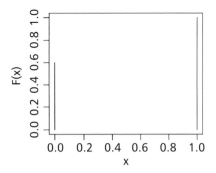

Taking a binomial with $p = 0.4$ and $n = 10$ trials we have

```
> #Binomial
> n<-10
> p<-0.4
> x<-0:n
> density<-dbinom(x,n,p)
> distrib<-pbinom(x,n,p)
> plot(x,density,"s",ylab="f(x)")
>> plot(x,distrib,"s",ylab="F(x)")
```

This produces plots

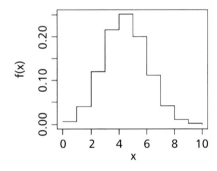

and

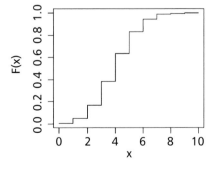

(*Continued*)

Quantiles at specified p-values are easy to produce using the qbinom() function, e.g.,

```
>n<-10
> p<-0.4
> #quantiles
> prob_levels<-c(0.001,.05,.25,.5,.75,.95,0.999)
> quants<-qbinom(prob_levels,n,p)
> quants
[1] 0 2 3 4 5 7 9
```

In Excel we have

Density

```
=BINOM.DIST(x,10,0.4,0), where x corresponds to the value
of x
```

Cumulative

```
=BINOM.DIST(x,10,0.4,1), where x corresponds to the value
of x
```

Quantile

```
=BINOM.INV(10,0.4,p), where p corresponds to the desired
quantile
```

Random number generation

Generating random number for the Bernoulli and binomial is quite easy and can be accomplished with either a simple random uniform number generator or with the built-in function rbinom(). The first approach computes the Bernoulli outcome by simply comparing a uniform(0,1) random number to p; if $U > p$ then $x = 0$, otherwise $x = 1$. The second directly invokes rbinom() in R.

```
> #generating Bernoulli random variables
> #specify p
> p<-0.35
> #specify n_reps
> n_reps<-1000
> #method 1
> x<-(runif(n_reps)<p)*1
> #method 2
> x<-rbinom(n_reps,1,p)
```

Generating binomial random variables can be accomplished by generating a series of *n* Bernoulli variables and then summing these.

```
>#generating binomial random variables
>#specify n
>n<-10
>#specify p
>p<-0.35
>#specify n_reps
>n_reps<-100
>#method 1
>x<-array(0,c(n_reps))
>for (i in 1:n_reps)
>{
>x[i]<-sum(runif(n)<=p)*1
>}
```

Alternatively, you can directly use the rbinom() function in R

```
#method 2
>x<-rbinom(n_reps,n,p)
```

The advantage of the first approach is that sometimes we will not want to assume that the parameter *p* remains constant, but instead allow it to vary from sample to sample (or even among Bernoulli trials within a sample). In such cases we can still simulate or model the data but no longer under binomial assumptions (which require *p* to be constant). We will look at an example of this below.

In Excel®, random outcomes for a binomial process using the above parameters can be generated using:

```
=BINOM.INV(10,0.35,RAND())
```

say between *a* and *b*. This probability is equal to the area under the pdf between *a* and *b* (Figure 5.5) and we could calculate it directly using integration

$$P(a \leq y \leq b) = \int_a^b f(t)dt$$

Alternatively, we can calculate this probability using the **cumulative distribution function (cdf)**. With the cdf, rather than thinking about the probability of a random event taking a particular value we calculate the value that is less than or equal to a specific value. The cdf (or sometimes distribution function) is defined as

$$F(x) = P(X \leq x)$$

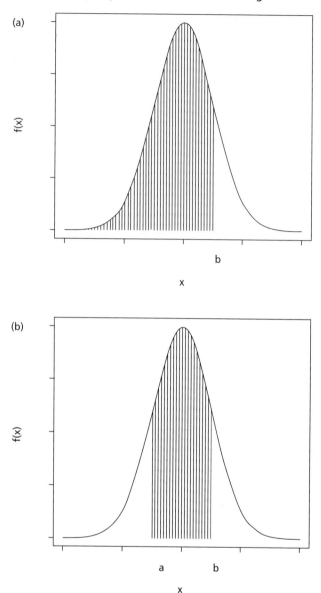

Figure 5.5 Area under a probability density function $f(x)$ representing (a) probability that the random variable X is $< b$; (b) probability that $a < X < b$.

cdfs have three simple rules:

1. $F(x)$ is non-decreasing.
2. $F(-\infty) = 0$.
3. $F(\infty) = 1$.

For example, continuing with the snowstorm example above, the Poisson distribution with a mean of 2.1 has support for $x = 0$, 1, 2, 3, 4, 5, 6, 7, 8, and 9 with $f(0) + f(1) + f(2) + f(3) + f(4) + f(5) + f(6) + f(7) + f(8) + f(9) = 0.122 +$

0.257 + 0.270 + 0.189 + 0.099 + 0.042 + 0.015 + 0.004 + 0.001 + 0.000 = 1.000. Note that the probability of any $x > 9$ are not zero but are very, very small and can be treated as zero. Therefore, the probability of getting 4 or more snowstorms in any given year is 0.099 + 0.042 + 0.015 + 0.004 + 0.001 + 0.000 = 0.161 or 16%, almost twice a decade. In general, cdfs are calculated for discrete distributions by

$$F(x) = \sum_{k \leq x: f(k) > 0} f(k)$$

Calculation of $F(x)$ for continuous distributions is more challenging and requires integration. By definition:

$$F(x) = \int_{-\infty}^{x} f(t) dt$$

where the lower limit will depend on the support of the distribution (may be zero). Usually these computations are done by computer functions (see Box 5.1) or looked up in standard tables. The advantage is that once $F(x)$ is available (discrete or continuous) we can ask questions like "what is the probability that x is between a and b?" since

$$Prob(a < x < b) = \int_{a}^{b} f(t) dt = \int_{-\infty}^{b} f(t) dt - \int_{-\infty}^{a} f(t) dt = F(b) - F(a)$$

To illustrate, let's assume that you have to choose between two the varieties of beans to plant in your garden and each has a different optimal range of temperatures for growth. The first variety grows best when daytime temperatures are between 24–30 °C and the second variety does best when temperatures range 28–33 °C. The weather almanac indicates that daytime temperatures during the growing season in your area average 28.5 °C with a standard deviation of 3.2 °C. Temperature is continuously valued and can take negative and positive values, so we use a normal distribution. Assuming a normal distribution with mean 28.5 °C and standard deviation 3.2 °C, for bean variety 1 we have $F(30) = 0.68038$, $F(24) = 0.07982$ and $Prob(1 < x < 2) = F(2) - F(1) = 0.600551$ or a 60% chance that temperatures will be in the optimal range. For bean variety 2, we have $F(33) = 0.92018$, $F(28) = 0.43792$ and $Prob(1 < x < 2) = F(2) - F(1) = 0.48226$ or a 48% chance that temperatures will be in the optimal range. Which variety would you choose?

We summarize the main features of several important probability distributions in Table 5.1; Appendix B provides a more complete description of these and several additional distributions.

We can reverse the idea of distributions and, for a given probability level of a distribution function obtain the value of x or **quantile** associated with that value. The quantiles are essentially found by inverting the distribution function and solving for x, though for discrete distributions they can easily obtained by examination and **interpolating** between values. Box 5.1 gives an example of calculating the quantiles of a normal distribution using the built-in functions in the software package R (http://www.r-project.org/) and Microsoft Excel ®. Boxes 5.2 and 5.3 give examples of calculating quantiles and other functions for two important discrete distributions: the Poisson and the binomial.

In addition to using probability distributions to model uncertain events, they are all also useful for summarizing information about a random variable, such as its mean or variance. The mean or **expected value** of a discrete random variable Y with pmf $f(y|\theta)$ is

$$E(Y) = \sum_y yf(y|\theta)$$

likewise the expected value of a continuous random variable Y is defined as

$$E(Y) = \int_{-\infty}^{\infty} yf(y|\theta)dy$$

We can return to the prescribed fire example from Chapter 2 where we had a simple probability model that said that following a burn we are 90% sure of the outcome, but have a 5% probability for the other two possible outcomes. We assigned each of the outcomes a **utility** $U(X = x_i)$, here it was the value obtained from a specific outcome (e.g., failure to burn is assumed to result in decline of habitat quality (value = −1), moderate burn is assumed to result in status quo habitat (value = 0), and intense burn is assumed to result in improvement of habitat (value = +1)), divided by the cost of each action (no burn = 1, moderate burn = 2, intense burn = 3). This information is then assembled to compute an expected value for each decision. The detailed calculations for this example are:

$$E(\text{No burn}) = (-1)0.9 + (0)0.05 + (1)0.05 = -0.85$$
$$E(\text{Moderate burn}) = (-1/2)0.05 + (0/2)0.9 + (1/2)0.05 = 0.0$$
$$E(\text{Intense burn}) = (-1/3)0.05 + (0/3)0.05 + (1/3)0.9 = 0.2833$$

Using data in statistical models for description and prediction

Why you should look at your data?

Although in many, if not most cases you will want to go on and conduct more sophisticated analyses, we consider it an important first step to summarize and visualize the data for several reasons. First, it is good to get familiar with the look and feel of the data – what are the types of variables (discrete, continuous, categorical, etc.)? What are the means, variances, ranges and other descriptive statistics for the variables? We will also want to be looking for obvious patterns in the data, such as data concentrated around certain values, missing in certain ranges. Additionally, we may want to explore for correlations among key variables: even though we discourage "data dredging" to build predictive models or test hypotheses, a certain amount of intelligent exploration of the data is perfectly fine. Last, but not least, an initial exploration of the data will often reveal errors, either in the original field data entry or in transcription to computer files, such as inadmissible values (impossibly low or high measurements, dates out of range of the study, locational coordinates out of the study area, etc.) or how missing values are handled (listed as 0s, NAs, −9999, . . .) that will need to be corrected or "filtered" before analysis. Box 5.4 illustrates how to use R and Excel for computing basic summary statistics and graphing.

Box 5.4   Summarizing and manipulating data using R

Here, we illustrate use of the free statistical program R to provide basic summary statistics, perform graphing, and conduct some initial, simple analyses of the data. Here, we use a dataset containing 101 samples of insect counts along an elevation gradient from 0 to 1000 m in 10-m intervals; the data are contained in a csv and Excel® file are available at the book website. We read the data into R using the commands

```
>insects<-read.table("insects.csv",header=T,sep=",")
>attach(insects)
```

First, we can obtain some simple summary statistics either for all the variables at once, or for a single variable at time. For example, the command

```
>summary(insects)
```

provides

```
       plot          elevation            count
 Min.   :  1    Min.   :   0    Min.   :  156
 1st Qu.: 26    1st Qu.: 250    1st Qu.:  528
 Median : 51    Median : 500    Median : 1856
 Mean   : 51    Mean   : 500    Mean   : 4452
 3rd Qu.: 76    3rd Qu.: 750    3rd Qu.: 6390
 Max.   :101    Max.   :1000    Max.   :21991
```

We can also produce some specific statistics for selected quantitative variables. For example, to obtain means, standard deviations, and ranges for elevation we use the commands:

```
> mean(elevation)
[1] 500
> sd(elevation)
[1] 293.0017
> range(elevation)
[1]    0 1000
```

(Continued)

To obtain the same summary statistics for "count" we use

```
> mean(count)
[1]  4451.911
> sd(count)
[1]  5572.118
> range(count)
[1]    156 21991
>
```

As we note in Appendix A, a **quantile** is defined as the value of a random variable that is associated with a specific cumulative probability level. For example, a 0.75 quantile (or 75th percentile) is a value that lies above 75% of the values of the distribution. A sample quantile based on the same idea applied to a random sample; for instance the 0.75 quantile from a sample of 1000 observations would be the 751st value (ranked from low to high). The 0.5 quantile (50th percentile) is also known as the **median**. We can easily obtain selected quantiles using R; for example

```
>quantile(count,c(.01,.5,.9))
```

provides the 1%, 50%, and 90% quantiles (percentiles) of the variable "count".

Likewise,
```
>median(count)
```

provides the median value for count, which you can confirm is the same as the 50% percentile.

All of the above commands are saved in an R script file located on the website.

Basics of graphing

Graphing is a very important feature of R, and one that is very useful at both the early exploratory stages of analysis and later when we need to do model diagnostics. While there are many types of graphs possible in R, 2 of the most common and useful ones are **histograms** and **scatterplots**. Histograms graphically show the relative frequencies of values of a variable. So for example, the insect data above has two variables that are recorded: elevation and count. We can produce histograms for this by the commands

```
>hist(elevation)
>hist(count)
```

which produces the graphs

Histogram of elevation

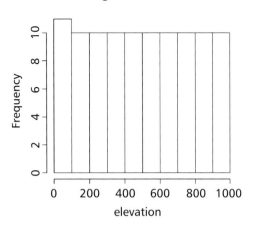

and

Histogram of count

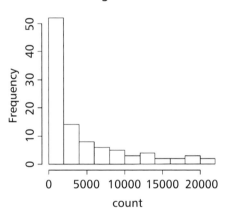

Thus, we can quickly see that the data is pretty evenly spread over the range of elevations, but that most of the data falls below counts of 5000, with a very small number >2000.

Scatterplots, by contrast, plot one variable against another in an attempt to discern patterns or correlations. Using the insect example

```
>plot(elevation,count)
```

produces

(*Continued*)

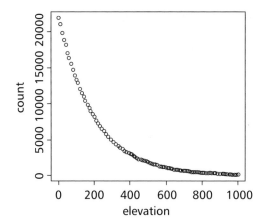

indicating a strong negative (but nonlinear) relation between elevation and insect abundance. Many times we will want to transform the data to better display relationships or meet model assumptions. For example for count data it is common to use a logarithmic transformation, so that

```
>plot(elevation,log(count))
```

produces a plot but this time with count on the natural log scale

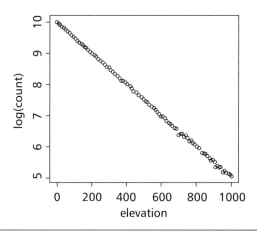

Linear models

The last foundational tool from our statistics toolbox that will be useful for conducting decision analysis is linear modeling. We use linear models for 2 main purposes: 1) to estimate relationships between variables using observed data and 2) to predict values of a response, assuming that the model is true or at least a reasonable approximation of the process. We will also touch on several other important topics that are addressed in more detail in the appendices: statistical distributions (Appendix B), estimation (Appendix C) and multi-model inference (Appendix D).

A linear model is simply a model in which a **response** (often written as Y) is modeled as a linear function of one or more **predictors** (or **explanatory factors**, often written as X) and **parameters** or **coefficients** (often written as b or β). The classic linear model has the following mathematical form:

$$\hat{Y}_i = \beta_o + \beta_1 X_i$$

Alternatively, this can be written as:

$$E(Y_i) = \beta_0 + \beta_1 X_i$$

which shows that we are modeling the average (expected) values of Y and how they vary as a function of the predictors (X). We can think of Y as the sum of a deterministic component $E(Y)$ and a random component ε_i. In this model, there are 3 parameters. The first 2 parameters are the intercept (β_0), and the slope (β_1) describing the linear relationship between Y and X. There is also a third parameter, σ_e^2, which describes the variation in Y_i left over or unexplained by the linear model. Under **normal regression** the residual errors

$$\varepsilon_i = Y_i - \hat{Y}_i$$

are assumed to follow a normal distribution with mean of zero and variance of σ_e^2. However, even in the absence of normal assumptions we can still fit a linear model by minimizing the sum of the squared residual errors under a procedure known as **least squares estimation** (Appendix C). If Y_i is a continuous variable and the explanatory variables (X_is) are continuous as well, we refer to this as simple linear regression (only one explanatory variable) or multiple linear regression (>1 explanatory variable). When the explanatory variables are only categorical or qualitative variables then we refer to this as analysis of variance or ANOVA models. Finally, we can have explanatory variables that are both quantitative and qualitative, while the response is still continuous; these models are referred to as analysis of covariance models or ANCOVA. Box 5.5 illustrates construction of linear models using the insect count example.

We can extend the linear model even further to allow for different types of response variables. The response, Y_i, can take many forms depending on the characteristics of the data; it does not have to be a continuous variable, but could be a count or integer, presence/absence data or a proportion. Therefore, we need something more flexible than the normal linear model; this larger class of models is called generalized linear models (GLM) and has the classic linear model (i.e., that assumes a normal distribution) as a subset or special case of GLMs. In allowing for different types of response variables, we have to change the assumptions about the statistical distribution being used to describe the data. For example, if the data were a count, say a population size or number of species, we would have to use a Poisson distribution to describe the process (see Appendix B). Remember for a decision analysis we are interested in describing the probability distribution of some response variable and how changes in explanatory variables affect that response variable (how the dependent variable "responds" to changes in the independent variables). In Box 5.6, we illustrate generalized linear models, with an example involving binary (binomial) responses, and extend the example to allow for the incorporation of multiple factors,

---

Box 5.5   Building a linear model in R

We illustrate construction linear models with counts of insects in light traps along an elevation gradient from 100–1500 m in 10-m intervals. At each elevation contour insects were removed from traps and counted after a 1-week sampling period (all sample times are the same within 12 h), elevation is recorded exactly, and average daily maximum temperature is recorded in degrees C using data loggers. The relationships of interest are the influence of elevation (insect density is hypothesized to increase with elevation) and temperature (thought to be related in a quadratic fashion to insect density, with an optimum temperature of approx. 20 °C). The data are available on the book website, and are the same data considered in Box 5.5, with the addition of maximum temperature measurements.

We start here with construction and evaluation a simple model relating counts to elevation. As before, we read the data in and create scatterplots of the data

```
> #read in length mass data
> data<-read.csv("insect_example.csv")
> attach(data)
>plot(elev,counts)
```

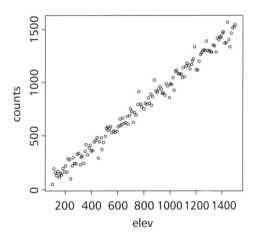

```
>plot(temp,counts)
```

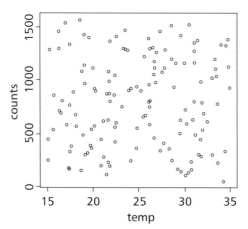

This seems to indicate a strong linear relationship between elevation and counts, but little suggestion of a relationship with temperature. We will proceed first with a model predicting counts as a function of elevation. We construct this using the lm() function, which produces estimates both under least-squares and normal assumptions:

```
>reg<-lm(counts~elev)
```

We then use the regression to produce predictions of the count response, and plot these against observed counts.

```
> ypred<-predict(reg,interval="p")
> plot(elev,counts)
> matlines(elev,ypred,col="red")
> summary(reg)
```

This produces summary output and graphics:

```
Call:
lm(formula = counts ~ elev)

Residuals:
     Min       1Q   Median       3Q      Max
-147.896  -31.384    6.619   32.942  157.602

Coefficients:
             Estimate Std. Error t value Pr(>|t|)
(Intercept) -12.63787    9.91870  -1.274    0.205
elev          1.01452    0.01105  91.809   <2e-16 ***
—
Signif. codes:  0 '***' 0.001 '**' 0.01 '*' 0.05 '.' 0.1
' ' 1
```

(Continued)

```
Residual standard error: 53.41 on 139 degrees of freedom
Multiple R-squared: 0.9838,     Adjusted R-squared: 0.9837
F-statistic:  8429 on 1 and 139 DF,  p-value: < 2.2e-16
```

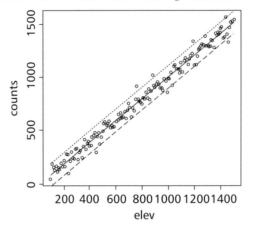

The graph produces a prediction line and a 95% prediction interval, which provides a confidence interval on the prediction of a future observation of count, given the specified elevation. We can also produce predictions at selected elevations by creating a new dataframe specifying just those elevations. For example

```
> #PREDICTION
> new<-data.frame(elev=c(100,500,1000,1500))
> y_pred<-predict(reg,new,interval="p")
> y_pred        fit       lwr        upr
1   88.8142  -18.25412   195.8825
2  494.6225  388.44961   600.7954
3 1001.8829  895.82244  1107.9433
4 1509.1432 1402.07492  1616.2116
```

Produces prediction intervals at 100, 500, 1000, and 1500 m.
    We can also directly calculate the R2 statistic as

```
> #Calculation of R2
> RSS<-deviance(reg)
> #total sum of squares (intercept only model)
> ss_total<-deviance(lm(counts~1))
> #regression sum of squares
> ssr<-ss_total-RSS
> R2<-ssr/ss_total
> R2
[1] 0.9837765
>
```

Overall, this analysis confirms that there is a strong predictive relationship between elevation and insect count. Finally, R has several built-in diagnostic graphics that can reveal problems with the data violating model assumptions:

```
> plot(elev,counts)
> matlines(elev,ypred,col="red")
```

```
> #diagnostic plots
> par(mfrow=c(2,2))
> plot(reg)
> savePlot(filename="Rbuilt_in_plots_reg",type="png",
device=dev.cur())
> par(mfrow=c(1,1))
>
```

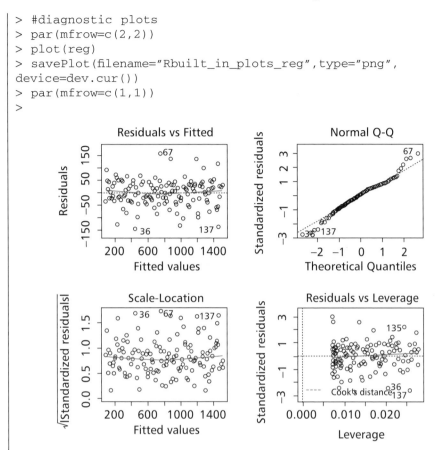

The results here all suggest that the data for this problem meet the assumptions of linear regression.

Box 5.6   Generalized linear models and multiple predictor models in R

As mentioned previously, the lm() function provided both least-squares estimates and MLEs (and associated log-likelihood values) under normal assumptions. A more general procedure that allows for specification of other likelihoods is provided by **generalized linear models** and the glm() procedure in R. Generalized linear models can be appropriate if normal assumptions either are thought *a priori* not to apply (e.g., with count or binary responses) or if the examination of residuals or other tests shows deviation from normality. The glm() procedure allows both specification of a distribution "family" (e.g., normal or Gaussian, Poisson, binomial) and a transformation or **link function** (identity [by default], log, logit, etc.).

As an example, let's consider a study of warbler nests at 150-m from 100 to 1500 m in the southern Appalachian Mountains. At each elevation, between

(Continued)

$n = 30$ and 40 nests were found and followed until success (fledging) or failure (destruction of nest, eggs, or nestlings). The response was thus the number y of successful nests in n trials at each elevation. The data for the study are available on the book website. We can build a linear (on the logit scale) model with binomial error assumption for these data like this:

```
> #GLM
> #nest success example
>
> nest_data<-read.csv("nest_success2.csv")
> attach(nest_data)
> glm1<-glm(y/n~elev,weights=n,family=binomial(link=logit))
> summary(glm1)
```

Generating confidence intervals on the predictions with glm() is somewhat trickier than lm due to the fact that we are not assuming a single distribution. Here, we will use a normal approximation for the predictions on the logit scale (which should by approximately true given maximum likelihood) along with the prediction and its standard error against the data:

```
> #normal assumption on logit scale, back transform to
probability scale
> pred_norm<-predict(glm1,se.fit=T)
> lower<-pred_norm$fit-1.96*pred_norm$se.fit
> upper<-pred_norm$fit+1.96*pred_norm$se.fit
> logitp<-cbind(lower,pred_norm$fit,upper)
> p_pred<-1./(1+exp(-logitp))
> #prediction equation
> plot(elev,pred_norm$fit,type="l",ylab="Predicted
Logit",xlab="Elevation")
> matlines(elev,lower,lty=2)
> matlines(elev,upper,lty=2)
```

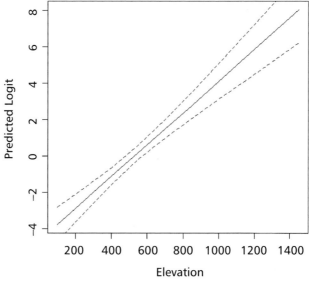

The back transformation to probability is accomplished by the code below, which also plots the predicted success vs. empirical success (y/n)

```
> expit<-function(x)
+ {
+ 1/(1+exp(-x))
+ }
>
>
>
>
> #prediction versus success
> success<-y/n
> plot(elev,success)
> matlines(elev,expit(pred_norm$fit),lty=1)
> matlines(elev,expit(lower),lty=2)
> matlines(elev,expit(upper),lty=2)
```

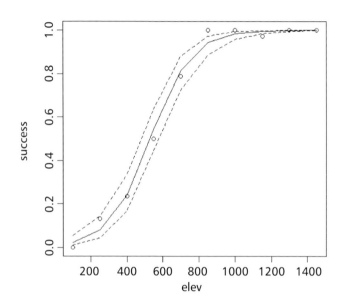

Finally, note that glm() provides an additional statistic called AIC or **Akaike's information criterion** (Appendix C). AIC is a metric, based on the likelihood computed for a specific model given the data, that we will use for comparing among alternative models later in this lab.

Multiple factor regression

In the above example, nests were grouped into levels of a single factor, and at each level n_i nests were observed with x_i successes. Often, we will be

(*Continued*)

interested in more than one factor, and typically factors vary in a continuous fashion along with the response, which now will be "success" or "failure" at each unique combination of factor values.

We illustrate this by returning to the warbler nest example, now with data on 70 nests located in a single season. The data file contains the response (in this case, 1 = the nest succeeded 0 = failure, elevation in meters, Julian date (days after 1 Jan) the nest was found, caterpillar abundance (number of cat-erpillars / 50 tree leaves), and shrub abundance (number of stems per 3 m × 3 m plot centered at the nest). We can start by reading the data in and building a model that predicts the response in terms of elevation and Julian date.

```
> #MULTIPLE REGRESSION
> nest_multiple<-read.csv("2011_nest_data.csv")
> attach(nest_multiple)
> glm1<-glm(success~jdate+elev,family=binomial(link=logit))
> summary(glm1)
```

Note that in this example we are not keeping track of number of nests, because at each location there is only 1 nest that succeeds (1) or fails (0). Therefore, we do not need a weighting factor as in the previous binomial example.

The summarized results for the glm are

Call:
glm(formula = success ~ elev + jdate, family = binomial(link = logit))

Deviance Residuals:

Min	1Q	Median	3Q	Max
-0.9235	-0.5698	-0.3120	-0.2000	2.4537

Coefficients:

	Estimate	Std. Error	z value	Pr(>|z|)	
(Intercept)	3.126836	5.270897	0.593	0.5530	
elev	0.004478	0.003330	1.345	0.1786	
jdate	-0.062751	0.028466	-2.204	0.0275	*

—

Signif. codes: 0 `***' 0.001 `**' 0.01 `*' 0.05 `.' 0.1 ` ' 1

(Dispersion parameter for binomial family taken to be 1)

    Null deviance: 49.754 on 69 degrees of freedom
Residual deviance: 42.065 on 67 degrees of freedom
AIC: 48.065

Number of Fisher Scoring iterations: 6

So, on the logit scale this model is predicting 1) an intercept (both predictors = 0) of 3.12, an increase of 0.004 with elevation, and a decrease of 0.06 with each Julian day.

Plotting the predicted response with the data is a bit tricky for this example, for 2 reasons. First, the response is a 1 (success) or 0 (failure) but we are really predicting the probability of success or failure, therefore we can't really expect the model predictions to line up with the data. The other issue is that if we wish to display the predictions in 2 dimensions (the usual scatterplot approach) we need to pick one predictor as the *x*-axis and then average over the second one. So for example, to plot versus elevation averaged over Julian date we can use code like this

```
> new <- data.frame(elev,jdate<-rep(mean(jdate),length
(elev)))
> ypred2<-predict(glm1, new, se.fit=T)
> lower<-ypred2$fit-1.96*ypred2$se.fit
> upper<-ypred2$fit+1.96*ypred2$se.fit
> y<-cbind(ypred2$fit,lower,upper)
> ypred<-1./(1+exp(-y))
> plot(elev,success)
> matlines(elev,ypred,col="red",lty=c(1,2,2))>
```

including polynomial terms. Finally, in Box 5.7 we show how to provide inference under multiple alternative models, using an information theoretic approach (Appendix D).

Readers who have followed the presentation to this point should have a basic grasp of statistical methods and will be able to perform many fairly sophisticated analyses. The remainder of the chapter delves into some areas that may not be

Box 5.7   Evaluating and comparing multiple models

The above example quickly leads to a problem (or opportunity, if you want to view it that way): we have more than one potential model that can be used to describe the data. So, above we built a model that predicted nest success as a linear function of elevation and Julian date:

```
>glm1<- glm(formula = success ~ elev + jdate, family =
binomial(link = logit))
```

However, we could just as easily have postulated models in which elevation alone, Julian date alone, or Julian date interacting with elevation, were predicting success, or neither. Also, we could have stipulated a quadratic response for a factor such as elevation, which would result in a maximum or minimum on the logit scale. We have thus created 6 models altogether

```
> glm0<-glm(success~1,family=binomial(link=logit))
> glm1<-glm(success~jdate,family=binomial(link=logit))
> glm2<-glm(success~elev,family=binomial(link=logit))
> glm3<-glm(success~elev+jdate,family=binomial(link=logit))
> glm4<-glm(success~elev+I(elev^2),family=binomial
(link=logit))
> glm5<-glm(success~elev+I(elev^2)+jdate,family=binomial
(link=logit))
```

Note that the code "I(eleve^2)" is R for "$elev^2$". Also note that glm0 is a "null" model: success~1 is simply a placeholder for "intercept only" ("1" here is "one" not "el").

So now we have 6 candidate models, and we haven't even talked about caterpillar abundance or shrub density (also measured) as potential predictors in replace of or in addition to elevation and Julian date. How do we decide which model to use? We discuss in detail various criteria to take into consideration in building and comparing models in Appendix D, and recommend the information theoretic approach based on AIC. We computed AIC, along with measures of fit (R^2, MSE) for the nest example, using code written in R. We focused specific interest in obtaining a prediction of nest survival at 1200 m elevation and Julian date of 176 days, both values near the averages observed for nests in the study.

The following code creates the 6 alternative models

```
> glm0<-glm(success~1,family=binomial(link=logit))
> glm1<-glm(success~jdate,family=binomial(link=logit))
> glm2<-glm(success~elev,family=binomial(link=logit))
> glm3<-glm(success~elev+jdate,family=binomial(link=logit))
> glm4<-glm(success~elev+I(elev^2),family=binomial(link=lo
git))
> glm5<-glm(success~elev+I(elev^2)+jdate,family=binomial(link
=logit))
```

Followed by this, we calculate predictions for each model at 1200 m elevation and Julian date of 176 days

```
> #predictions
> p0<-predict(glm0,se.fit=TRUE)
> p1<-predict(glm1,list(elev=1200,jdate=176),se.fit=TRUE)
> p2<-predict(glm2,list(elev=1200,jdate=176),se.fit=TRUE)
> p3<-predict(glm3,list(elev=1200,jdate=176),se.fit=TRUE)
> p4<-predict(glm4,list(elev=1200,jdate=176),se.fit=TRUE)
> p5<-predict(glm5,list(elev=1200,jdate=176),se.fit=TRUE)
```

The predictions, standard errors, and desired summary statistics are then formed into a table

```
> pred<-c(p0$fit[1],p1$fit,p2$fit,p3$fit,p4$fit,p5$fit)
> se_pred<-c(p0$se.fit[1],p1$se.fit,p2$se.fit,p3$se.fit,p4$se.
fit,p5$se.fit)
> mod<-c("null","jdate","elev","elev+jdate","elev+elev**2",
"elev+elev**2+jdate")
> npar<-c(summary(glm0)$df[1],summary(glm1)$df[1],
summary(glm2)$df[1],summary(glm3)$df[1],summary(glm4)$df[1],
summary(glm5)$df[1])
> RSS<-c(deviance(glm0),deviance(glm1),deviance(glm2),
deviance(glm3),deviance(glm4),deviance(glm5))
> sigma<-c(summary(glm0)$sigma,summary(glm1)$sigma,
summary(glm2)$sigma,summary(glm3)$sigma,summary(glm4)$sigma,
summary(glm5)$sigma)
> ss_total<-rep(deviance(glm0),length(mod))
> #regression sum of squares
> ssr<-ss_total-RSS
> R2<-ssr/ss_total
> lik<-c(logLik(glm0),logLik(glm1),logLik(glm2),
logLik(glm3),logLik(glm4),logLik(glm5))
> aic<-c(AIC(glm0),AIC(glm1),AIC(glm2),AIC(glm3),
AIC(glm4),AIC(glm5))
> N<-sum(n)
> aic_c=aic+2*npar*(npar+1)/(N-npar-1)
> delta<-aic_c-min(aic_c)
> Z<-exp(-0.5*delta)
> wt<-Z/sum(Z)
>
> models1<-data.frame(model=mod, R_squared=R2,RSS=RSS,
lik=lik, n_par=npar,pred=pred,se=se_pred,AICc=aic_c,wt=wt)
> attach(models1)
```

Finally, the table sorted by AIC is displayed

```
models1[order(AICc),]
write.csv(models1,"mult_models.csv")
```

(*Continued*)

Model	R_squared	RSS	Lik	n_par	pred	Se	AICc	wt
Null	0.000	49.75	−24.877	1	−2.048	0.376	51.812	0.066
Jdate	0.116	43.98	−21.992	2	−2.391	0.507	48.164	0.411
Elev	0.012	49.13	−24.566	2	−2.092	0.391	53.311	0.031
elev+jdate	0.155	42.06	−21.032	3	−2.543	0.562	48.428	0.360
elev+elev^2	0.014	49.05	−24.528	3	−2.233	0.656	55.419	0.011
elev+elev^2 +jdate	0.156	42.01	−21.006	4	−2.654	0.749	50.527	0.120

This indicates that the model with Julian date alone has the best AIC support, followed by the model with elevation and Julian date. Finally, a model-averaged prediction and unconditional standard error and confidence intervals is produced.

```
> #model averaging
> model_avg_pred<-sum(wt*pred)
> uncond_se<-sum(wt*sqrt(se_pred^2+
(pred-model_avg_pred)^2))
> model_avg_pred
[1] -2.44351
> uncond_se
[1] 0.5688231
> print("predicted probabilty")
[1] "predicted probabilty"
> 1./(1+exp(-model_avg_pred))
[1] 0.07991443
> lower<-model_avg_pred-1.96*uncond_se
> upper<-model_avg_pred+1.96*uncond_se
> print("CI")
[1] "CI"
> c(1./(1+exp(-lower)), 1./(1+exp(-upper)))
[1] 0.02769538 0.20938823
>
```

Mathematical details are provided in Appendix D, and the R code to perform this analysis is provided in Appendix G.

familiar to many readers, and that can be somewhat advanced. Therefore, some readers may wish to skip these sections for now, with the understanding that we may refer to the material at other points in the book.

Hierarchical models

Hierarchical models arise because nature – and the data that we observed from it – is constructed of hierarchies. There are numerous types of hierarchies, and

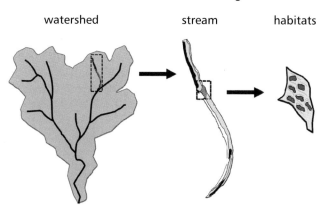

watershed   stream   habitats

Figure 5.6 A graphical example of a spatial hierarchy where stream reaches are nested within watersheds and habitats are nested within stream reaches.

they frequently overlap one another. In **systematic hierarchies**, individual animals are member of species, which in turn are nested in genera, families, orders, classes and so on. In **ecological hierarchies**, we may have individuals organized into populations, communities, ecosystems, biomes, etc., or they may be organized functionally, as in a food web (predators, herbivores, primary producers, detritivores, etc.). Finally, we often encounter **spatial hierarchies**, in which for example a landscape is composed of watersheds, streams, stream habitats, etc. (Figure 5.6). A common feature of hierarchies is that the different hierarchical levels relate to each other in predictable ways, and that information is "shared" across levels. Thus lower levels of the hierarchy tend to inherent features of the upper levels, while upper levels tend to "organize" features of the lower ones. For example, dogs and pumas are both animals of the order Carnivora, but belong to different families within that order. Both inherit features of the order (skull and dentition, feeding habits), as well as those of the higher levels (Mammalia: hair; vertebrata: bony skeletons) that they share with very dissimilar organisms (pigs; fish). Likewise, a watershed is "organized" as a collection of interconnected streams; in turn, individual streams inherit properties that are defined at the watershed level. Finally, some types of hierarchical, nested relationships come about principally because of the way that we acquire our field measurements. For instance, we may be interested in measuring the mass of a frog, but do so by sequential measurements on individuals, repeated through time. The individual frogs are properly viewed as the subjects, within which the measurements over time are hierarchically nested (Figure 5.7a). Likewise, interest may center on measuring the mean albumen content of bird eggs. However, the sample of eggs is obtained by first obtaining a sample of nests, and then sampling eggs within nests (Figure 5.7b).

These last two examples in particular bring to our attention the issue of **nonindependent errors**, one of the principal motivations of hierarchical modeling. That is, measurements on sampling units in the same hierarchical unit will tend to be correlated with one another, if for no other reason than they share the

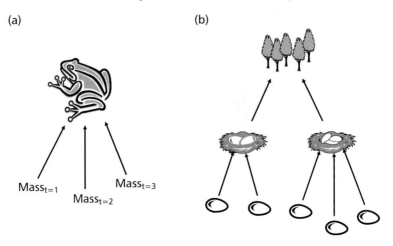

(a)   (b)

Mass$_{t=1}$ Mass$_{t=2}$ Mass$_{t=3}$

Figure 5.7 Examples of hierarchically structured data with (a) multiple measures on a single specimen and (b) multiple eggs from individual clutches.

attributes of that hierarchical unit. To take an extreme case, suppose that we are trying to characterize species occurrence at the landscape level, and are doing so by samples from streams within 50 watersheds. Suppose further that all streams within a watershed are perfectly correlated with respect to species occupancy, that is, all streams in a watershed are either occupied or all unoccupied. A random sample of $n = 500$ that is spread evenly over the 50 watersheds will as a result contain 50, not 500 independent measures. Further complicating the problem is that a completely random sample could easily over- or undersample some watersheds, resulting in a biased estimate of overall occupancy. More generally, non-independent errors can lead to estimates of variance that are biased low; biased measures of model fit and comparison; biased parameter estimates in predictive models; or all of the above. Other factors, such as missing or unknown (so unmeasured) covariates, shared histories of the subjects (e.g., animals that travel in pairs or other social groups), and overdispersion in count models (Appendix D) all lead to non-independence of errors being a common (even the usual) situation in ecology.

The problem of non-independent errors sometimes can be solved by modifying the sampling design to break dependencies, but in other cases this is not possible – or the hierarchical dependencies themselves are of interest. In either case, it is important to properly consider non-independent errors in modeling error structure, and we start by motivating the problem by means of a basic linear predictive model, now applied to a hierarchical system. The response we wish to consider, Y, is the response in terms of a population attribute, say abundance, of stream fishes, to an attribute X measured at the stream level. A simple linear model

$$Y_i = \beta_0 + \beta_1 X_i + \varepsilon_i$$

(a)

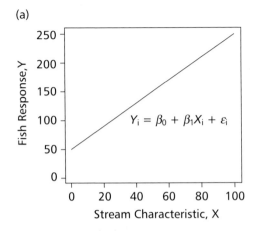

(b)

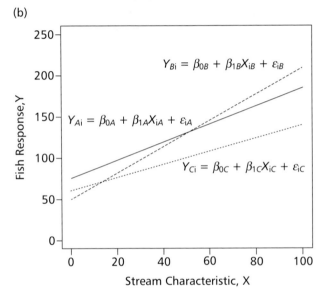

Figure 5.8 Hierarchical linear models (a) non-hierarchical response, (b) hierarchical response for 3 watersheds. β_{0i} refers to intercept and β_{1i} to slope for each watershed, $i = A, B,$ and C.

where the ε_i are assumed to be independent across the n replicated streams (Figure 5.8a). If we wanted to invoke normal regression (Appendix C) we would also assume that the errors follow a common normal distribution, that is

$$\varepsilon_i \sim Normal(0, \sigma^2)$$

we could invoke some other error distribution using generalized linear models (Appendix C), but we would need to assume independence in the error structure.

Now consider the situation depicted in Figure 5.8b, where we have three different watersheds ("A", "B", and "C"), each with different, but linear responses to the factor X measured in their streams. We could simply treat these as three different linear regression models

$$Y_{Ai} = \beta_{0A} + \beta_{1A}X_{iA} + \varepsilon_{iA}$$
$$Y_{Bi} = \beta_{0B} + \beta_{1B}X_{iB} + \varepsilon_{iB}$$
$$Y_{Ci} = \beta_{0C} + \beta_{1C}X_{iC} + \varepsilon_{iC}$$

This modeling approach obviously ignores the hierarchical relationships involved, namely that whereas the individual watersheds may have certain behaviors in common, others may be operating at a landscape scale. One way to capture this idea is to think of the parameters β_{jk}, $j = 0,1$, $k =$ A,B,C as *themselves* arising from linear models. Thus, we can model the intercepts for each watershed β_{0k} as linear responses to watershed-level attributes (e.g., various landscape metrics W_{1k}):

$$\beta_{0k} = \gamma_{00} + \gamma_{01}W_{1k} + u_{0k}$$

Similarly, we can model the slopes for the effects of the stream-level measurements (the site characteristics X_{iA}) as linear responses to the watershed-level metrics W_{1k}:

$$\beta_{1k} = \gamma_{10} + \gamma_{11}W_{1k} + u_{1k}$$

We can combine these models into a single overall model as:

$$Y_{ki} = \beta_{0k} + \beta_{1k}X_{ik} + \varepsilon_{ik}$$

where the β_{0k} and β_{1k} are defined hierarchically by the above relationships. This combined model can be expanded by replacing β_{0k} and β_{1k} by their respective predictive models

$$Y_{ki} = \gamma_{00} + \gamma_{01}W_{1k} + u_{0k} + (\gamma_{10} + \gamma_{11}W_{1k} + u_{1k})X_{ik} + \varepsilon_{ik}$$

Among other benefits, this hierarchical representation makes clear that there are three components or levels to model error:

- γ_{0k} – average watershed level, after accounting for watershed factors;
- γ_{1k} – average effect of the site-level attributes at the watershed level, after accounting for watershed factors;
- u_{ik} – unique effect of the sampling units (ik) that is unexplained in the model.

Consequently, there are many ways that these error components can be modeled, and by which the components may be partitioned. One of the simplest error models would assume that each of the components is independently distributed (e.g., normal). Assuming that the models are unbiased predictions this would give

$$u_{0k} \sim Normal(0, \sigma_{00}^2)$$
$$u_{1k} \sim Normal(0, \sigma_{11}^2)$$
$$\varepsilon_{ik} \sim Normal(0, \sigma^2)$$

More complex models arise by assuming a covariance structure (Appendix C). For example,

$$\underline{u}_k \sim MultiNormal(\underline{0}, \Sigma)$$

where

$$\underline{u}_k = \begin{bmatrix} u_{0k} & u_{1k} \end{bmatrix}$$

and

$$\Sigma = \begin{bmatrix} \sigma_{00}^2 & \sigma_{01} \\ \sigma_{10} & \sigma_{11}^2 \end{bmatrix}$$

Other, special cases arise from this model by simplifying constraints. For example, stipulating $\gamma_{01} = \gamma_{10} = \gamma_{11} = u_{1k} = 0$ produces:

$$Y_{ki} = \gamma_{00} + u_{0k} + \varepsilon_{ik}$$

which is a 1-way analysis of variance (ANOVA) model describing a grand mean γ_{00}, watershed effects u_{0k}, and unique site-level effects ε_{ik}. Another special case arises from assuming that $u_{0k} = u_{1k} = 0$; this produces:

$$Y_{ki} = \gamma_{00} + \gamma_{01}W_{1k} + \gamma_{10}X_{ik} + \gamma_{11}X_{ik}W_{1k} + \varepsilon_{ik}$$

which is a two-way ANOVA with interaction. Similarly, random effects and random coefficients models arise as special cases, by eliminating the fixed effects of the watershed-level factors. Then, the site-level coefficient models simplify to

$$\beta_{0k} = \gamma_{00} + u_{0k}$$
$$\beta_{1k} = \gamma_{10} + u_{1k}$$

which is a pure random coefficients model. Further simplification by $u_{1k} = 0$ simplifies the above to a random effects (random intercept) model. In Box 5.8, we illustrate hierarchical linear models, with an example involving continuous (normal) and binary (binomial) responses.

Shrinkage estimators, sharing information

One of the key features of hierarchical modeling is that it allows information to be "shared" among different sampling units, because of the linkages provided by the hierarchical model structure. This becomes important when some units of the system are sampled with better precision than others. If a portion of the variability of all units can be explained by the hierarchical model, then information from the more precise units can be "borrowed" by the less precise ones, to increase overall precision. An empirical Bayes **shrinkage estimator** (EB; Appendix C) will tend to "shrink" the model estimates from less precise units towards the mean relationship. The EB estimator computes a reliability measure

---

Box 5.8   Linear hierarchical modeling and partitioning of variance components

Hierarchical linear models can be fit in R using the lmer() function available in the lme4 library. Note that lme4 does not come with the base R package and must be installed. The lmer() procedure allows both specification of a distribution "family" (e.g., normal or Gaussian, Poisson, binomial) and a transformation or **link function** (identity [by default], log, logit, etc.).

To illustrate normal linear hierarchical modeling, we use a pooled set of sampling and data collected principally by the Idaho Department of Fish and Game and collaborating agencies. A total of 236 samples were collected from a 100-m long stream reaches within 23 watersheds within the Interior Columbia River Basin. The number of samples per subwatershed ranged from 8–16. The dataset includes the estimated density (no./m^2) of westslope cutthroat trout (*Oncorhynchus clarki lewisi*), a stream reach predictor, wetted stream width, and a watershed predictor, percent of watershed with productive soil type. The data for the study are available on the book website.

We begin the modeling procedure by evaluating whether or not a hierarchical model is appropriate by fitting our global model (all predictors) and examining the residuals ordered by the group that we suspect may have dependence, here watershed.

```
>
> REG_REGRESS<-glm(DENS~Soil_p+MN_WDTH+Soil_p*MN_
WDTH,data,family = gaussian)
>
> #### here we evaluate dependence by plotting the global
model residuals
> #### by watershed YOU ALWAYS DO THIS BEFORE ANY
HEIRARCHICAL MODELING
> boxplot(residuals(REG_REGRESS)~watershed)
>
```

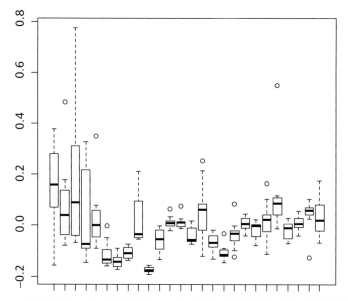

170102051503 170103011102 170602071807 170603050703

This plot indicated the predictions from the model are consistently higher and lower for some watersheds and suggests dependence among observations collected within individual watersheds. To account for the dependence, we could model the intercept and slope (for the site level predictor, mean wetted width) as randomly varying among watersheds. Before we do that, it is always useful to fit a random effects ANOVA to partition the variation within and among watersheds.

```
> ## Random effects ANOVA
> ## Note you need to specify REML = FALSE to get true
likeklihood and AIC
> rnd_anova<-lmer(DENS~1+(1|watershed),data, REML = FALSE)
> list(rnd_anova)
[[1]]
Linear mixed model fit by maximum likelihood
Formula: DENS ~ 1 + (1 | watershed)
   Data: data
   AIC  BIC  logLik deviance REMLdev
 -391.7 -381  198.8   -397.7  -391.9
Random effects:
 Groups    Name        Variance  Std.Dev.
 watershed (Intercept) 0.0119588 0.109357
 Residual              0.0098898 0.099447
Number of obs: 261, groups: watershed, 26
```

Fixed effects:

```
            Estimate Std. Error t value
(Intercept)  0.08543    0.02234   3.824
```

(Continued)

West-slope cutthroat density varied by $0.0119588/(0.0119588 + 0.0098898)$ = 0.547 or 54.7% among watersheds. Also, notice that the output includes AIC, which we can use to evaluate the relative support of candidate models. Below we fit several models with random effects with and without a covariance.

```
> ## Evaluate the support for multiple models below
> ## Model density function of soil productivity random
intercept
> model<-lmer(DENS~Soil_p+(1|watershed),data, REML = FALSE)
> list(model)
[[1]]
```

Linear mixed model fit by maximum likelihood
```
Formula: DENS ~ Soil_p + (1 | watershed)
   Data: data
   AIC    BIC  logLik deviance REMLdev
 -391.5 -377.3 199.8  -399.5  -380.6
Random effects:
 Groups    Name        Variance  Std.Dev.
 watershed (Intercept) 0.0110668 0.105199
 Residual              0.0098896 0.099446
Number of obs: 261, groups: watershed, 26
```

Fixed effects:
```
            Estimate Std. Error t value
(Intercept) 0.0499920 0.0334280   1.496
Soil_p      0.0007977 0.0005753   1.387
```

```
>
> ## Model density function of random intercept and random
slope for width
> ## model includes covariance between random intercept and
slope
> model<-lmer(DENS~MN_WDTH+(1 + MN_WDTH|watershed),data,
REML = FALSE)
> list(model)
[[1]]
```

Linear mixed model fit by maximum likelihood
```
Formula: DENS ~ MN_WDTH + (1 + MN_WDTH | watershed)
   Data: data
   AIC    BIC  logLik deviance REMLdev
 -422.9 -401.5 217.5  -434.9  -418.2
Random effects:
 Groups    Name        Variance   Std.Dev. Corr
 watershed (Intercept) 0.02813003 0.167720
           MN_WDTH     0.00021969 0.014822 -0.996
 Residual              0.00891584 0.094424
Number of obs: 261, groups: watershed, 26
```

Fixed effects:

```
              Estimate Std. Error t value
(Intercept)  0.137412   0.036213  3.795
MN_WDTH     -0.010380   0.003457 -3.002
>
> ## Model density function of soil productivity random
intercept and
> ## and random slope for width. The interaction between
soil productivity and width
> ## means that we are trying to account for the
variability on the slopes with a
> ## watershed predictor
> ## model includes covariance between random intercept and
slope
> model<-lmer(DENS~Soil_p+MN_WDTH+Soil_p*MN_WDTH+(1 + MN_
WDTH|watershed),data, REML = FALSE)
> list(model)
[[1]]
```

Linear mixed model fit by maximum likelihood

```
Formula: DENS ~ Soil_p + MN_WDTH + Soil_p * MN_WDTH + (1 +
MN_WDTH | watershed)
   Data: data
   AIC    BIC logLik deviance REMLdev
 -426.4 -397.9  221.2  -442.4  -394.1
```

Random effects:

```
 Groups    Name        Variance   Std.Dev. Corr
 watershed (Intercept) 0.02444508 0.156349
           MN_WDTH     0.00022085 0.014861 -1.000
 Residual              0.00885732 0.094113
Number of obs: 261, groups: watershed, 26
```

Fixed effects:

```
              Estimate Std. Error t value
(Intercept)   8.542e-02 5.419e-02  1.576
Soil_p        1.404e-03 9.021e-04  1.557
MN_WDTH      -8.206e-03 5.553e-03 -1.478
Soil_p:MN_WDTH -6.434e-05 9.101e-05 -0.707
>
> ## Model density function of soil productivity random
intercept and
> ## and random slope for width
> ## model DOES NOT includes covariance between random
intercept and slope
> model<-lmer(DENS~Soil_p+MN_WDTH+Soil_p*MN_
WDTH+(1|watershed)+(0+MN_WDTH|watershed),data, REML = FALSE)
> list(model)
[[1]]
```

(Continued)

Linear mixed model fit by maximum likelihood

```
Formula: DENS ~ Soil_p + MN_WDTH + Soil_p * MN_WDTH +
(1 | watershed) +      (0 + MN_WDTH | watershed)
   Data: data
   AIC    BIC logLik deviance REMLdev
 -399.3 -374.3  206.6   -413.3  -366.1
```

Random effects:
```
 Groups     Name         Variance  Std.Dev.
 watershed (Intercept) 0.0070001 0.083667
 watershed MN_WDTH     0.0000000 0.000000
 Residual              0.0097620 0.098803
Number of obs: 261, groups: watershed, 26
```

Fixed effects:
```
             & Estimate Std. Error  t value
(Intercept)    1.079e-01  4.014e-02   2.689
Soil_p         1.160e-03  6.877e-04   1.687
MN_WDTH       -8.941e-03  4.335e-03  -2.063
Soil_p:MN_WDTH -2.631e-05  6.773e-05  -0.388
>
```

The third model was the best approximating. We should check goodness-of-fit of that model by ordering the residuals by watershed. The plot below suggests that the dependence was accounted for.

```
> ## Check goodness of fit of best approximating model
> model<-lmer(DENS~Soil_p+MN_WDTH+Soil_p*MN_WDTH+(1 +
MN_WDTH|watershed),data, REML = FALSE)
>
> #### here we evaluate if the dependence was accounted for
> boxplot(residuals(model)~watershed)
>
```

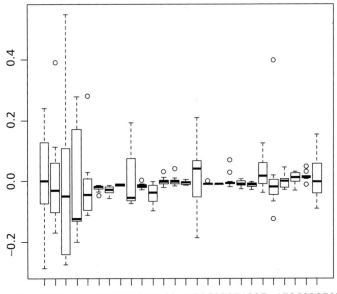

Logit linear model

Below is the R code and output for fitting a logit-normal hierarchical model to the fish survival example data in box 5.10.

```
> data <- read.table("fish_hierch_data.csv",skip=0,head=T,
sep=",")
> attach(data)
> names(data)
[1] "Site.no"  "Elev"  "Flow"  "Fish"  "Lived"
>
> ## Site = sample site ID
> ## Elev = elevation (site level measure)
> ## Flow = average stream discharge
> ## Fish = number of fish
> ## Lived = number of fish living
>
> library(lme4)
> Y<-cbind(Lived,Fish-Lived)
> ## Random effects logit-linear model
> ## Note you need to specify family = binomial,
> Logit_HLM<-lmer(Y ~1+Elev + Flow+(1|Site.no),data, family
= binomial, REML = FALSE)
> list(Logit_HLM)
Generalized linear mixed model fit by the Laplace
approximation
Formula: Y ~ 1 + Elev + Flow + (1 | Site.no)
   Data: data
   AIC   BIC logLik deviance
 114.0 124.4 -52.99    106.0
Random effects:
 Groups Name        Variance Std.Dev.
 Site.no  (Intercept) 2.6372   1.6240
Number of obs: 100, groups: Site, 10
```

Fixed effects:

	Estimate	Std. Error	z value	Pr(>|z|)	
(Intercept)	-1.040806	1.047723	-0.993	0.320516	
Elev	0.007125	0.002137	3.334	0.000855	***
Flow	0.082067	0.040543	2.024	0.042947	*

```
―
Signif. codes:  0 '***' 0.001 '**' 0.01 '*' 0.05 '.' 0.1
' ' 1
```

for a model coefficient (β_{0k} in this case) as $\hat{\lambda}_k = \dfrac{\hat{\tau}_{00}}{\hat{\tau}_{00} + \hat{\sigma}_k^2 / n_k}$, and computes the shrinkage estimator as

$$\hat{\beta}_{0k}^{EB} = \hat{\lambda}_k \hat{\beta}_{0k} + (1 - \hat{\lambda}_k)\hat{\gamma}_{00}$$

where $\hat{\tau}_{00} = 1 / \hat{\sigma}_{00}^2$ and $\hat{\sigma}_k^2$ is an empirical estimate of variance based on a sample of n_k on the kth unit, and $\hat{\gamma}_{00}$ is an estimate of the grand mean. We illustrate this with the earlier fish example (Figure 5.8). Suppose for simplicity that we have two watersheds, both having the same or similar slopes in the relationship with X, but different intercepts, and different precisions for the individual watershed regressions, with one relatively precise but the second imprecise. The EB estimator "shrinks" the prediction line for second toward the average, but has small influence on the other estimated line, since it is more precise (Figure 5.9).

The idea of borrowing information and shrinkage has potentially important implications for parameterizing natural resource decision models. Natural resource management decisions often involve at risk species or ecosystems for which there is little empirical data. For example, rare species are infrequently encountered during sampling, leading to information-poor datasets with few observations. In these instances, there often are insufficient data for fitting models with species-specific or system-specific data. One useful approach is to

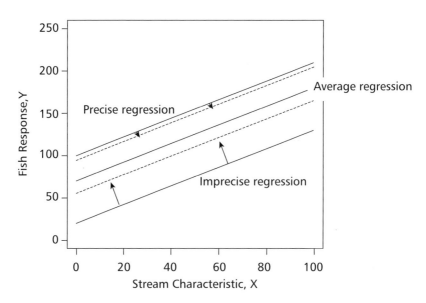

Figure 5.9 Shrinkage estimation with linear hierarchical models. Shrinkage of imprecision regression line to average regression is greatest. Dotted lines represent shrinkage toward average regression (middle solid line) from each separate regression (outer lines).

include data from similar species (e.g., within the same genera) or systems and fit a hierarchical model that models the parameters randomly varying among species or systems. The EB parameter estimates would integrate the information contained in the data on the species or system of interest and borrow information from the others. Alternatively, if data from the species or system of interest are completely lacking, the γ_{1k} and σ^2_{1k} could be used to estimate the expected effect of predictor k on the response and the associated uncertainty on the effect, respectively. When taking this approach to estimating parameters when data are sparse or lacking, you should be careful to include data from only those species or systems that are believed to be similar to those of interest.

Bayesian inference

The procedures to this point are based approaches such as least square or maximum likelihood estimation (Appendix C). In either case, we were operating implicitly under a "paradigm", called the **frequentist paradigm**, that (1) treats parameters as fixed (but unknown) constants, and (2) bases probabilistic inference on the notion of repeated experiments (see the earlier description in this chapter about the different ways of viewing probability). That is, when we compute a variance or confidence interval on a parameter estimate, we are *not* making a probability statement about the parameter. Rather, we are making a statement about what proportion of repeated experiments (similar to the one we just conducted) would contain the true parameter value. So if we repeated our study 10 000 times, we would expect a 95% confidence interval to contain the parameter approximately 9500 of those times. Finally, (3) we would be basing all of our conclusions about confidence on the results of this one study, without regard to previous studies.

By contrast, the **Bayesian paradigm** applies the rules of probability to inference about unknown quantities (e.g., parameters, predictions, missing values), conditional on known quantities (e.g., observed data). Under the Bayesian paradigm, parameters may have a fixed value, but uncertainty in that value is expressed by probability distributions; thus in contrast to the frequentist paradigm, it is permitted (or even necessary) to make probability statements about parameters. With Bayes, we update information about the parameters or other unknown quantities using data, using the rules of conditional probability and Bayes' theorem (Appendix A). Under the Bayesian paradigm we can incorporate information from previous knowledge (e.g., another study) to inform the distribution of the parameter; thus, under Bayes our posterior (after data collection) inference about the parameter is a blend of previous knowledge (if any) and our current study. Finally, Bayes does not require a hypothetical repetition of the study for inference: probability statements about the parameter (not some statistic) are possible from a single study (or even before the collection of data). However, replication of studies is valued, in that each additional bit of empirical knowledge can be folded into our current understanding of the system.

The basic theory behind Bayesian inference is covered in Appendix C, but the core is the application of Bayes' theorem to the relationship between the data and the unknowns. First, the joint distribution of unknowns (parameter, predictions, etc., denoted as θ) and the data is given by

$$f(\theta, x)$$

which in turn can be written in terms of conditional distributions (Appendix A) as:

$$f(\theta, x) = f_x(x \mid \theta)f_\theta(\theta) = f_\theta(\theta \mid x)f_x(x)$$

providing us Bayes' theorem, one form of which is:

$$f_\theta(\theta \mid x) = \frac{f_x(x \mid \theta)f_\theta(\theta)}{f_x(x)}$$

Because the denominator of the above expression represents the probability of the data, which have been collected, it is treated as a constant, leading to:

$$f_\theta(\theta \mid x) \propto f_x(x \mid \theta)f_\theta(\theta)$$

where "$\propto$" denotes "is proportional to". Thus, our knowledge of the unknown quantities after collection of data is changed proportionally by the multiplier $f_x(x \mid \theta)$ which turns out to be the sampling model or **likelihood** for the data (Appendix C).

Example – Binomial likelihood with a beta prior on *p*

We illustrate with an example where interest focuses on updating knowledge about the binomial parameter p. We will initially start by assuming that we have no prior information about p, apart from the obvious fact that $0 < p < 1$. In this case a non-informative ("vague") prior would be captured by a uniform distribution (Appendix A) between 0 and 1

$$f(p) = U(0, 1)$$

which is equivalent to

$$f(p) = Beta(1, 1)$$

(Appendix B). The sampling distribution for the binomial is

$$f(x \mid p) = Binomial(x \mid n, p)$$

Thus, it follows that

$$f(p \mid x) = Beta(x + 1, n - x + 1)$$

That is, the posterior distribution of p is simply another beta distribution, with parameters that contain the original prior parameters (1 and 1, known as **hyper-parameters**) and a summary of the data result (x, $n - x$). More generally, if the prior of p given by a beta(α,β) (so, with hyper-parameters α and β), and our sample experiment has n trials with x success, we obtain the posterior of p by

$$f(p \mid x) = Beta(x + a, n - x + b).$$

Thus, for example, if we took a "vague" prior on p from beta(1,1) and then observed $x = 6$ in $n = 10$ trials, the posterior distribution of p would be

$$f(p \mid x = 6, n = 10) = Beta(6 + 1, 10 - 6 + 1) = Beta(7, 5)$$

We show how to implement this example (and more general binomial–beta examples showing how to incorporate data from previous studies) in Box 5.9.

In Appendix C we provide a more general discussion of Bayesian approaches and discussion of some of the computational approaches to Bayes. A summary is that the binomial–beta example above is an example of a situation where the prior and posterior distributions form a **conjugate pair**, meaning that the posterior and prior distributions are the same (in this case, both are betas). Conjugate pairings are handy and should be used when available, but many if most of our more complex models will not be amenable to this approach. For more general problems we will almost certainly need specialized programs such as WinBUGS (Appendix G) to conduct the necessary computations using Markov chain Monte Carlo (MCMC) procedures (Appendix C).

Hierarchical and random effects models

Hierarchical and random effects models both naturally fit into the Bayesian paradigm, where now some of the "unknowns" are hierarchically dependent on other unknowns. In a simple example, we have 10 experimental units, say streams, and for each one we conduct a binomial experiment, where we mark $n = 10$ fish in each stream and determine the number of x_i successes (number that survive), $i = 1, \ldots, 10$. Let $p =$ the probability of survival in the marked sample, with survival rates potentially different in the 10 streams, p_i, $i = 1, \ldots,$ 10. A random effects model might stipulate

$$f_y(x_i) = Normal(\mu, \sigma^2)$$

and p_i is predicted by an inverse logit transform

$$p_i = 1 / (1 + \exp(-y_i))$$

Thus, y comes from a normal distribution, and p is a transformation of that result (to place the value of p on the (0,1) interval). The upshot is that we are now modeling p with the parameters μ and σ^2 that in turn determine (on the logit scale) the mean and variance of p among the 10 units. The data

Box 5.9 Bayesian example: binomial likelihood with a beta prior

We can obtain exact posterior inference under the binomial-beta model, once we have specified a prior and then obtained data. Let's return to the example of the stream fishes. Suppose that we mark a sample of 10 fish with radio tags, follow them for 1 year, and determine that 6 of the 10 survive. Given our vague (uniform) prior on p, the posterior distribution can be gotten exactly by

$$f(p \mid x = 6, n = 10) \sim Beta(6 + 1, 4 + 1) = Beta(7, 5)$$

R easily computes and plots the posterior distribution:

```
> #BINOMIAL-BETA

> # UNIFORM prior on p
> a<-1
> b<-1
> n<-10
> x<-6
> p<-seq(0,1,0.001)
> prior<-dbeta(p,a,b)
> post<-dbeta(p,a+x,b+n-x)
> plot(p,post,type="l")
> lines(p,prior,col="red",lty=2,lwd=3)
```

The posterior (solid line) is plotted overlaid with the prior (dashed) line. We can see that the observations have now given us some confidence that p is around 0.6 (and not just somewhere between 0 and 1).

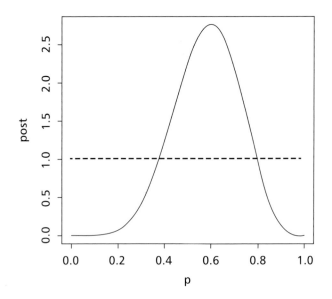

We can obtain more formal statistics easily enough from the posterior. To calculate and plot the median and the lower and upper 2.5% quantiles:

```
> #statistics from the posterior
> #median
> med<-qbeta(.5,a+x,b+n-x)
med
[1]  0.5881096
> #lower, upper
> interval<-c(qbeta(.025,a+x,b+n-x),qbeta(.975,a+x,b+n-x))
>interval
[1]  0.3079047 0.8325119
> #plot median, upper and lower
> plot(p,post,type="l")
> lines(c(med,med),c(-.20,dbeta(med,a+x,b+n-x)),lty=2)
> lines(c(interval[1],interval[1]),c(-.20,dbeta(interval[1],
a+x,b+n-x)),lty=2)
> lines(c(interval[2],interval[2]),c(-.20,dbeta(interval[2],
a+x,b+n-x)),lty=2)
>
```

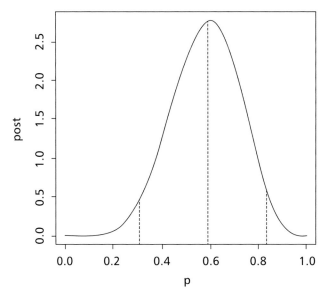

The median (or sometimes mean) value can be used as a Bayesian estimate of p (=0.588 in this example). The upper and lower 2.5% quantiles (0.307, 0.832) form a 95% probability interval under the posterior distribution also known as a **Bayesian credible interval** (BCI). The BCI is analogous to the confidence interval (CI) in frequentist statistics. A notable difference is that the BCI is directly interpretable in terms of the probability that the parameter takes on a value in the interval. By contrast, the CI only has meaning in the context of repeated samples, in x% of which the CI is predicted to include the parameter value.

(Continued)

We can also simulate data from the posterior distribution, treat these as samples of the parameter value, and compute descriptive statistics. For example, to get descriptive statistics and a histogram from 10 000 samples from the above posterior:

```
> #posterior statistics by simulation
> x_post<-rbeta(10000,a+x,b+n-x)
> quantile(x_post,c(0.025,0.5,0.975))
     2.5%       50%      97.5%
0.3091363 0.5858421 0.8364010
> hist(x_post,xlim=c(0.4,0.8))
```

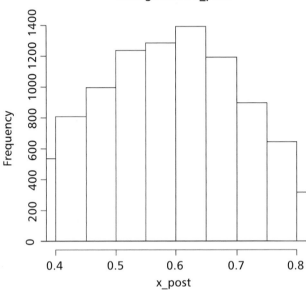

Histogram of x_post

The results are (as they should be) very close to those provided by directly using the distribution, so you may be asking "why bother"? For cases like this in which the posterior distribution takes on the form of a known distribution, there is no advantage. However, many more complex models cannot be expressed as known distributions, but it may be possible still to sample from these distributions using simulation methods such as **Markov chain Monte Carlo (MCMC)**.

Sometimes information about the parameter will be based on multiple sources or studies. In some cases there will be enough information to directly build a prior distribution, but often there will be only summary information such as means, ranges, and perhaps variances on the prior estimates. If we have a prior estimate of the mean of p and its variance we can use the method of moments for the beta distribution to get a reasonable estimate of the parameters a and b. To refresh your memory, we built a small function in R to do this.

```
beta.mom<-function(mean,sd){
v<-sd**2
```

```
x<-mean
a<-x*(x*(1-x)/v-1)
b<-(1-x)*(x*(1-x)/v-1)
c(a,b)
}
```

The inputs are the mean and standard deviation of *p*, and the output will be a list containing the MOM estimates of *a* and *b*. So for example, if we have prior information that leads us to believe that the mean of *p* is approximately 0.8 and the sd is 0.1, we can use this function:

```
> beta.mom(.8,.1)
[1] 12   3
```

We would then use beta(12,3) as our informative prior, together with whatever new data we collect.

The posterior (solid) and prior (dashed) distributions are plotted by

```
parms<-beta.mom(0.8,0.1)
a<-parms[1]
b<-parms[2]
n<-10
x<-6
p<-seq(0,1,0.001)
prior<-dbeta(p,a,b)
post<-dbeta(p,a+x,b+n-x)
plot(p,post,lty=1,type="l")
lines(p,prior,lty=2)
```

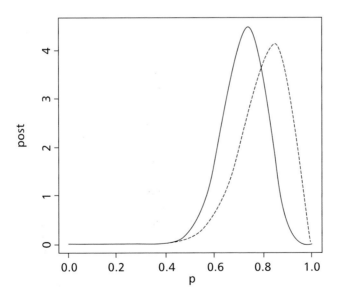

(Continued)

```
> #statistics from the posterior
> #median
> med<-qbeta(.5,a+x,b+n-x)
> #lower, upper
> interval<-c(qbeta(.025,a+x,b+n-x),qbeta(.975,a+x,b+n-x))
> med
[1] 0.7259438

> interval

[1] 0.5328872 0.8738479
>
> #plot median, upper and lower
> plot(p,post,type="l")
> lines(c(med,med),c(-.20,dbeta(med,a+x,b+n-x)),type="l")
> lines(c(interval[1],interval[1]),c(-.20,dbeta(interval[1],
a+x,b+n-x)),lty=2)
> lines(c(interval[2],interval[2]),c(-.20,dbeta(interval[2],
a+x,b+n-x)),lty=2))
```

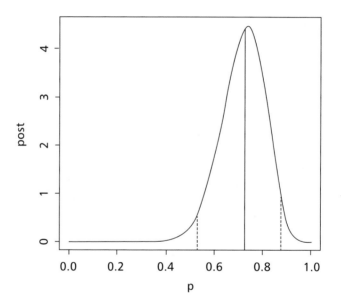

A couple of words of caution are in order here. First, we are now potentially giving a lot of weight to prior information, and maybe not much to our data, and we will have to think carefully about whether that is a good idea. The second is a technical issue that the MOM approach for the beta sometimes fails (produces negative values for the parameters for example); this can happen if we use particularly large sd values (try the above with beta.mom(0.8,0.8) to see what we mean). In such cases we are going to have to use other approaches to get a reasonable prior, for example describing the prior distribution of p by a discrete distribution in which we specify the probability of p being in certain intervals (e.g., the first, second, third and fourth quartiles).

likelihood is a product of conditionally independent (Appendix A) binomial distributions:

$$f_x(x_i) = Binomial(n, p_i)$$

where

$$p_i = 1 / [1 + \exp(-y_i)]$$

See Box 5.10 for an example implemented in WinBUGS. More complex models, e.g. involving hierarchical predictors, are likewise easy to construct in the Bayesian paradigm; we give an example in Box 5.11.

Box 5.10 Modeling a random effect using WinBUGS: Binomial success with random variation

Here, we illustrate how a random effect can be incorporated into a statistical model using WinBUGS. In the example, we have 10 sites (e.g., streams) and at each site we take a binomial sample to estimate a success parameter. For example, we mark 10 fish with radio transmitters and return in 30 days and determine x, the number of fish that have survived. We could simply pool the 100 marked fish and determine x overall, but we suspect that survival varies from stream to stream in a random fashion. So, we will instead allow p the success parameter to vary by drawing it from a normal distribution on the logit scale.

```
#model  code for BUGS
{
mu~dnorm(0,0.0001)
sd~dunif(0.0001,100)
tau<-pow(sd,-2)
for(i in 1:nsites)
{
beta[i]~dnorm(mu,tau)
logit(p[i])<- beta[i]
x[i]~dbin(p[i],n[i])
}
}

library(R2OpenBUGS)
data<-list("x","n","nsites","n")
parameters<-c("mu","sd","p")
model.file<-"binomial_re_mod.txt"
```

(Continued)

```
inits<-list(list(mu=1,sd=1),list(mu=-1,sd=2),list(mu=0,sd=1))
temp=bugs(data,inits,parameters,model.file,n.chains=3,n.
iter=50000,OpenBUGS.pgm="C:/OPENBUGS/OpenBUGS.
exe",codaPkg=FALSE,debug=TRUE)
plot(temp)
```

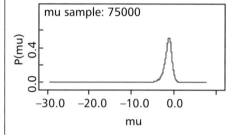
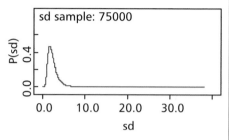

	mean	sd	MC_error	val2.5pc	median	val97.5pc	start	sample
deviance	28.8	4.719	0.02525	21.66	28.11	39.84	25001	75000
mu	−1.422	1.048	0.006315	−3.763	−1.304	0.2529	25001	75000
p[1]	0.2079	0.1152	4.369E-4	0.03808	0.1903	0.4737	25001	75000
p[2]	0.05436	0.0637	3.736E-4	7.722E-5	0.03138	0.2293	25001	75000
p[3]	0.6484	0.1406	6.173E-4	0.3545	0.6576	0.8908	25001	75000
p[4]	0.5597	0.1448	5.746E-4	0.2753	0.5623	0.8276	25001	75000
p[5]	0.5591	0.1449	5.987E-4	0.2718	0.5618	0.8281	25001	75000
p[6]	0.05462	0.06436	3.916E-4	7.423E-5	0.03157	0.234	25001	75000
p[7]	0.2929	0.13	4.967E-4	0.08061	0.2803	0.5765	25001	75000
p[8]	0.2069	0.1147	4.298E-4	0.03841	0.19	0.4717	25001	75000
p[9]	0.5589	0.1452	5.771E-4	0.2722	0.562	0.8278	25001	75000
p[10]	0.05478	0.06403	3.744E-4	7.208E-5	0.03201	0.2307	25001	75000
sd	2.474	1.357	0.01416	0.9517	2.177	5.767	25001	75000

Lunn, D.J., A. Thomas, N. Best, and D. Spiegelhalter. (2000) WinBUGS – a Bayesian modelling framework: concepts, structure, and extensibility. *Statistics and Computing* **10**, 325–337.

Box 5.11   Hierarchical model with random effects in WinBUGS

We illustrate hierarchical modeling involving hierarchical predictors with another fish survival example. This time, 10 streams were sampled along an elevation gradient (100–1000 m). Within each stream, sites were selected at 10 flow levels (m³/s). At each site, 10 fish were equipped with radio transmitters and 30-day survival monitored. We modeled the stream-level intercepts (b0) as a function of elevation (g1); site level logit survival probability took the stream-level intercept and modeled the additional effect of flow rates (b1). The model code and R code to read the data and interface with OpenBUGS are summarized below (the data and complete R script file and model code are on the book website).

```
#model
{
## site level intercept and slope
g0~dnorm(0,0.37)
g1~dnorm(0,0.37)


## Flow slope
b1~dnorm(0,0.37)


sd~dunif(0.0001,n_obs)
tau<-pow(sd,-2)


# unique effect for each site
for(j in 1:10) { r[j]~dnorm(0,tau) }


for(i in 1:100)
{
logit(p[i])<- g0 + g1*Elev[i] + b1*Flow[i] + r[Site.no[i]]
p2[i]<-max(0.00001,min(0.99999, p[i]))
Lived[i]~dbin(p2[i],Fish[i])
}
}


fish_hier<-read.csv("fish_hierch_data.csv")
attach(fish_hier)
x<-lived
n_sites<-10
n_streams<-10
n<-10
n_obs<-n_sites*n_streams

library(R2OpenBUGS)
data<-list("Lived","Fish","n_obs","Elev","Flow")

parameters<-c("b1","g0","g1","sd")

model.file<-"binomial_hier_mod.txt"

inits<-list(list(g0=0,g1=0,b1=0,sd=1),list(g0=0,g1=0,b1=0,sd=
1),list(g0=0,g1=0,b1=0,sd=1))
#inits<-list(g0=0,g1=0,b1=0,sd=1)


temp=bugs(data,inits,parameters,model.file,n.chains=3,n.
iter=50000,OpenBUGS.pgm="C:/OPENBUGS/OpenBUGS.
exe",codaPkg=FALSE,debug=TRUE)
plot(temp)
```

(Continued)

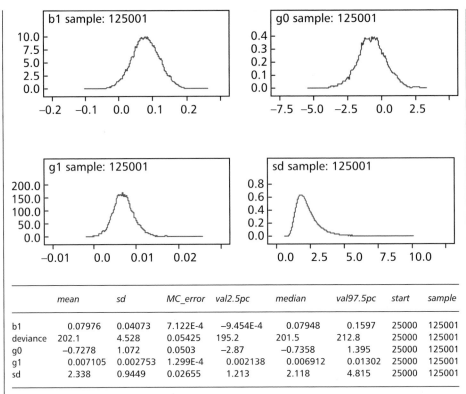

	mean	sd	MC_error	val2.5pc	median	val97.5pc	start	sample
b1	0.07976	0.04073	7.122E-4	−9.454E-4	0.07948	0.1597	25000	125001
deviance	202.1	4.528	0.05425	195.2	201.5	212.8	25000	125001
g0	−0.7278	1.072	0.0503	−2.87	−0.7358	1.395	25000	125001
g1	0.007105	0.002753	1.299E-4	0.002138	0.006912	0.01302	25000	125001
sd	2.338	0.9449	0.02655	1.213	2.118	4.815	25000	125001

Lunn, D.J., A. Thomas, N. Best, and D. Spiegelhalter. (2000) WinBUGS – a Bayesian modelling framework: concepts, structure, and extensibility. *Statistics and Computing* **10**, 325–337.

Resampling and simulation methods

Here, we discuss two important types of procedures for estimating variances and other distribution attributes that employ computationally intensive approaches. These procedures can be very useful in decision making when, for example, it is very difficult to compute variances by standard methods such as maximum likelihood, due to the complex mathematical form of the quantity being estimated, or because of small sample size or non-random sampling. The first of these, **jackknife estimation** (covered in more detail in Appendix C) is a data resampling procedure that allows for the estimation of variance and evaluation of bias. The basic jackknife involves passing through the data, sequentially leaving out observations and computing the estimate on the remaining observation. If we have an observed sample $\underline{x} = [x_1, x_2, .. x_n]$, the ith jackknife sample is given by

$$\underline{x}(i) = [x_1, x_2, .. x_{i=1}, x_{i+1}, \ldots, x_n]$$

The ith jackknife replicate of $\hat{\theta}$ is then given by

$$\hat{\theta}(i) = \hat{\theta}(\underline{x}(i))$$

A jackknife estimate of variance is then provided by

$$\text{var}\left(\hat{\theta}\right)_{jack} = \frac{n-1}{n}\sum_{i=1}^{n}\left(\hat{\theta}(i)-\hat{\theta}(.)\right)^2$$

We can also use **bootstrap sampling** to estimate bias and standard deviation. Standard (nonparametric) bootstrapping involves resampling with replacement from the sample data, under the assumption the data itself represents the underlying probability distribution. Bootstrapping allows us to build up a large "sample" of estimates, by repeatedly re-estimating the parameter from permutations of the data. We illustrate jackknife and bootstrap variance estimation in Box 5.12.

Box 5.12   Jackknife and bootstrap estimation of variance and bias

We illustrate jackknife variance estimation with an example of the computation of the ratio between y and x, where y and x are sample counts or other measures. Interest focuses on R, the estimate of y/x. A simple (but slightly biased) estimate of R is provided by

$$\hat{R} = \frac{\bar{y}}{\bar{x}}$$

The variance and standard deviation of this estimate can be approximated based on the variances of y and x. We will instead use the jackknife to get estimates.

We will assume that we have sample data for y and x, which we will read into a dataframe in R

```
x<-c(2,5,6,7,9,14)
y<-c(3,5,7,8,9,15)
dat<-data.frame(cbind(x,y))
```

We wrote a small function in R to compute estimates of bias and standard deviation using the jackknife.

```
#JACKKNIFE SAMPLING OF RATIO ESTIMATOR
jack<-function(x,y)
{

theta.hat<-mean(y)/mean(x)
n<-length(x)
theta.jack <- numeric(n)
for (i in 1:n){
theta.jack[i] <- mean(y[-i]) / mean(x[-i])
}
xx<-theta.jack-mean(theta.jack)
```

(Continued)

```
var_jack<-(n-1)/n*sum(xx**2)
bias <- (n - 1) * (mean(theta.jack) - mean(theta.hat))
return(c(bias,sqrt(var_jack)))
}
jack(dat$x,dat$y)
```

The program produces an estimate of standard deviation of 0.035; it also produces an estimate of bias of 0.0048, suggesting that the estimator is slightly positively biased.

We also illustrate bootstrapping with the ratio example.

```
#BOOTRAP SAMPLING OF RATIO ESTIMATOR
>ratio <- function(d,w )
>sum(d$y * w)/sum(d$x * w)
```

Here we use the built-in R function boot() to perform bootstrapping. First, we define the ratio estimator with a user-defined function:

```
>ratio <- function(d,w )
>sum(d$y * w)/sum(d$x * w)
```

Then, we apply bootstrapping to this function to take 1000 samples:

```
>boot(dat, ratio, R=1000, stype="w")
```

In general, we can apply the bootstrap to any estimating function, drawing a large number of (1000, 10 000 or more) of bootstrap samples from the data with replacement and calculating the estimate for each sample. For the ratio example, this results in an estimated standard deviation of 0.0376 and bias of 0.004, very similar to the values obtained under jackknifing.

A special case of bootstrapping is termed **parametric bootstrapping**, where the parameters of the simulating distribution are estimates (e.g., by maximum likelihood) and then values are simulated, conditional on these as fixed values. Parametric bootstrapping is essentially a form of Monte Carlo simulation (Chapter 6, Chapter 8, and Appendix E) where instead of assuming fixed values for the parameters we estimate them from data. Parametric bootstrapping can be very useful when decisions depend on a complex array of inputs but each of the inputs can be characterized by a parametric distribution. We can then use parametric bootstrapping to simulate output values and characterize the statistical distribution of the output without necessarily knowing its parametric form. We give an example of parametric bootstrapping in Box 5.13.

Box 5.13 Parametric bootstrap estimation of confidence intervals

Parametric bootstrapping (and more generally Monte Carlo simulation) can be very useful for generating samples for functions of random variables such as estimators, but also for output from models where components of the model may have in turn been simulated from parametric distributions.

We illustrate parametric bootstrapping for estimating confidence intervals for a simulation–optimization problem that we will visit in more detail in Chapter 8. In this problem we are optimizing resource allocation to 2 species in order to meet an objective, but there is a constraint on the total resources available. However, we are uncertain about the constraint value (but have an estimate of its mean (280) and sd (5), and we assume that the realized constraint has a normal distribution with these parameters. Because the constraint affects the objective outcome in a nonlinear manner, it is not immediately obvious how to get a confidence interval on the objective value. Our approach is to

- Make an allocation based on the mean constraint value.
- Draw random values of the constraint from a normal (280,5).
- Recalculate the objective value for each constraint value.
- Calculate summary statistics including a CI from the simulated objective value.

The R code to accomplish this, along with example results, is shown below.

```
> #PARAMETRIC BOOSTRAPPING EXAMPLE
>
> #optimal value for mean constraint, then simulate value
with normal with that mean and specified sd
> #find optimal for mean constraint
> coef_mean<-280
> coef_sd<-20
> n_reps<-10000
```

Here, we define the objective function, using a Lagrangian multiplier (Appendix E) to incorporate the constraint.

```
> obj_c<-function(param,coef)
+ {x1<-param[1]
+ x2<-param[2]
+ lambda<-param[3]
+ L<-224*x1+84*x2+x1*x2-2*x1^2-x2^2+lambda*(coef-3*x1-x2)
+ L
+ }
>
```

(Continued)

The value of L (the constrained objective) is then passed to the linear solver in R

```
> #use linear solver
> a<-matrix(data<-c(4,-1,3,-1,2,1,3,1,0),nrow=3,ncol=3,
byrow=T)
> #b<-c(224,84,280)
> b<-c(224,84,coef_mean)
> print("optimal decision")
[1] "optimal decision"
> dec<-solve(a,b)
> dec
[1] 69 73  7
>
```

We then simulate random values of the constraint and see how these affect the objective value.

```
> #generate samples of the constraint value
> constraint<-rnorm(n_reps,coef_mean,5)
>
> print("sd =5")
[1] "sd =5"
> obj_val<-obj_c(dec,constraint)
> mean(obj_val)
[1]  11774.07
> sd(obj_val)
[1]  34.65471
> range(obj_val)
[1]  11657.80 11919.07
>
>
> hist(obj_val, xlab="Objective",breaks=50)
> #95 % bootstrap CI
> quantile(obj_val,c(.025,0.975))
     2.5%     97.5%
11705.87 11842.1612
```

In summary, we first solved the optimization problem under deterministic conditions using standard optimization approaches described in Appendix E in more detail. We then simulated the of that solution in terms of objective value allowing some of the model inputs to vary at random (in this case, constraint coefficients). Finally, we summarize the simulated output by computing means, standard deviations, and quantiles and plotting a histogram:

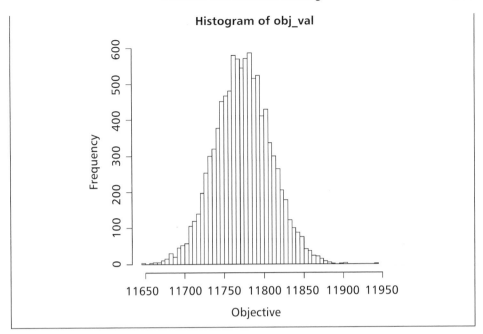

Histogram of obj_val

Statistical significance

If you have some familiarity with statistical analyses, you may have noticed that we do not include null hypotheses testing and associated concepts such as statistical significance in this chapter. Statistical null hypothesis testing has increasingly come under criticisms for use in natural resource-related fields (Yoccoz 1991, Johnson 1999, Anderson et al. 2000). Null hypotheses of "no difference" or "no effect" are usually known to be false, a priori. In addition, statistical hypotheses tests are strongly affected by sample size, so that a small effect (e.g., Pearson correlation $r = 0.1$) can be significant when the sample size is large, and a moderate effect size can be insignificant when the sample size is small. More importantly to decision making, statistical null hypothesis tests provide little, if any, information about the magnitude of the differences. Indeed, the magnitude of an effect and measure of the uncertainty are more useful to decision makers because they provide the basis for action. To illustrate, consider a situation where a fish hatchery manager has to select one of two types of fish food additives. Experiments indicated that expected growth increase with the first additive is 5 g/day with a standard error (SE) of 2.75, whereas the expected growth of the second additive is 2 g/day (SE = 0.75). A null hypotheses indicated that the growth under the first additive was not statistically significant ($p > 0.05$) from zero, whereas the second additive was significant. The hypotheses tests suggest that the manager should choose the second additive, but if we use the normal cumulative probability distribution a different conclusion can be reached. Using the estimated growth rate and SE, we estimate the probability that the growth will be greater than 2, 4, and 6 g/day for additive 1 as: 0.86, 0.64, and 0.36, respectively; and additive 2 as: 0.50, 0.00, and 0.00, respectively. Clearly, the null hypotheses tests were not as useful to the manager for evaluating the relative benefits of each additive. Structured decision-making practitioners should always focus statistical analysis on obtaining estimates of the magnitude and precision of the estimate, and on resolving decisions, rather than achieving arbitrary significance levels.

References

Anderson, D.R., K.P. Burnham, and W.L. Thompson. (2000) Null hypothesis testing: problems, prevalence, and an alternative. *Journal of Wildlife Management* **64**, 912–923.

Johnson, D.H. (1999) The insignificance of statistical significance testing. *Journal of Wildlife Management* **63**, 763–772.

Lunn, D.J., A. Thomas, N. Best, and D. Spiegelhalter. (2000) WinBUGS – a Bayesian modelling framework: concepts, structure, and extensibility. *Statistics and Computing* **10**, 325–337.

Yoccoz, N.G. (1991) Use, overuse, and misuse of significance tests in evolutionary biology and ecology. *Bulletin of the Ecological Society of America.* **72**, 106–111.

Additional reading

Link, W.A. and R.J. Barker. (2010) *Bayesian Inference with Ecological Applications.* Elsevier-Academic, London, UK.

Royle, J.A. and R. M. Dorazio. (2008) *Hierarchical Modeling and Inference in Ecology.* Elsevier-Academic. London.

Williams, B.K., J. D. Nichols, and M.J. Conroy. (2002) *Analysis and Management of Animal Populations.* Elsevier-Academic.

6

Modeling the Influence of Decisions

So far, we have clearly identified the problem and the decision alternatives. We've also identified and structures objectives – being careful to separate our means and fundamental objectives – and quantified one or more objectives in a utility function. We now need a means to connect alternative management actions to objectives. In other words, we need a **model** to estimate the value of management actions. This chapter builds on the previous chapter where we showed how to use familiar statistical techniques to parameterize models than can be used for decision making.

Structuring decisions

Up to this point (Chapters 2 and 3) we have emphasized the importance of casting decision making in a structured manner, in which we develop a clear statement of the problem, develop fundamental and means objectives, explicitly identify the relation between fundamental and means objectives in an objectives network, and connect our available decision options to outcomes that relate to our objectives in a **decision network**. Here, we will describe in some detail two common approaches used to depict decision models: **influence diagrams** and **decision trees** and demonstrate how these models are used to estimate the value of alternative decisions. We illustrate techniques for creating and parameterization of influence diagrams using Netica software (http://norsys.com/) in Boxes 6.1–6.3.

There are number of other ways to model the influence of decisions, including analytical approaches, Monte Carlo simulation, and formal optimization

Decision Making in Natural Resource Management: A Structured, Adaptive Approach,
First Edition. Michael J. Conroy and James T. Peterson.
© 2013 John Wiley & Sons, Ltd. Published 2013 by John Wiley & Sons, Ltd.

procedures considered in more detail in Chapter 8. Regardless of the specific approach used, it is important that the method captures the essential structure of our decision-making problem, and involves the connection between our candidate decisions and our fundamental and means objectives (Chapter 3). Thus, all decision modeling should begin with the creation of a graphical representation of the decision that explicitly links objectives and decisions and represents the decision makers' ideas about the ecological and physical dynamics of the system being managed.

Influence diagrams

Influence diagrams are graphical representations of decision models. They provide explicit representations of the individual components of the decision: the decision, uncertain events, and the consequence or outcome of the decision and the causal relationship among them. Individual model components are represented by geometrical shapes referred to as **nodes**. The shape of the node is different for each type of model code component. **Decision nodes** are represented by rectangles; chance or **uncertainty nodes**, by ovals; and **consequence nodes**, by diamonds or hexagons. Figure 6.1 is an influence diagram of a habitat improvement decision with four nodes: the decision node, *improve habitat*; two chance nodes, *current population size* and *future population size*; and one consequence node, *conservation value*. **Directed arcs** are used to indicate dependencies between model components. For example, here future population size depends on the improving habitat decision and current population size, whereas conservation value depends on future population size. Influence diagrams may resemble flowcharts, but they are fundamentally different. An influence diagram represents an instantaneous moment in time. Therefore, the arcs usually do not represent the timing or sequence of events (but see **value of information** below) but are used to represent causality or flow of information. Influence diagrams also do not contain feedbacks or loops, i.e., influence diagrams are **acyclic**. A node that has no other node pointing into it is defined as a **root node**. For example, the influence diagram in Figure 6.1 has two root nodes: the decision node – improve habitat and the uncertain node – current population size. This means that current population size is not known with certainty when the decision is made. In many natural resource applications, root nodes are used to represent the current **state** of the system and events that are not under the control of the decision maker, such as future weather events.

  We mentioned earlier that influence diagrams are acyclic (i.e., they do not contain loops) and that the arcs connecting nodes usually represent causality. The only instance where arcs do not represent causality is when they point from one node into a decision node. In these instances, they are used to represent sequence (timing) or the flow of information. In the improve habitat influence diagram (Figure 6.2a), an arrow pointing from current population into the improve habitat decision node means that the current population size is known with certainty when the decision is made. This also implies **state-dependent decision making**. That is, the state (condition) of the population is known when the

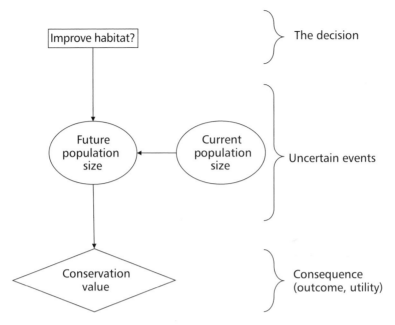

Figure 6.1 An example influence diagram of a habitat improvement decision.

decision is made. Later, we will expand on this concept when discussing sequential dynamic (Markov) decision making.

In the habitat improvement influence diagram, the arc connecting the current population size node to the decision (Figure 6.2a) implied that the population size is known with certainty when the decision is made. This is known as the **value of perfect information** (more on this in Chapter 7). In reality, the current system state is never known with certainty. For example, abundance is usually estimated using sample data and, perhaps, a population estimator, such as mark-recapture. Because estimates are made using sample data, they will have some error associated with them, usually expressed as a standard error, and maybe even biases. In other words, true abundance will not be known with absolute certainty. To express the fact that information is imperfect, we add another chance node to the network defined as **field sampling results** and connect an arc from current population size to it (Figure 6.2b). This means that the sampling results depend on the current population size. We then connect an arc from the field sampling results to the decision. This link implies that the field-sampling results will be known with certainty, which they should be, and that the sampling results are imperfect measures of true population size. These types of links represent imperfect information. In Chapter 7, we will explore the uses of the concept of **imperfect information** to estimate the **value of imperfect information** and use it to evaluate the relative benefits of sampling or conducting additional studies.

Arcs also are used to indicate a sequence of decisions. A decision node pointing into another decision node means that the first decision (first node) is made

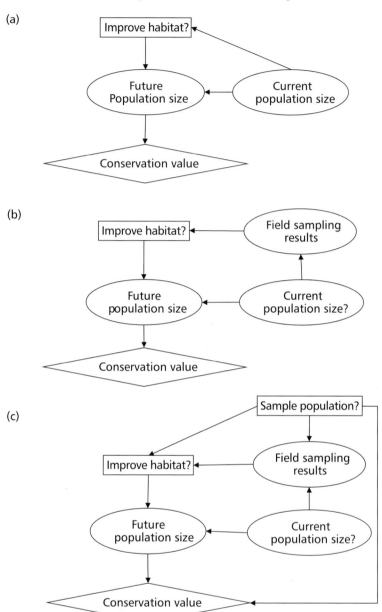

Figure 6.2 A habitat improvement decision influence diagram with (a) arc connecting current population size with the habitat improvement decision representing prefect information; (b) arc connecting sampling results to habitat improvement representing imperfect information; and (c) an additional decision to sample the population that is made before the habitat improvement decision.

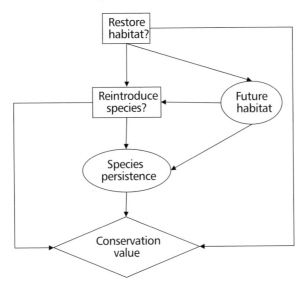

Figure 6.3 An influence diagram of two linked decisions, the first involving a habitat restoration decision and the second, a species reintroduction decision.

before the second decision is made. For example, the habitat improvement influence diagram can be expanded to include the decision to sample the population before deciding to improve the habitat (Figure 6.2c). This influence diagram implies that the first decision is to sample the population because the arc points from *sample population* to the *improve habitat* decision indicating the sequence of events. The sample population arc pointing into sampling results implies that the sample population decision can change sampling results implying that the decision involves not just the decision to sample (yes, no) but perhaps the level of sampling effort. For example, more sampling effort (samples) will probably result in lower variance, which can affect the value of information (more in Chapter 7). The sample population decision also points into conservation value implying some direct cost associated with the action, more on this later in the chapter. Linked decisions are similarly represented in an influence diagram with the first decision and its consequence pointing into the second decision. For example, Figure 6.3 represents two linked decisions; the first decision involves a decision to restore habitat prior to reintroducing a species to an area. The decision to reintroduce the species is influenced by the structure of the habitat after the restoration decision. In other words, the decision on whether to reintroduce the species depends on whether the habitat restoration was successful.

Sequential dynamic decisions are represented in an influence diagram as a series of identical subdiagrams that represent the dynamics that occur after a decision is made. For example, the influence diagram in Figure 6.4 represents an animal population harvest decision over a 3-year time period. Here, the arcs connecting the harvest decisions indicate temporal sequencing of the decisions.

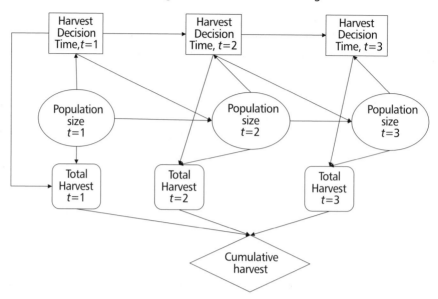

Figure 6.4   Influence diagram of a sequential dynamic decision of animal population harvest for three time periods.

The population size at t node points into the harvest at time t indicating that the size of the population is known before the decision is made. That is, the decision is **state dependent**. The total harvest for a particular year is influenced by the harvest decision and the population size. The population size in the next time period is influenced by the population size and the harvest decision in the previous time period. This is a feature of dynamic decision problems. The decision at a particular time period affects the decisions in succeeding time periods because that decision potentially changed the status of the population. Notice too that the total harvest in a year is represented by a rectangle with rounded corners. You have not seen this type of node before. It is defined as a calculated or constant node. It implies that the total harvest can be calculated with certainty. Do you think this is possible? Constant (aka calculated) nodes are fairly rare in ecological modeling applications and we will use them sparingly in this book. The utility for this model is the cumulative harvest over the entire time horizon, that is, the sum of the harvest from time 1, 2, and 3. In most real-world population harvest decisions, the time horizon will be long (typically 100 years) and the cumulative harvest over long time periods is one of the fundamental objectives. This long run perspective helps ensure sustainability (Chapter 8). Finally, notice that Figure 6.4 is a very complicated diagram for a relatively simple decision over a 3-year time period and it would be impossible to create a similar diagram for an infinite time period. This is why most practitioners generally show only a single time step in influence diagrams created to represent a dynamic decision.

As stated earlier, to be useful in structuring decisions influence diagrams should closely relate to our fundamental and means objectives. Objectives at the top of the objectives network should be the fundamental objectives or attributes of fundamental objectives (Chapter 3) that are used to calculate the utility of the decision. The means objectives relating to each fundamental objective often reveal ideas about the dynamics of the system being managed and, quite often, objectives at the bottom of the objectives network consist of potential management actions that can help the decision makers achieve their fundamental objectives. Thus, fundamental and means objectives networks can serve as a valuable starting point when structuring decisions. Objectives networks, however, generally do not have spatial and temporal components, which are important for modeling decisions. It is important then to explicitly define the spatial and temporal dimensions of the decision, in particular the grain and extant before proceeding with model building. These will help focus the model building at the proper level of resolution and will help facilitate communication among decision makers and members of the stakeholder working group (see Chapter 4).

Frequent mistakes when structuring decisions

Several mistakes can be made when creating influence diagrams and developing decision models. As we discussed in Chapter 3, one of the most common mistakes is failing to incorporate the cost of management actions in the objectives network and consequently, the subsequent influence diagram. For example, the improve habitat influence diagram (Figure 6.1) suggests that there is no cost associated with implementing the management action because the decision only affects the future population size, which in turn is the only factor that affecting conservation value. To incorporate the management costs into the utility, we connect the decision node to the consequence node – conservation value (Figure 6.5a). This linkage implies that the management action is not cost free and that the decision takes into account both the effect of the action on future population size and the cost of the action.

Another common mistake when building influence diagrams is failing to include important uncertainties or direct causal linkages. Because transparency is an important objective of structured decision making (Chapter 2), the influence diagram should faithfully represent the (hypothesized) dynamics of the system being managed. For example, the habitat improvement influence diagram (Figure 6.5a) implies that the future population only depends on the management action and the current size of the population. This suggests the management action will involve the direct manipulation of the population, such as harvest or supplementation (stocking), and that the population is relatively unaffected by other factors, such as the quality of the habitat. Most animal populations are influenced to some extent by physical factors, such as weather and habitat features, and most natural resource management influence diagram should include these components. In addition, a habitat improvement decision should only affect the population through its effect on the habitat. Thus, the arc from the decision should be pointing into *future habitat* and *future habitat* should be pointing into *future population size* (Figure 6.5b). We interpret the resulting influence diagram as: the habitat improvement decision affects the structure of the future habitat and the conservation value (i.e., presumably

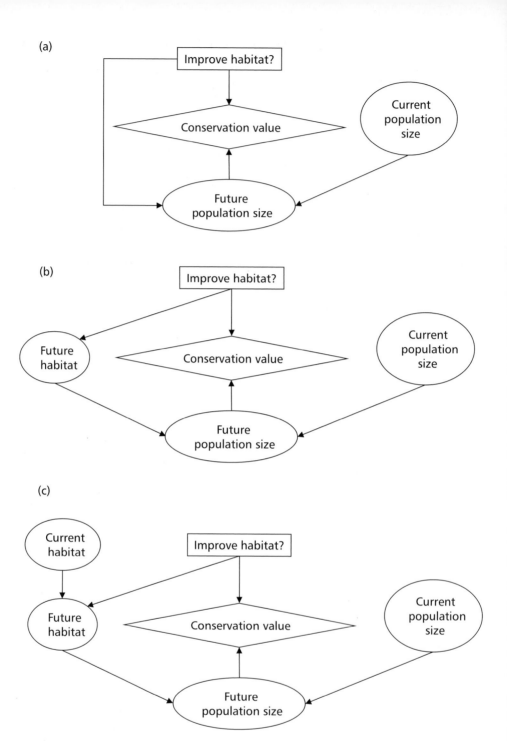

Figure 6.5 A habitat improvement decision influence diagram with (a) arc connecting the decision to conservation value representing the cost of the management action affects the utility; (b) arc connecting improve habitat into to future habitat improvement indicating that the decision will affect the future population by potentially modifying the habitat; and (c) current habitat node connected to future population indicating that future habitat is influenced by the current state of habitat and the decision.

because the action has costs); future population size is the result of current population size and future habitat; and the conservation value of a decision depends on the future population size and the management costs associated with the decision.

The habitat improvement influence diagram in Figure 6.5b also illustrates another common mistake – the failure to incorporate the current state of the system. This version of the habitat improvement influence diagram implies that future habitat was only affected by the management action and did not depend on the current habitat structure. In some instances, this may be a reasonable assumption. For example, the future structure of a wildlife food plot only depends on what is planted in that year (i.e., the management action) and not on the types of crops that were planted last year. However, in many natural resource applications, the current state of the system (e.g., population size, habitat quality) will have a substantial effect of the future system state and are usually considered when making a decision. These are defined as **state-dependent decisions** and are an important characteristic of dynamic decision making. To incorporate the effect of the current state of habitat in the habitat improvement decision, we add a new node *current habitat* and connect it to *future habitat* (Figure 6.5c).

Another difficulty when building influence diagrams is properly specifying the relationships among the model components. This is illustrated by two models that consist of the nodes: A, B, and C (Figure 6.6). The vectored (straight) version of the model implies that the value of node B depends on A and the value of node C depends on B. At this point, the nature of the relationships is unspecified and could be linear or nonlinear. However, the structure of this model implies that node A only affects node C through its effect on node B. That is, if the value of B were known with certainty, node A no longer affects C and node C is **conditionally independent** of A (Chapter 5). This idea of conditional independence is important and we will discuss it in detail later. Node B in the vectored version of the model is defined as an intermediate node because only it is used to translate how node A affects node C. The branched version of the three-node model in Figure 6.6 implies that A and B jointly affect node C. We emphasize that the links between nodes indicate causality and do not specify the nature of the relationships among components. For example, the branched version of the three-node model could indicate that the effects of A and B on node C are additive or that they interact. Similarly, the directed arcs can represent linear or nonlinear relationships. Thus, influence diagrams are not useful for conveying information on the nature of the relations among components, but are primarily used to indicate causality.

The key to avoiding mistakes and building a faithful graphical representation of a decision is to know what the linkages mean in an influence diagram. Arcs connecting two nodes indicate a direct effect or influence of one model component on another, while two nodes that are linked through an intermediate node indicate an indirect effect. Combining both types of connections in a single model can imply relatively complex processes. For example in Figure 6.7a, the value of node Y depends on node X and the value of node Z depends on the value of nodes X and Y. Ecologically, this might mean that node X has an indirect

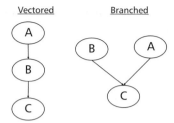

Figure 6.6 Vectored and branched versions of three node diagrams. The vectored version implies that factor A only affects C through its effect on factor B, while the branched version indicates that factors A and B jointly affect C.

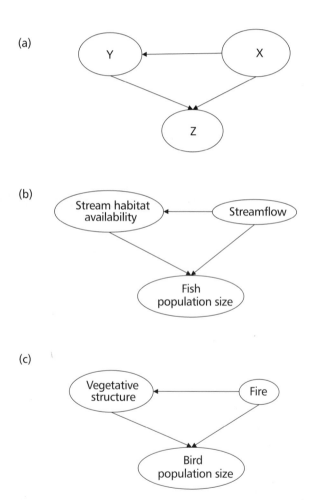

Figure 6.7 Three node models representing direct and indirect effects of (a) node X on node Z, (b) streamflow on fish populations, and (c) fire on bird populations.

effect on node Z through its effect on node Y and a direct effect on node Z. That is, there are two mechanisms through which node X affects node Z. To illustrate, assume that the model represents the effects of streamflows on stream habitat and fishes (Figure 6.7b). Streamflows can affect fish populations indirectly by altering stream habitat availability and directly by flushing fishes downstream during floods causing mortality. Similarly, this type of model could represent the effect of fire on terrestrial habitats and ground nesting bird populations (Figure 6.7c). Fire can indirectly affect ground nesting birds by changing the vegetative structure and can directly affect birds through fire-related mortality (i.e., due to smoke inhalation, burns). As you might suspect, linkages like these can lead to confusion and misinterpretation because they represent unstated mechanisms. For example, the effect of streamflow on habitat availability could represent the effect of habitat availability on the carrying capacity of a stream or could represent the effect of the availability of refugial habitat on survival. To avoid confusion, influence diagrams should be accompanied by descriptions with explicit definitions of the nature of the relationships among model components.

Defining node states

Thus far, we have concentrated on properly defining the relationships among nodes in influence diagrams. We now shift our focus to characterizing nodes (i.e., model components). Each node in an influence diagram consists of a set of **mutually exclusive** and **collectively exhaustive states**. For example, the previous habitat improvement influence diagram (Figure 6.5c) could consist of two states: *Yes*, improve habitat and *No*, do not improve habitat or alternatively, the states could represent discrete levels of habitat improvement: 0, 10, and 100 hectares (ha); Figure 6.8). The states are mutually exclusive in that there is no overlap in values that define each class and are collectively exhaustive in that they represent the complete range of possible values. For example, consider the node describing available habitat available in a management unit as a percentage of the area of the management unit. To be collectively exhaustive, the node states should be bounded by 0% (the lowest possible value) and 100% (the highest possible value) so a potential discretization of values for the range of 0–100% is 0–32.999%, 33–66.999%, 67–100%. The 0.999 is used here to reinforce the idea that the states are mutually exclusive (no overlap). For simplicity in this book, we will use discretization of 0–33%, 33–67%, etc. with the assumption that the upper bound for a state does not include that value. For example, the state 0–33% includes values that range from zero up to but not including 33%.

Node and node state definitions are a potential source of confusion and uncertainty (i.e., **linguistic uncertainty,** Chapter 7) that can be minimized or eliminated by explicitly defining each state. We find that defining states based on numeric values or ranges is better than defining them based on a narrative. For example, consider a hypothetical node defined as biotic integrity with two

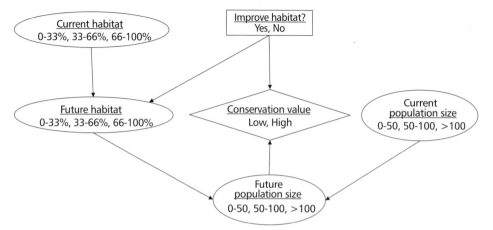

Figure 6.8 Influence diagram of habitat improvement decision with node states listed below the node names.

states: *Good* and *Poor*. These states can be defined using a narrative or numeric criteria.

Narrative:
Good: comparable to good reference ecosystem that supports the full range of ecological functions possible.
Poor: significant deviation from reference ecosystem condition with less than half of ecological functions supported.

Numeric criteria:
Good: More than 50 native species (list provided) present.
Poor: Less than 50 native species (list provided) present.

The narrative illustrates two common mistakes when building influence diagrams and decision models. First, the state definition depends on something external to the system of interest, i.e.., the reference condition. This means that the decision maker needs to know something about the reference condition before an observation can be assigned to a particular state. In addition, the narrative does not explicitly define the ecological function and leaves it to the reader to interpret the meaning. These interpretations will likely vary from reader to reader and can lead to confusion and greater uncertainty. In contrast, the numeric criteria are unambiguous once the list of native species is known. Remember that an important objective of structured decision making is transparency, so states should be defined as unambiguously as possible.

   As we stated earlier, states do not have to be discrete values but can be discrete ranges of continuous values. Various approaches to discretizing continuous variables are presented in Chapter 5. Other factors to consider when defining states can include ecological and statistical considerations. The latter of these will be of particular concern when we discuss optimization methods for solving dynamic decision problems (Chapter 8), where the number of states should be kept as small as possible provided they are sufficient for capturing the dynamics of the system.

Decision trees

For practical reasons that will become apparent, we primarily focus on the use of influence diagrams for depicting management-decision models. **Decision trees** can also be used to display the decision in greater detail. Similar to influence diagrams, geometric shapes are used to represent the various components of the model. The branches leading out of the geometric shapes represent the possible decisions, outcomes, or chance events. For example, Figure 6.9 depicts a simple habitat improvement decision as 3 node influence diagram and a decision-tree version of the same decision. The *improve habitat* decision has two states, *Yes* and *No* and the decision tree has 2 branches that correspond to each decision. Decisions are generally the first branches of the tree and outcomes (i.e., utility) are the terminal branches. The consequence of each choice or chance event also is displayed at the ends of the branches. For example, future population size can

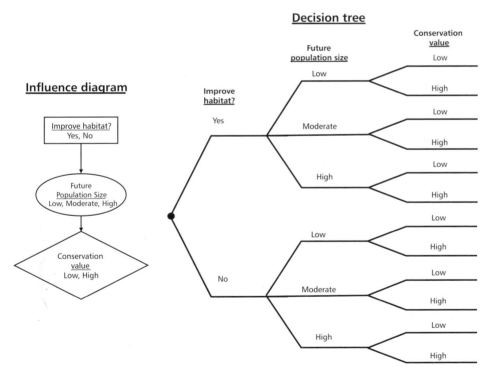

Figure 6.9 Influence diagram and corresponding decision tree of simple habitat improvement decision. States for each model components are listed below the node titles in the influence diagram and are above the branches in the decision tree.

take one of three states after a habitat improvement decision is made: *Low*, *Moderate*, and *High* and each of these states can have *Low* or *High* conservation value. Unlike influence diagrams, decision trees are very useful for explicitly displaying the alternative combinations of events that can lead to an outcome. This is accomplished by following the branches of the tree that correspond to each combination of events. For example in Figure 6.9, we can identify the 6 ways of achieving *High* conservation value.

>   Habitat is improved (*Yes*) and:
>   1)   future population size is *Low*; or
>   2)   future population size is *Moderate*; or
>   3)   future population size is *High*; or
>
>   Habitat is not improved (*No*) and:
>   4)   future population size is *Low*; or
>   5)   future population size is *Moderate*; or
>   6)   and future population size is *High*.

Thus, there are multiple pathways that can lead to a particular outcome and there are multiple outcomes that can result from a single decision. As we shall see, not every pathway or outcome is equally likely. Next, we will learn how to apply the statistical concepts from Chapter 5 to estimate the likelihood of a particular outcome and find the **optimal decision**.

Solving a decision model

The ultimate goal of the structured decision-making process is to identify the **optimal decision**, that is the decision or combination of decisions that will result in the best outcome in terms of the objectives. This is usually the decision that results in the greatest value of the utility, but depending on objectives it could also be the smallest value (e.g., minimizing costs). Although we generally strive to identify the optimal decision, it can frequently be the case that several decision alternatives (i.e., combinations of choices for our decision controls) provide nearly the same objective result. In such cases it may not be deemed worth the trouble to select the optimal decision, and the stakeholders may be satisfied with an outcome that is technically suboptimal, often referred to as a **satisficing decision** (Byron 1998). In Chapter 8, we provide an example of a satisficing decision as applied to harvest management.

Solving a decision model involves estimating the objective value or utility of each management alternative and choosing the decision that maximizes (or minimizes, depending on the management objective) the utility. For example, consider a stream management decision to add large wood habitat (e.g., snags, logs) to a stream section where none currently exists. The management objective is to maximize the size of a fish population in the stream section and the three decision alternatives under consideration are to increase available large wood habitat by: 0 (none), 10, and 20 m^2. To find the decision that increases the fish population the most, we have the following linear equation that was obtained by fitting fish abundance and large wood habitat availability data with linear regression (Chapter 5):

$$N_i = 20 \times H_i - 0.75 \times H_i^2$$

where N is the estimated number of fish and H the large wood habitat availability for each decision i. Using the above equation, we estimate that fish abundance under the 0 (none), 10, and 20 m^2 habitat increase decisions are 0, 125, and 100 fish, respectively. Given the three decision alternatives, the best decision is to increase large wood habitat by 10 m^2 because that is predicted to result in the greatest number of fish, 125. This is the essence of solving a decision model; we predict the outcome of alternative management actions and choose the management alternative that provides the best result. This procedure does not change with the complexity of the model. However, incorporating and propagating uncertainty in the decision model is a bit more complicated than the above example and requires the use of probability.

To illustrate how to solve a decision problem using probability, we begin with a timber harvest influence diagram that consists of: the timber harvest decision with two states, *yes* and *no*; the **consequence** or outcome following the decision (in this case, large or small future population); the **utility** (in this example, socioeconomic value) of the outcome; and the uncertain events that represent the dynamics of system of interest: *stream habitat* and *current* and *future fish population status* (Figure 6.10). Stream habitat consists of two states: *good* and *poor* and current and future population status also consists of two states: *small* and *large*. Here stream habitat is influenced by the harvest decision; future population size by current population size and stream habitat; and the utility, socioeconomic value by timber harvest decision and the future fish population size. In influence diagrams, uncertainty in the value of root nodes (i.e., nodes with no connections into them) is expressed using probability. For example, the timber harvest influence diagram has one root node, current population size. Uncertainty in current population size could be due to sampling error because population size is estimated using sample data. This uncertainty is expressed by assigning probabilities that the fish population is *small*, 0.8, or *large*, 0.2 (Figure 6.10). Notice that these two probabilities sum to one. Node states are collectively exhaustive (i.e., they comprise all of the possible states of the system), so the state-specific probabilities for a node need to sum to one. Later, we will discuss where we get the probabilities, but for now we focus on solving the model.

The relationship among the model components (nodes) are modeled using conditional probability. In Chapter 5, we introduced the concept of conditional probability and defined it as the probability of an event occurring given that another event has occurred. That is, the first event is conditional on the second event. In a timber harvest influence diagram, stream habitat is conditional on the timber harvest decision. Thus, we use conditional probability to express the

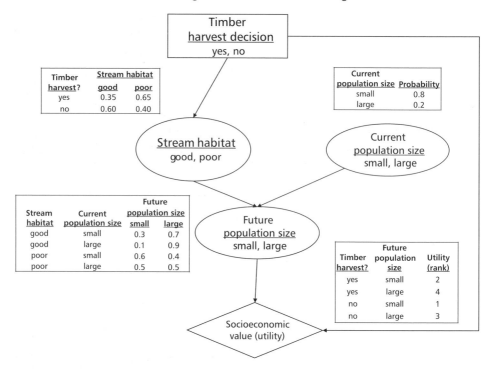

Figure 6.10   Timber harvest decision influence diagram with probability, conditional probability, and utility tables displayed for illustration.

uncertainty in effect of timber harvest on stream habitat. If timber is harvested (yes) then the probability that stream habitat is *good* is 0.35 and *poor* is 0.65. Under no timber harvest, the probability that stream habitat is *good* is 0.60 and *poor* is 0.40. As discussed above, each pair of probabilities sum to one because the outcomes are collectively exhaustive. Similarly, future population size was influenced by stream habitat and current population size. That is, future population size is conditional on stream habitat and current population size. The probability that the future population size is *small* or *large* then depends on if the stream habitat is *good* or *poor* and if the current population size is *small* or *large*. Given the conditions that stream habitat is *good* and current population size is *small* (i.e., these conditions are known with certainty), the probability the future population size is small is 0.3 and the probability it is *large* is 0.7. For the timber harvest model, we have 4 possible combinations of stream habitat and current population size. Thus, the conditional probabilities of future population for each combination of current population and stream habitat are contained in a **conditional probability table**. In some applications, these probability tables may be defined as **transition probabilities** or **transition probability matrices** (Chapter 8). Finally, the value (utility) of the each possible outcome is based on the ranking the combinations of timber harvest (economic gain) and future population size with the best outcome, yes harvest timber and large future population, receiving

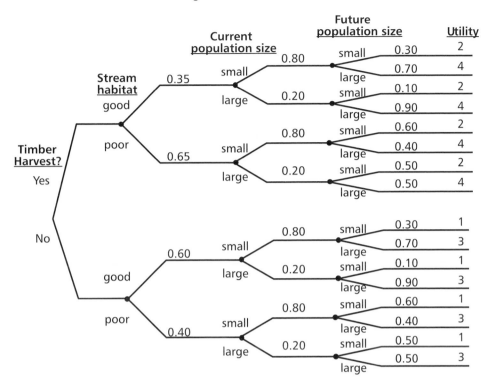

Figure 6.11 Decision tree for timber harvest decision with branches labeled with corresponding probability and utility values.

a rank of four and the worst outcome, no timber harvest and small future population, receiving a rank of 1 (Figure 6.10).

Calculating the value if each decision is best shown using the decision tree (Figure 6.11). Notice that each decision has a complete branch that includes habitat, current population, future population, and the utility. The only difference in the *Yes* and *No* branches are the probabilities for stream habitat. If timber is harvested, the probabilities for *good* and *poor* habitat are 0.35 and 0.65, respectively. If there is no timber harvest, the probabilities for *good* and *poor* habitat are 0.6 and 0.4, respectively. The conservation value (utility) is at the far right-hand side of the tree. To estimate the value of a decision, the probabilities and the conservation value on each branch that correspond to a decision are multiplied and added. For example, consider the timber harvest yes decision. One potential combination of events of a *Yes* decision is that stream habitat is *good* (0.35 probability), the current population size is *small* (0.8), and the future population size is *small* (0.3). The utility of harvesting timber and the future population being small is 2. These values are multiplied together as: $0.35 \times 0.80 \times 0.30 \times 2$. Remember that a "Yes" decision could potentially lead to other outcomes represented by the various branches in the tree, stream habitat could end up *poor*, current population size *small*, and future population size *small* or *large*. To take these outcomes into account, we multiply the probabilities and values along each branch as:

| Stream _habitat_ | Population size | | Tree branch product |
	Current	_Future_	
good	small	small	0.35 × 0.80 × 0.30 × 2
good	small	large	0.35 × 0.80 × 0.70 × 4
good	large	small	0.35 × 0.20 × 0.10 × 2
good	large	large	0.35 × 0.20 × 0.90 × 4
poor	small	small	0.65 × 0.80 × 0.60 × 2
poor	small	large	0.65 × 0.80 × 0.40 × 4
poor	large	small	0.65 × 0.20 × 0.50 × 2
poor	large	large	0.65 × 0.20 × 0.50 × 4
		sum	3.064

The expected value of the yes timber harvest is 3.064. When we repeat the process for the no timber harvest decision, we obtain 2.224. The optimal decision then is to harvest timber because it has the greatest utility.

As demonstrated above, the expected value of relatively simple decisions can be calculated in a spreadsheet with the rows corresponding to all of the combinations of node states and objective values. Solving more complicated decision models, particularly approaches that involve sequential dynamic decision making, require the use of specialized graphical or statistical modeling software. In Chapter 8, we describe the different approaches used to obtain optimal decisions and provide detailed examples.

Conditional independence and modularity

Earlier, we introduced the concept of **conditional independence** and said that it was useful to decision making. Conditional independence is important to the decision-model building process because it allows two or more models to be integrated through common elements. For example, a land-use dynamics model that predicts the amount of fine sediment delivered to a stream following a management action can be combined with a fish population dynamics model that uses estimates of fine sediment in streams to predict fish abundance (Figure 6.12). Most natural resource decisions involve complex physical and biological processes and rarely, if ever, are decision models built using the results from single study. More often than not model builders must combine multiple elements from unrelated research with new or existing data to create a decision model. The important thing to keep in mind is that the common elements (e.g., nodes, components) must be expressed in a common currency. For example, a

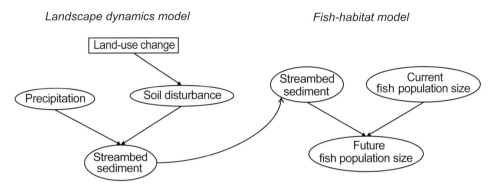

Figure 6.12 Conditional independence allows us to integrate two or more models through common elements.

landscape dynamic model that predicts streambed sediment in terms of tons per hectare is of little use to the fish biologist who needs sediment estimated in terms of percent streambed composition. The idea of combining models or elements into a single model also brings to mind the importance of communication among technical advisors, especially when the team of advisors comes from different disciplines. All too often, decision-model development is delayed or worse, stopped because team members did not communicate effectively and failed to properly integrate their respective elements. The take-home message here is that conditional independence allows us to integrate very different elements and that successful model integration requires an integrated and communicating team from the very beginning of the model-building process.

Parameterizing decision models

Decision models can be parameterized using a wide variety of information ranging from empirical data to expert opinion and quite often, multiple types of information are used to parameterize a single decision model. In Chapter 5, we discussed the use of statistics in decision making and provided examples of how to use empirical data and common statistical procedures, such as linear regression, to estimate variables that can be used to parameterize decision models. We illustrate the use of these techniques by parameterizing an influence diagram using empirical data and models in Box 6.2.

Dynamic processes, such as animal population dynamics, are difficult to model with graphical techniques (e.g., influence diagrams and decision trees) because the models cannot contain feedback loops. One useful approach for parameterizing graphical decision models that include dynamic processes is to simulate the process, summarize the results of the simulations (e.g., construct frequency distributions), and use the summaries to parameterize the model. Simulations are usually performed with computer programs, although some simple simulation modeling can be accomplished by hand or in spreadsheets. Inputs to the simulation model (e.g., survival probabilities, starting population sizes) are pre-assigned to correspond to the inputs in the decision model (i.e., decision and the root nodes) or are drawn from statistical distributions (Chapter 5). The simulations are then run using the inputs and the simulated outcomes (e.g., species persistence, population size) are output to a dataset and summarized. This was the approach taken in the largemouth bass fishing regulations case study (Chapter 9). Largemouth bass population parameters, such as fecundity, growth rate, natural mortality, and fishing mortality were randomly drawn from statistical distributions that corresponded to the values in root nodes of the influence diagram. The population dynamics were simulated on an annual time step and population characteristics: fish density, large fish density, and annual variability in population size were output at 20 years from the present. This was repeated 100 000 times and conditional probabilities were calculated using frequency distributions. We illustrate the use of simulation for parameterizing an influence diagram in Box 6.3. Next, we discuss situations where no data or models are available.

Box 6.1   Creating an influence diagram with Netica ® software

Netica software (http://norsys.com/) can be used to create and parameterize influence diagrams and Bayesian belief networks (BBN; i.e., influence diagrams without decision and utility nodes). The program is user friendly and is available in licensed (requires payment) and limited (no cost) versions. We find that the limited version is ideal for instruction and for demonstrating concepts and techniques to beginners. It has all of the abilities of the licensed version but restricts the size of the model to 15 nodes (version 4.16). This enforces parsimony in model development for beginners.

Models can be created relatively easily in Netica using the graphical user interface (GUI). Symbol shapes and colors also are used to represent different types of nodes, ovals for chance or uncertainty nodes (Netica defines these as *Nature* nodes), squares for decision nodes, and pentagons for utility nodes. Causal links between nodes are defined using directed arcs or arrows. These various symbols are located at the top of the Netica GUI. To create an influence diagram, users can click on the symbol corresponding to the type of model component. For example, clicking on the rectangle (decision node) and then clicking in the body of the GUI will produce:

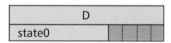

This is in the default "belief bar" format. Here the name or title of the node is displayed at the top and the decision alternatives (or state names for nature nodes) are displayed on the left-hand side below the title. To change the name or title of the node and add or change decision alternatives, double click on the node and the node box will appear. The node box has a place to add a node name and a node title. The node name should be simple and not contain and spaces or special characters (e.g., @ # $ +). The title can be longer (more descriptive) and can contain spaces, etc. To add a name and title, simply move the cursor in the corresponding box and type the name. Here, we'll use the name "mgnt" and title "Pest management decision". You will notice pull-down menus in the node box where you can specify or change the type of node from decision node to nature (aka uncertainty), utility or constant node. Users can also specify if a node is a discrete node or a continuous node. Note that continuous nodes need to be discretized (more below). Because this is a decision node, we will leave it as a discrete node. There are also two additional boxes and a pull-down menu for naming the note states or the value of a state. To create two management decisions, we'll add the decision alternative name, "No_action" (note no spaces or special characters) in the state box, click the new button and add the decision alternative "Control_pests" in the state box. Before we approve the changes, notice the pull-down menu toward the middle left of the table box. The default is *Description* and below the menu there is a space for entering a detailed description of the model (always a good idea). The other choices on the pull-down menu include: *Equation*, where equations can be used to define relations among nodes (more in Box 6.2); *State numbers*, where discrete values of node states can be added; *States*, where state names can be added or deleted (it is an alternative method for naming states); *State titles*, where the titles of states can be added or removed (state titles can include spades or special characters unlike state names); and several other options for adding comments or names and dates. We will leave the Description box empty because this is only an example. To approve the changes, click the "Okay" button and you will have the following:

Pest management decision			
No action			
Control pests			

We now need to add additional nodes to complete our example diagram. Let's add an uncertainty node by clicking on the oval node symbol at the top of the Netica GUI and then in the body of the GUI. Open the node box and enter "pest" as the name and "Number of pests" as the title. Here, we would like the node to be continuous, so select Continuous from the pull-down menu in the node box. You should now notice that the node box has two boxes labeled *Interval*. These are locations for entering values for the lower and upper bounds for defining state ranges. For example, let's assume that we would like to use discretize the number of pests node into 3 evenly divided states over the range of 0–75 pests: 0–25, 25–50, and 50–75 pests. Enter the lower (zero) and upper (25) bounds for the first state into the Interval boxes and click the *New* button and add the upper and lower bounds for the next

(Continued)

state (25–50) and repeat the process for the final state (50–75). Select Okay
and you should have the following:

Number of pests		
0 to 25	33.3	
25 to 50	33.3	
50 to 75	33.3	

Open the number of pests node box again. An alternative (and easier) way to
enter the ranges for continuous nodes is to select *Discretization* from the
pull-down menu toward the middle left of the table box (the one usually
showing *Description*). You should now see the 4 values defining the 3 state
ranges: 0, 25, 50, and 75. These values could have been entered manually or
using a shorthand function used by the software:

$$[\text{minimum, maximum}] / \#\text{states.}$$

For example, to obtain 3 equally sized states over the range of 0–75 pests.
You could add the following to the *Discretization* box: [0,75]/3 and choose
Okay.

Add an additional uncertainty node, open its node box and enter the name
"freeze" and the title "Number of subzero days". This is a discrete node that
will have states: 0, 1, and 2. You can enter these values in the *Value* box,
being sure to click *New* after entering each value or alternatively, you could
choose *State Numbers* from the pull-down menu toward the middle left of
the table box and enter the values with one value per line. You should then
have the following:

Number of subzero days	
0	33.3
1	33.3
2	33.3

Finally, we need to add a utility node by choosing the clicking on the pentagon
node symbol at the top of the Netica GUI and then in the body of the GUI.
Open the node box and name it "utility" with no title. Click Okay.

We now have four nodes with no causal relations defined. Let's assume
that the number of pests in influenced by the decision and the number
of subzero days. Choose the arrow at the top of the diagram click on
the decision node and then on the number of pests node. Repeat the process
but this time click first on the number of subzero days node then on the
number of pests node. Let's also assume that the utility is a function of
the decision and the number of pests so select the arrow at the top of the
GUI and connect these nodes to the utility node. We should have the
following:

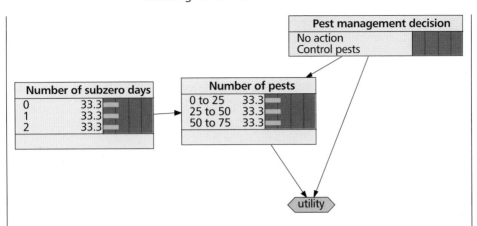

We now need to parameterize the influence diagram. Let's start with the root node number of subzero days. Notice that this consists of discrete values. In Chapter 5, we learned that the Poisson distribution is useful for representing the uncertainty in discrete-valued model components. Netica has several statistical distributions functions for parameterizing nodes and relations that can be found using the help feature. To parameterize the node, open the number of subzero days node box and choose *Equation* from the pull-down menu toward the middle left of the table box. The probability distribution function for the Poisson in Netica is:

$$PoissonDist(name, lambda)$$

where *name* is the name (not title) of the node and *lambda* is the desired mean of the Poisson distribution. To estimate the probability that the number of subzero days is 0, 1, or 2 assuming the mean number (lambda) is 0.25, we enter the following in the Equation box:

$$p(freeze \,|\,) = PoissonDist(freeze, 0.25)$$

and click Okay. You will notice that nothing has changed. This is because the equation needs to be used to create the probability table. To do this, select the node by clicking on it once and at the top of the GUI choose the *Table* pull-down menu and choose the option *Equation to table*. To see that the probabilities have been added, open the number of subzero days node box and select the *Table* button and you should see the following:

0	1	2
78.0488	19.5122	2.43902

(*Continued*)

These are the probabilities (expressed as percentages) that the number subzero days is 0, 1, and 2 assuming a Poisson distribution with a mean of 0.25. Close the table and the node box and let's parameterize the relation between the decision, number of subzero days and the number of pests.

Open the number pests node box and select the *Table* button and you should see the following conditional probability table (CPT):

Pest management decision	Number of subzero days	0 to 25	25 to 50	50 to 75
No action	0			
No action	1			
No action	2			
Control pests	0			
Control pests	1			
Control pests	2			

We need to add the values to the CPT tables expressed as percentages. These values could come from eliciting experts or from a model. Remember that the rows need to sum to 100%. Let's assume that we have the following values:

Pest management decision	Number of subzero days	0 to 25	25 to 50	50 to 75
No action	0	10.000	30.000	60.000
No action	1	20.000	40.000	40.000
No action	2	35.000	45.000	20.000
Control pests	0	30.000	40.000	30.000
Control pests	1	40.000	50.000	10.000
Control pests	2	55.000	40.000	5.000

Netica has an interesting feature that allows you to copy values from a spreadsheet and paste into the CPT table. After entering that values, select the Okay button. We now need to parameterize the utility, so open the utility node box and select the *Table* button and you should see the following table:

Number of pests	Pest management decision	utility
0 to 25	No action	
0 to 25	Control pests	
25 to 50	No action	
25 to 50	Control pests	
50 to 75	No action	
25 to 50	Control pests	

Here, we need to enter the utility for each combination of the number of pests and the management decision based on the values of the stakeholders. Let's assume that we have the following ranked values with 6 being the best outcome and 1 the worst.

Number of pests	Pest management decision	utility
0 to 25	No action	6
0 to 25	Control pests	4
25 to 50	No action	5
25 to 50	Control pests	3
50 to 75	No action	2
25 to 50	Control pests	1

After entering these values, select the Okay button. We almost have a working model. In Netica, the final step in the model-building process is to compile the model. This is accomplished by selecting the compile button (the lightning bolt) at the top of the GUI. Doing so will get you the following influence diagram.

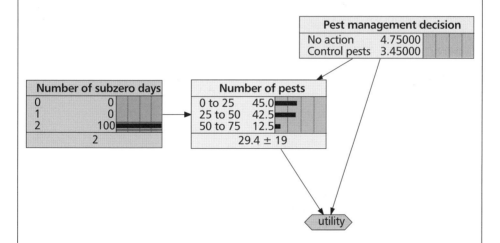

Netica displays the expected values of each decision alternative in the decision node to the right of the alternative. Here, the optimal decision is no action with a value of 3.47. In the uncertainty (nature) nodes, the state-specific probabilities are displayed to the right of each state and the length of the bar next to the value represents the relative magnitude of the probability. The mean and standard deviation are also displayed at the bottom of nodes with quantitative values, such as the two uncertainty nodes in our example diagram. Netica allows users to change the values of each node by clicking on the state

(*Continued*)

in the node. For example, to fix the number of subzero days at 2, place the cursor over the 2, click once and you will get the following:

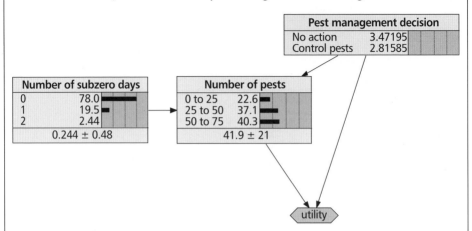

This shows the expected value of each decision alternative when the number of subzero days is known with absolute certainty to be 2. We will use this feature in Chapter 7 to conduct sensitivity analysis and estimate the value of information. Netica has many more useful features and we encourage you to explore these by reading the manual and examining the example networks provided with the documentation.

Box 6.2   Parameterizing an influence diagram with empirical data and regression models

Many dabbling ducks in North American require temporary wetlands to successfully reproduce. Management of these temporary wetlands on waterfowl refuges in the northern prairie region typically includes burning or thinning vegetation to maintain open water in the temporary wetlands. Because of limited management resources, management of individual wetlands occurs every 3 years. Vegetation removal is usually accomplished by either burning or mowing the wetlands after the breeding season when the water has dried up. Of these two options, burning typically prevents vegetative encroachment the longest but is the most expensive. To aid in the decision-making process, refuge managers developed models for predicting the effectiveness of alternative vegetation management actions on the degree of openness (%) of the wetland for the years following treatment. The fundamental objectives of refuge managers were to maximize duck production (i.e., the duckling chick density) while minimizing the cost of management. The corresponding influence diagram is shown below.

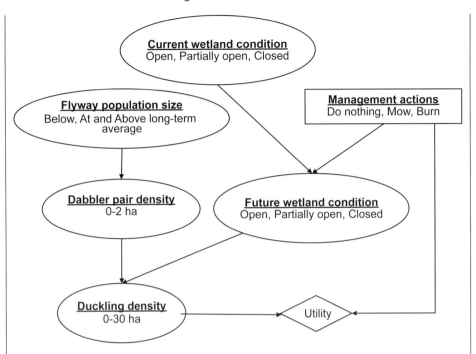

Here, managers are considering three management actions for an individual wetland: *Do nothing, Mow,* and *Burn* and three states for current and future wetland condition: open, with less than 33% of the surface vegetated; partially open, 33–67% vegetated; and closed, more than 67% vegetated. Managers assumed that the effectiveness of the alternative management actions would differ the most in year 2 following the treatment, so the model estimated wetland conditions in year 2. Future wetland condition is influenced by current wetland condition and the management action. Dabbling duck pair density is influenced by future wetland condition and the size of the flyway population (in year 2), while ducking density is influenced by dabbler pair density and future wetland condition. The states for dabbler pair and duckling density were discretized continuous values that encompassed the range of values reported in the literature.

Managers at several refuges have been collecting monitoring data on the status of wetlands before and after management actions. These data were used to create the frequency table shown below. Each row of the table corresponds to a combination of management action and wetlands conditions before treatment and the columns correspond to the wetland condition 2 years after treatment. The numbers in each column are the total number of wetlands for each combination of management action and wetland condition before treatment. For example, the first row indicates that managers did not treat 16 wetlands that were *open*. Of these, 5 (31%) were *open*; 9 (56%), *partially open*; and 2 (13%), closed after 2 years. These row frequencies were used to estimate the conditional probabilities for future wetland condition by calculating the proportion of observations shown in parentheses.

(Continued)

Management action	Wetland condition before treatment	Wetland condition 2 years after			
		Open	Partially open	Closed	Row total
Do nothing	Open	5 (0.31)	9 (0.56)	2 (0.13)	16 (1.0)
	Partially open	0 (0.00)	3 (0.14)	19 (0.86)	22 (1.0)
	Closed	0 (0.00)	1 (0.12)	7 (0.88)	8 (1.0)
Mow	Open	10 (0.77)	2 (0.15)	1 (0.08)	13 (1.0)
	Partially open	4 (0.18)	16 (0.73)	2 (0.09)	22 (1.0)
	Closed	1 (0.06)	11 (0.61)	6 (0.33)	18 (1.0)
Burn	Open	3 (0.75)	1 (0.25)	0 (0.00)	4 (1.0)
	Partially open	9 (0.69)	3 (0.23)	1 (0.08)	13 (1.0)
	Closed	11 (0.65)	5 (0.29)	1 (0.06)	17 (1.0)

Field personnel also recorded the dabbler pair density during monitoring. The relations between the density of dabbler pairs on a wetland and wetland condition and flyway population size were evaluated using an analysis of variance. The estimated number of pairs of dabblers for each combination of wetland condition and flyway population size is shown below.

Wetland condition	Flyway population size	Estimated pairs dabblers*	Dabbler pair state conditional probabilities			
			0–0.5	0.5–1.0	1.0–1.5	1.5–2.0
Open	Above	1.38	0.00	0.00	0.81	0.19
	Average	0.95	0.00	0.64	0.36	0.00
	Below	0.64	0.16	0.84	0.00	0.00
Partially open	Above	1.62	0.00	0.00	0.19	0.81
	Average	1.18	0.00	0.09	0.90	0.01
	Below	0.87	0.00	0.83	0.17	0.00
Closed	Above	0.9	0.00	0.77	0.23	0.00
	Average	0.46	0.61	0.39	0.00	0.00
	Below	0.16	0.98	0.02	0.00	0.00

*Residual standard error = 0.137.

The conditional probabilities for dabbler pair density are calculated using the cumulative distribution function for the normal distribution and estimated dabbler pair density and the residual standard error from the ANOVA. In R code, the probability of 1.0–1.5 dabbler pairs for open wetland condition and above average flyway population is estimated using:

```
estimate <- 1.38
SE <- 0.137
pnorm(1.5,estimate,SE)- pnorm(1.0,estimate,SE)
```

and in Excel:

NORMDIST(1.5,1.38,0.137,1) - NORMDIST(1.0,1.38,0.137,1).

Both functions return 0.81, which is the value included in the table above.

Conditional probability tables can also be created using functions and equations included in most graphical modeling software. For instance, the dabbler pair conditional probabilities for each combination of future wetland conditions (future) and flyway population are estimated in Netica software Version 4.02 using the following code in the "Equation" box.

```
p (pairs | future, Flyway) =
(future== Open && Flyway==Above)?
NormalDist(pairs,1.38,0.137):
(future== Open && Flyway==Average)?
NormalDist(pairs,0.95,0.137):
(future== Open && Flyway==Below)?
NormalDist(pairs,0.64,0.137):
(future== Partially_open && Flyway==Above)?
NormalDist(pairs,1.62,0.137):
(future== Partially_open && Flyway==Average)?
NormalDist(pairs,1.18,0.137):
(future== Partially_open && Flyway==Below)?
NormalDist(pairs,0.87,0.137):
(future== Closed && Flyway==Above)?
NormalDist(pairs,0.9,0.137):
(future== Closed && Flyway==Average)?
NormalDist(pairs,0.46,0.137):
(future== Closed && Flyway==Below)?
NormalDist(pairs,0.16,0.137)
```

Field personnel did not estimate the number or density of ducklings on wetlands. However, a previous investigation modeled the relation between dabbler ducking density, the density of hen dabblers, and the condition of the wetland using linear regression. The resulting model was:

$$\text{duckling} = 9.75 \times \text{hens} + 2.40 \times \text{partial} \times \text{hens} - 0.48 \times \text{closed} \times \text{hens}$$

where *duckling* is the estimated density of ducklings, *hens* is the density of hens, *partial* is 1 when the wetland is partially closed otherwise 0, and *closed* is 1 when the wetland is closed, otherwise 0. The residual standard error of the regression was reported as 0.417. Assuming that a dabbler hen is equivalent to a dabbler pair, we can use the above equation to estimate duckling density. Calculating the conditional probabilities for duckling density is a bit more complicated because it requires incorporating the variation in dabbler pair density within a state. For example, the dabbler pair densities range from 0–0.5 pairs per hectare in the smallest density state. To include this variability, (1) we randomly generate dabbler densities for values within

(Continued)

the range of each state from a uniform distribution, (2) randomly generate values for the residual error from a normal distribution with a mean of 0 and standard deviation equal to the residual standard error of the regression (0.417), (3) use regression equation and randomly generated values to estimate duckling density, and (4) plot the cumulative distribution of the estimated duckling densities. The R code for this is shown below.

```
# randomly select 10,000 pair density from range 0-0.5
pairs<-runif(10000, min=0, max=0.5)

# randomly select 10,000 residual error values from normal
distribution
err<- rnorm(10000, mean = 0, sd = 0.417)

# calculate duckling density for open wetlands
ducklings <- 9.95*pairs + err

# plot empirical cumulative distribution of densities
plot(ecdf(ducklings),xlab = "duckling density", xlim=c(0,7))
minor.tick(nx=5, ny=5, tick.ratio=.5)
```

The conditional probabilities are then obtained using the cumulative distribution plot. For example using the plot created with the above code, the probability that the duckling density is 0–5 ha is 0.955 and the probability that the duckling density is 5–10 ha is 0.045 given that the pair density is 0–0.5 and the wetland condition is open.

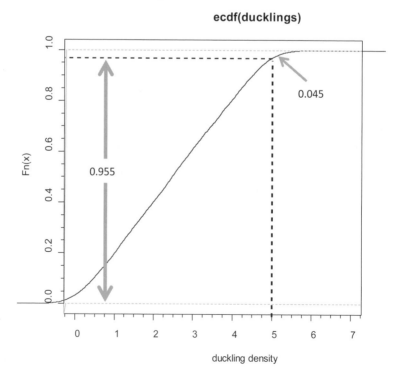

Conditional probability tables also can be created using functions and equations included in most graphical modeling software. Below, the duckling density conditional probabilities for each combination of future wetland conditions (future) and dabbler pair density are estimated in Netica software Version 4.02 using the following code in the "Equation" box.

```
p (duckling | future, pairs) =
(future == Partially_open)? NormalDist(duckling,
pairs*(9.95 + 2.39), 0.417):
(future == Closed)? NormalDist(duckling, pairs*(9.95
- 0.482), 0.417):
NormalDist(duckling, pairs*(9.95), 0.417)
```

Box 6.3   Parameterizing an influence diagram with output from a simulation model

Below is an influence diagram of a hypothetical management decision for an isolate metapopulation of a rare amphibian species. Small isolated wetlands are the only suitable habitats for the amphibian species and are assumed to be suitable habitat patches. The fundamental management objective is to maximize the persistence of the metapopulation. Managers have a fixed budget and can implement various combinations of two management options: improve connectivity among wetlands by removing migration barriers or increase the number of wetlands through restoration activities. To identify the optimal combination of management activities, refuge managers developed a model for predicting the probability of species persistence to 100 years in response to management activities.

Given the relatively small number of suitable wetland patches, the risk of metapopulation extinction (and the decision) is likely to be influenced by demographic stochasticity. To estimate the probability of persistence to 100 years of such a small metapopulation, we can simulate the metapopulation dynamics using various combinations of values for the three nodes pointing into the probability of species persistence node. Below is the R code for simulating metapopulation dynamics using randomly selected combinations of number of wetlands, colonization probability and persistence probability. The program creates 2 files. The first is a summary of each combination of simulation inputs discretized to the range of values in the corresponding node. For each combination, the proportion of simulations that the population persisted is calculated and the values written to a conditional probability table (CPT). The second creates a file with each simulation run that can be read into Netica as a case file for parameterizing the CPT tables.

(Continued)

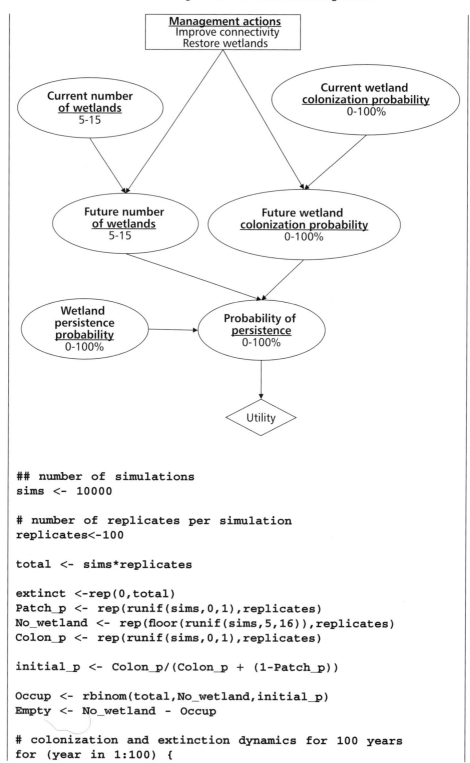

```
## number of simulations
sims <- 10000

# number of replicates per simulation
replicates<-100

total <- sims*replicates

extinct <-rep(0,total)
Patch_p <- rep(runif(sims,0,1),replicates)
No_wetland <- rep(floor(runif(sims,5,16)),replicates)
Colon_p <- rep(runif(sims,0,1),replicates)

initial_p <- Colon_p/(Colon_p + (1-Patch_p))

Occup <- rbinom(total,No_wetland,initial_p)
Empty <- No_wetland - Occup

# colonization and extinction dynamics for 100 years
for (year in 1:100) {
```

```
        Occup <-rbinom(total,Occup,Patch_p) +
rbinom(total,Empty,Colon_p)
        Empty <- No_wetland - Occup

        for (i in 1:total){ if (Occup[i] == 0)
extinct[i] <-1 }
        }
Persist <- 1-extinct

# First alternative for creating conditional probabilities
# create CPT table using summary of persistence for each
combination
# of colonization, persistence, and number of patches
# create groups corresponding to discretization of
colonization and
# extinction
# group is delimited by lower limit of discretized range
col_grp <- floor((Colon_p*10)/2)*.2
pers_grp <- floor((Patch_p*10)/2)*.2
CPT<- as.data.frame.table(tapply(Persist,list(col_grp,
No_wetland,pers_grp), mean))
names(CPT)<- c("Colon_p","No_wetland","Patch_p",
"Pr_persist")

write.table(CPT,"C:\\<your path> \\CPT.txt",sep = "\t",
row.names = FALSE)

# Second alternative create an input file to be read into
Netica
# note that column names must match node names in Netica
IDnum<- 1:total
result<-data.frame(IDnum,Patch_p,No_wetland,Colon_p,Persist)
write.table(result,"C:\\<your path> \\Data4_Netica.cas",sep
= "\t", row.names = FALSE)
```

The simulation output is only used to parameterize the relations among the persistence probability and the number of wetlands, wetland persistence probability, and wetland colonization probability. The relations among the remaining components (i.e., the probabilities) need to be specified using empirical data, models, or expert opinion.

Elicitation of expert judgment

When information is completely lacking, decision models can be parameterized using expert judgment. The use of expert judgment involves asking one or more experts that are familiar with the phenomenon being modeled (e.g., the effects of disease on wildlife populations) to parameterize the relationship between two or more model components. For example, there is often little empirical information available on the population-level effects of low levels of contaminants on wildlife survival. To parameterize the relationship between contaminants and survival, we could ask wildlife disease experts to use their best judgment to estimate survival under the different exposures levels of contaminants. As a measure of uncertainty, the experts could also be asked to provide the lowest and highest possible values for survival for each level of contaminants. These subjective estimates can then be combined when two or more experts are employed and the variation from expert to expert used as a measure of uncertainty.

Eliciting useful expert information can be a difficult task and several approaches have been developed to help analysts elicit information. These approaches can be categorized by the means of elicitation: **direct elicitation** and **indirect elicitation**, and the type of measure provided by the experts: **quantitative** or **qualitative** (O'Hagan et al. 2006). In direct elicitation, experts are provided with detailed description of the model components and are asked to provide the probabilities or response values. For example, a direct elicitation of the conditional probabilities in the timber harvest decision (Figure 6.10) involves asking an expert the probability that a fish population will be *large* the following year given that the habitat is *good* and the current population size is *small*. Indirect elicitation involves asking an expert a question, such as: how many streams out of 10 that have *good* habitat would you expect a population to be *large*? The probabilities are the calculated using the answer. For example, if the expert answered that 2 out of 10 streams would have large populations; the probability of *large* would be 20% (2 divided by 10). Quantitative measures are numerical values, such as 10% probability or 1 in 5 chance of survival, whereas qualitative measures are generally nominal or ordinal values, such as low, medium and high or increase and decrease. The best elicitation method to use depends on a variety factors including:

- the type of information desired, i.e., point estimates or estimates with measures of uncertainty;
- the experts' familiarity with probability and statistics;
- the necessity for multiple experts;
- the amount of time and resources available to conduct the elicitation.

Of these considerations, we've found that the experts' familiarity with statistical concepts is often the chief concern when eliciting information from most natural resource managers and ecologists. Thus, we tend to favor indirect approaches over direct approaches under these circumstances. Model development for most natural resource applications also occurs with limited budgets over relatively short timeframes, so we generally prefer approaches that can be used remotely (e.g., over the phone or via web-conferencing). Although there are numerous approaches to eliciting expert information, such practical considerations often dictate which methods are most appropriate for the application of interest. Below, we describe four quantitative approaches that are relative quick and inexpensive and can be used by experts with varying degrees of statistical expertise: probability, frequency, value, and function elicitation. Examples of each elicitation method are provided in Box 6.4.

Box 6.4   Elicitation of expert information and the construction of probability distributions

Probability elicitation

Probability elicitation is a direct, quantitative method that requires experts that are familiar and comfortable with statistical concepts, such as probability. A typical type of question for probability elicitation is:

Given that the maximum summer water temperature is 33°C or greater, what is the probability that an amphibian species will survive?

Assume that we asked 6 experts this question and received the following, values 0.5, 0.6, 0.45, 0.65, 0.45, 0.7. We could combine the values across experts by treating these values as data and calculating a mean. Uncertainty among experts can be estimated by calculating the variance.

```
> Surv<- c(0.5, 0.6, 0.45, 0.65, 0.45, 0.7)
> s.mean<-mean(Surv)
> s.var<- var(Surv)
>
```

As we learned earlier (Chapter 5, Appendix A), uncertainty in a probability can be modeled using a beta distribution. Thus, we could use the method of moments to estimate the parameters of the beta.

```
> ### beta method of moments
> beta.mom<-function(mean,v){
+   x<-mean
+   a<-x*(x*(1-x)/v-1)
+   b<-(1-x)*(x*(1-x)/v-1)
+   c(a,b)
+ }
> beta.mom(s.mean,s.var)
[1] 11.501531  9.098226
```

Often, one or more experts have greater experience than others and hence, we have greater faith in the values they provide. Weights are represented using positive numbers with larger numbers representing greater weight. The absolute values of the weights do not matter. Rather it is the relative differences among the weights that are used to weight the values provided by the experts.

For example, the weights 1,1,2,1,2,4 are equal to 0.091,0.091,0.182,0.091,0.182,0.364 because the relative proportions are equal. To demonstrate:

```
> # weights as whole numbers
> wt1<-c(1,1,2,1,2,4)
> # weights as proportions
> wt2<- c(0.091,0.091,0.182,0.091,0.182,0.364)
```

(Continued)

```
> #means
> weighted.mean(Surv,wt1)
[1] 0.5772727
> weighted.mean(Surv,wt2)
[1] 0.5772727
> #variance
> wt.var(Surv,wt1)
[1] 0.01601064
> wt.var(Surv,wt2)
[1] 0.01601064
```

In some instances, you will need to ask experts to fill out conditional probability tables that are used in influence diagrams and combine values across experts. For example, assume that we asked 3 experts to fill out the following tables for estimating species status:

	Snag density	Forest canopy	Species status		
			Absent	Rare	Abundant
expert 1	Many	Open	0.36	0.36	0.29
	Many	Closed	0.15	0.35	0.50
	Few	Open	0.67	0.25	0.08
	Few	Closed	0.44	0.38	0.19
expert2	Many	Open	0.37	0.45	0.18
	Many	Closed	0.17	0.41	0.41
	Few	Open	0.69	0.28	0.03
	Few	Closed	0.51	0.47	0.01
expert3	Many	Open	0.36	0.38	0.26
	Many	Closed	0.17	0.43	0.41
	Few	Open	0.66	0.30	0.03
	Few	Closed	0.56	0.42	0.03

Here as before, we can calculate the means and weighted means of the probabilities in corresponding cells. However, be careful that the same cells are selected as below or that you read each expert table as a matrix in R.

```
> Many.Open<-c(0.36,0.37,0.36)
> Many.Closed<-c(0.15,0.17,0.17)
> Few.Open<-c(0.67,0.69,0.66)
> Few.Closed<-c(0.44,0.51,0.56)
> #means
> mean(Many.Open)
[1] 0.3633333
> mean(Many.Closed)
[1] 0.1633333
> mean(Few.Open)
[1] 0.6733333
> mean(Few.Closed)
[1] 0.5033333
```

```
> #expert weights
> wt<-c(5,10,100)
> #weighted means
> weighted.mean(Many.Open,wt)
[1] 0.3608696
> weighted.mean(Many.Closed,wt)
[1] 0.1691304
> weighted.mean(Few.Open,wt)
[1] 0.6630435
> weighted.mean(Few.Closed,wt)
[1] 0.5504348
```

Frequency elicitation

Frequency elicitation is an indirect, quantitative method that does not require experts that are familiar and comfortable with statistical concepts.

A typical type of question for frequency elicitation is: 100 individuals from an amphibian species are exposed to temperatures $> 33\,°C$, how many will survive? These values can be used to calculate probabilities and average across experts. For example assume, that 4 experts answered the above question by providing the following values: 50, 30, 25, and 40.

```
> n.surv<-c(50, 30, 25, 40)
#remember 100 individuals
> p.surv<-n.surv/100
> mean(p.surv)
[1] 0.3625
> var(p.surv)
[1] 0.01229167
```

Here, we also could use the weighting scheme as demonstrated above and use the method of moment or maximum likelihood approach to estimate the parameters of a beta distribution, if we desired. Alternatively, we could use the values directly to estimate the parameters of a beta distribution. Remember that the parameters of a beta distribution represent: a = number of successes (e.g., survivors) and b = the number of losses (e.g. deaths), so we have

```
> n.surv<-c(50, 30, 25, 40)
> n.die<-100- n.surv
> #let's see what that gets us in terms of means and
variance for probability
> p.s<-rbeta(1000,n.surv,n.die)
> mean(p.s)
[1] 0.3631856
> var(p.s)
[1] 0.01114131
```

The mean value was fairly close as was the variance. If we wanted to weight experts differently, we could rescale the values. For example, assume that

(*Continued*)

expert 1 and 2, should receive half the weight of experts 3 and 4. We could do the following:

```
> n.surv<-c(25, 15, 25, 40)
> n.die<-c(25, 35, 75, 60)
> ## notice that the sum of survive and die for expert 1
and 2 is 50
> n.surv+n.die
[1]   50  50 100 100
> p.s<-rbeta(1000,n.surv,n.die)
> mean(p.s)
[1]  0.361134
> var(p.s)
[1]  0.01268896
```

Keep in mind that the variance on the beta is influenced by the size of the parameters, for example:

```
> p.s<-rbeta(1000,1,1)
> mean(p.s)
[1]  0.4963211
> var(p.s)
[1]  0.0871831
> #expected value the same, variance different
> p.s<-rbeta(1000,100,100)
> mean(p.s)
[1]  0.4995342
> var(p.s)
[1]  0.001307685
```

Value elicitation

Value elicitation is a direct, quantitative method that does not require experts that are familiar and comfortable with statistical concepts. A typical type of question for value elicitation is:

What is the LD50 temperature for an amphibian species? (LD50 is the temperature where 50% of animals die).

Assume that we received the following answer from 6 different experts: 30, 31, 35, 37, 32, 34.5

```
> temp<-c(30, 31, 35, 37, 32, 34.5)
> mean(temp)
[1]  33.25
> sd(temp)
[1]  2.678619
```

These parameters can then be used to parameterize relations among model components. For example, to estimate the probability that 50% of the amphibians will die if temperatures reach $34\,^{\circ}C$:

```
> pnorm(34,mean(temp),sd(temp))
[1]  0.6102593
```

Function elicitation

Function elicitation is an indirect, quantitative method that does not require experts that are familiar and comfortable with statistical concepts. Here, experts define the functional relationship between two or more model components using a graphical representation. For example, an expert is asked to define the relation between amphibian survival and summer temperatures by drawing a line on paper or in a spreadsheet. For example, the figure below was drawn by an expert:

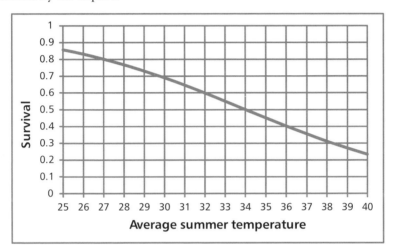

This drawn relationship needs to be turned into a function. The first step is to estimate pairs of survival and summer temperature values on the line. Because we are modeling a probability, logit-linear regression model seems appropriate. However, the R glm function requires the response to be binary and in live, dead format. For convenience, assume that we began with 100 individuals and calculate the number surviving and dying based on the probabilities estimates from the figure drawn by the expert. We then have:

```
> surv<-round(Estimated.survival*100)
> died<-100-surv
>
> response<-cbind(surv,died)
> glm(response ~ Ave.summer.temp, fam=binomial)
Call: glm(formula = response ~ Ave.summer.temp, family =
binomial)
Coefficients:
   (Intercept)   Ave.summer.temp
       6.6889          -0.1971
```

The function for defining this relationship is a logit linear model 6.689–0.197* temperature. To incorporate the variability among experts, the slopes and intercepts would be averaged and the averages used to parameterize the model.

(*Continued*)

Quantifying uncertainty in expert judgment

To minimize the effect of overconfidence on expert, we recommend the four-step process to uncertainty elicitation proposed by Speirs-Bridge et al. (2010). In the 4-step process, experts are asked the following 4 questions:

1) What do you think the lowest value could be? ____
2) What do you think the highest value could be? ____
3) What is your most likely estimate?___
4) How confident (%) are you that the interval you created, from lowest highest, will capture the true value? ___

For example, 100 individuals from an amphibian species are exposed to 33 °C temperatures.

1) What are the fewest number that would survive? ____
2) What are the most that would survive? ____
3) What is your most likely estimate of the number surviving?___
4) How confident (%) are you that the interval you created, from lowest highest, will capture the true value? ___

Note that the 4-step process can be used with any of the types of elicitation covered above.
   Let's assume that we asked an expert the first question above:

Given that the maximum summer water temperature is 33 °C or greater, what is the probability that an amphibian species will survive

1) What do you think the lowest value could be? _0.25_
2) What do you think the highest value could be? _0.9_
3) What is your most likely estimate? _0.55_
4) How confident (%) are you that the interval you created, from lowest highest, will capture the true value? _0.90_

To estimate the parameters, we need to find the beta distribution that best fits those conditions, i.e., a median value of 0.5 and upper and lower 90% confidence limits of 0.25 to 0.9. We can fit these parameters in R with the following code:

```
> ## lower, median, and upper values
> w<-c(0.25, 0.55, 0.9)
> ### load library first
> library(fitdistrplus)
> # 90% confidence so specify lower 5 and upper 95
percentiles
```

```
> est<-qmedist(w, "beta", probs=c(0.05, 0.95))
> parms<-est$estimate
> parms
  shape1    shape2
3.906427  2.729426
> ## let's see how close we got
> test<-rbeta(1000,parms[1],parms[2])
> quantile(test,c(0.05,0.5,0.95))
      5%        50%        95%
0.2855605  0.5938664  0.8611368
```

This is not too bad considering expert's judgments don't always follow the laws of probability. Note that shape1 is alpha and shape2 is beta for the beta distribution. We could combine these across experts using the Bayesian methods we used earlier.

Let's try the same thing, but this time asking:

What is the LD50 temperature for an amphibian species?

1) What do you think the lowest value could be? __33__
2) What do you think the highest value could be? __39__
3) What is your most likely estimate?__35__
4) How confident (%) are you that the interval you created, from lowest highest, will capture the true value? __0.80__

The corresponding R code and output are:

```
> ## lower, mean, and upper values
> w<-c(33, 35, 39)
> library(fitdistrplus)
> # 80% confidence so specify lower 10 and upper 90
percentiles
> est<-qmedist(w, "norm", probs=c(0.10, 0.90))
> parms<-est$estimate
> parms
     mean         sd
35.799954  1.872723
> ## let's see how close we got
> test<-rnorm(1000,parms[1],parms[2])
> quantile(test,c(0.10,0.5,0.90))
     10%        50%        90%
33.42040  35.71296  37.98570
```

As before, the estimates are not too bad and they can be combined across experts using the Bayesian techniques we learned in Chapter 5.

Probability elicitation is a direct, quantitative method that requires experts that are familiar and comfortable with statistical concepts, such as probability. The benefits of probability elicitation are that it is relatively quick and easy and the information can be incorporated directly into models. A typical type of question for probability elicitation is: Given that the maximum summer water temperature is 33 °C, what is the probability that an amphibian species will survive?

Frequency elicitation is an indirect, quantitative method that does not require experts that are familiar and comfortable with statistical concepts. The benefits of the frequency elicitation are that it is relatively quick and is one of the easiest methods for getting probabilities from experts that are uncomfortable with statistical concepts. Typical type of question for frequency elicitation is: 100 individuals from an amphibian species are exposed 33 °C temperatures, how many will survive?

Value elicitation is a direct, quantitative method that does not require experts that are familiar and comfortable with statistical concepts. The benefits of the value elicitation are that it is relatively quick and easy. However, some time may be required to process the values provided by the experts. A typical type of question for value elicitation is: What is the lethal maximum temperature for an amphibian species?

Function elicitation is an indirect, quantitative method that does not require experts that are familiar and comfortable with statistical concepts. Function elicitation is unlike the three approaches described above in that experts are not asked to provide values. Rather, experts define the functional relationship between two or more model components using a graphical representation. For example, an expert is asked to define the relation between amphibian survival and summer temperatures by drawing a line on paper or in a spreadsheet. It is a relatively quick and easy method for eliciting complex relationships among model components from experts. However, it can be time consuming for the analyst to process the information to obtain parameter estimates that define the graphical representation.

Quantifying uncertainty in expert judgment

Thus far, we have discussed probability, frequency, and value elicitation in terms of single point estimates where uncertainty is not explicitly quantified by a single expert. Obtaining estimates of uncertainty requires multiple experts, for example, estimating the mean and standard error of probabilities across experts. Alternatively, uncertainty can be elicited from individual experts. For example, experts could be asked to provide the minimum and maximum value for a model parameter. Most approaches to uncertainty elicitation, however, are subject to **overconfidence** by the experts. Overconfidence is when an expert underestimates the amount of uncertainty and tends to result in biased estimates. To minimize the effect of overconfidence, we recommend the four-step process to uncertainty elicitation proposed by Speirs-Bridge et al. (2010). In the 4-step process, experts are asked the following 4 questions:

1) What do you think the lowest value could be? ____
2) What do you think the highest value could be? ____
3) What is your most likely estimate?___
4) How confident (%) are you that the interval you created, from lowest highest, will capture the true value? ___

For example, 100 individuals from an amphibian species are exposed to 33 °C temperatures.

1) What are the fewest number that would survive? ____
2) What are the most that would survive? ____
3) What is your most likely estimate of the number surviving?___
4) How confident (%) are you that the interval you created, from lowest highest, will capture the true value? ___

These values, then, are used to parameterize the model or if two or more experts are used, the values are combined.

Quantifying uncertainty in function elicitation can also involve averaging model parameters across experts and treat the resulting function similar to a random effects linear regression model (Chapter 5). Alternatively, individual experts could be asked to create (draw) confidence limits around the function relationship and estimate their degree of confidence that the true relationship is contained within the limits in a manner analogous to the 4-step process outlined above. The limits can then be used to quantify the uncertainty and incorporate it into the decision model.

Group elicitation

The expert judgment process can benefit for the input of multiple experts. Multiple experts can provide the basis for quantifying uncertainty. The interaction and discussion among experts also can lead to an improved understanding of the physical or biological processes under question and better estimates of the relations between two or more model components. The Delphi method (Delbecq et al. 1975) is among the most widely used **group elicitation** techniques in natural resources management. The objective of the Delphi technique is to reach consensus among experts regarding the value of parameters (e.g., ecological thresholds) or the relations among factors. Such group judgments, however, can be influenced by human behaviors that dictate group dynamics. For example, strong personalities can attempt to dictate the outcome of the process and lead to biased estimates. Conversely, group members may not communicate their differences of opinion because they want to be viewed as "getting along" with the group. Groups also tend to be much more overconfident than individuals when making group judgments. To avoid these problems, we recommend the use of a modified Delphi technique (Clark et al. 2006). Here, individual experts are furnished with the background and description of the model components and are asked to provide their expert judgment independently from other group members. The judgments of the experts are summarized. Anonymous summaries are provided to and discussed by the experts during a review session. The anonymity of the each expert judgment is important during the review phase because it minimizes the potential of group members to defer to the opinion of others they may view and more competent or alternatively, discount the opinions of others that are perceived as less competent.

During the review phase, the discussion should focus on clarifying the model components and their relations to ensure that all the experts are addressing the same problem and minimize linguistic uncertainty. In addition, experts should be encouraged to provide information on new or unknown studies that may provide the rationale for modifying the experts' initial judgments. However, care must be taken to ensure that the discussion does not turn into an argument of who is right and who is wrong. It is important to represent and quantify genuine differences of opinion in the parameter estimates. Following the review and discussion, the experts should be provided with the opportunity to revisit and revise their initial judgments.

The care and handling of experts

As you may have guessed, eliciting expert information requires preparation and skills that are more suited to social sciences and cognitive psychology. The elicitation process should begin with a complete and clearly defined model. Linguistic uncertainty, the uncertainty due to the imprecision of everyday language, is a potentially significant source of uncertainty when eliciting expert information that can greatly inflate uncertainty in the decision model and may bias estimates (Regan et al. 2002; Carey and Burgman 2008). Therefore, every attempt should be made to clearly define the model components and their component states, including the reasoning and justification behind specific relationships. Whenever possible, node states should be defined by unambiguous numeric values. When experts are asked to provide a measure of uncertainty, such as a range or degree of confidence in a value, it should be made clear that the uncertainty should be epistemic uncertainty (i.e., incomplete knowledge of ecological processes) rather than aleatory uncertainty (i.e., unpredictable variation of the system). Remember that you are asking experts for information on phenomenon that have never been studied (otherwise you wouldn't need the expert) and no one, not even an expert wants to be wrong. Experts can be tentative or reluctant to provide estimates that may be reviewed or commented on by others. Therefore, you should develop and implement procedures for ensuring confidentiality and anonymity.

The drawbacks of using expert judgment approaches are that use of expert judgment places a heavy burden of proof on the decision maker. Expert judgment should not be used when empirical data or models are available because it is less defensible. There are several problems that can arise from the use of experts to parameterize decision models. Many of these arise from miscommunication or the failure of the expert to understand (or the analyst explain) the nature of the component they are being asked about. As we discussed, these factors can be minimized by being as clear and concise as possible. However, experts are human and they undoubtedly have biases that could affect their judgment and ultimately the decision. Therefore, it is important to know these potential biases in advance and choose experts accordingly.

References

M. Byron. (1998) Satisficing and optimality. *Ethics* **109**, 67–93.

Carey, J. and M.A. Burgman. (2008) Linguistic uncertainty in qualitative risk analysis and how to minimize it. *Annals of the New York Academy of Science* **1128**, 13–17.

Clark, K.E., J.E. Applegate, L.J. Niles, and D.S. Dobkin. (2006) An objective means of species status assessment: adapting the Delphi technique. *Wildlife Society Bulletin* **34(2)**, 419–425.

Delbecq, A.L., A.H. Van de Ven, and D.H. Gustafson. (1975) *Group Techniques for Program Planning: A Guide to Nominal Group and Delphi Processes*. Scott Foresman and Company, Glenview, Illinois.

O'Hagan, A., Buck, C.E., Daneshkhah, A., Eiser, J.R., Garthwaite, P.H., Jenkinson, D.J., Oakley, J.E., and Rakow, T. (2006) *Uncertain Judgments: Eliciting Expert Probabilities*. Wiley, Chichester, UK.

Regan, H.M., M. Colyvan, and M.A., Burgman. (2002) A taxonomy and treatment of uncertainty for ecology and conservation biology. *Ecological Applications* **12**, 618–628.

Speirs-Bridge, A., F. Fidler, M. McBride, L. Flander, G. Cumming, and M.A. Burgman. (2010) Reducing overconfidence in the interval judgments of experts. *Risk Analysis* **30**, 512–523.

Additional reading

Kuhnert, P.M., T.G. Martin, and S.P. Griffths. (2010) A guide to eliciting and using expert knowledge in Bayesian ecological models. *Ecology Letters* **13**, 900–914.

Kuhnert, P.M., T.G. Martin, K. Mengersen, and H.P. Possingham. (2005) Assessing the impacts of grazing levels on bird density in woodland habitat: a Bayesian approach using expert opinion. *Environmetrics* **16**, 717–747.

Yamada, K., J. Elith, M. McCarthy, and A. Zerger. (2003) Eliciting and integrating expert knowledge for wildlife habitat modelling. *Ecological Modelling* **165**, 251–264.

7

Identifying and Reducing Uncertainty in Decision Making

In the previous chapter, we built a decision model to estimate the outcomes of alternative management actions and used it to identify the optimal or satisficing alternative. Natural resource management, however, is complicated by uncertainty and not all uncertainties have an equal effect on decision making. In this chapter, we discuss uncertainty in natural resources decision making. First, we discuss important types of uncertainty, the effects of uncertainty on decision making, and introduce approaches for identifying the key or important uncertainties. Next, we discuss way to reduce uncertainty and introduce adaptive resource management as a special case of structured decision making that reduces uncertainty through time and improves decision making.

Types of uncertainty

We begin by defining **uncertainty** using **certainty** as our benchmark. An event is considered **certain** if it will happen with a 100% probability. Certain events are generally predicted using deterministic models and are displayed graphically in influence diagrams as constant nodes. Uncertainty then is anything that falls short of absolute certainty; there is uncertainty even if an event has a 99.9% probability of occurring. Later in the chapter we will discuss methods for identifying the important uncertainties and reducing them through the use of monitoring data in processes known as **adaptive management**.

There are several different types of uncertainty that can be broadly broken down into two categories: **irreducible** and **reducible uncertainty**. We use these terms to reinforce the idea that some sources of uncertainty (i.e., those that are reducible) can be decreased through additional efforts by decision makers (e.g.,

Decision Making in Natural Resource Management: A Structured, Adaptive Approach,
First Edition. Michael J. Conroy and James T. Peterson.
© 2013 John Wiley & Sons, Ltd. Published 2013 by John Wiley & Sons, Ltd.

studies), whereas other sources of uncertainty are irreducible and cannot be decreased. Here we focus on two types of reducible uncertainty, **linguistic** and **epistemic uncertainty**. Linguistic uncertainty is the uncertainty due to the imprecision of language. Epistemic uncertainty reflects the incomplete knowledge and the inherent limits of using data to estimate parameters and understand ecological processes. The type of irreducible uncertainty important to natural resources decision making is defined as **aleatory uncertainty**. It represents the natural, unpredictable variation of the system that is being managed. The knowledge of experts and additional data cannot be used to reduce aleatory uncertainty although they may be useful in quantifying aleatory uncertainty. We first address aleatory or irreducible uncertainty, since this type of uncertainty by definition cannot be reduced by conducting additional studies or collecting more data. We will then return to reducible forms of uncertainty, and deal at length with methods for quantifying and reducing this type of uncertainty.

Irreducible uncertainty

The two most common forms of aleatory uncertainty in natural resource management decision making are **environmental** and **demographic stochasticity**. Environmental stochasticity arises from variation in environmental factors, such as the weather, and disturbances, such as earthquakes and fires. Demographic stochasticity is the unpredictable variation in the survival and reproduction among individuals in a population. This variation occurs even if all individuals in a population have the same expected survival and reproductive rates and has important implications for managing small populations. Environmental and demographic stochasticity can be incorporated in decision models using statistical distributions.

Another type of uncertainty that may be irreducible is **partial controllability**, which is the inability to perfectly control the system of interest. For example, a waterfowl refuge manager may take an action that is predicted to result in a specific density of moist soil plants. The actual density of plants is likely to differ from the intended amount due to random events such as weather patterns. Partial controllability is similar to environmental stochasticity because much of the uncertainty is often due to environmental stochasticity. Partial controllability can also arise from the behaviors of individual people. For example, wildlife park managers may set park-access regulations based on the number of people that are predicted to visit the park. The actual number of visitors, however, depends on the choices of individual people, so it is likely to differ from the predicted number. Partial controllability can also be incorporated in decision models using statistical distributions. Other examples of partial controllability can be, to some extent, reducible. For example, uncertainty in the burn resulting from a prescribed fire (intensity, area extent) may be reduced by confining fire application to certain environmental conditions (ideal wind speed and humidity). Nonetheless, decision makers need to be aware that the intended result(s) of management actions can, and quite often do, differ from the actual results and partial controllability should be incorporated into decision models.

Reducible uncertainty

In contrast to the above forms of uncertainty, we now consider uncertainty that *can* be reduced. One of the most familiar of these is **linguistic uncertainty,** which you have undoubtedly encountered in everyday life. When people write or talk, the language they use is often vague and filled with ambiguous terms and phrases or words that can be interpreted in multiple ways. For example, consider the phrase "species diversity" as it is commonly used in the biological disciplines. Species diversity can mean multiple things. For example, species diversity can mean the total number of species; the number of native species; or the relative distribution of individuals among species, also known as evenness. Similarly, other commonly used terms such as population, community, ecosystem, viability, and sustainability can each be defined in several ways. Because we use language to communicate, linguistic uncertainty is a potential problem throughout the structured decision-making process. However, it is particularly problematic when identifying stakeholder objectives. Words can have multiple meanings; hence a single stakeholder objective can also mean something different to each stakeholder. This can lead to potential conflict among stakeholders and the failure to properly define the decision situation and management alternatives. For example, the management alternative "improve stream habitat" could include any number of actions ranging from installing large wood structures to changing the shape of the stream channel. Linguistic uncertainty can also be particularly problematic in interdisciplinary stakeholder working groups. Each discipline has its own terminology and definitions and quite often the same term can mean very different things to different disciplines. For example in biological disciplines, spatial scale generally refers to the grain or extent of a system being managed, whereas spatial scale means the ratio of a distance on the map to the corresponding real-world distance in geographical disciplines. Minimizing linguistic uncertainty is important in every step of the process and can be minimized by thoroughly defining objectives, decision alternatives, and the components of the decision model and their relationships. When possible, use unambiguous quantitative values and mathematical functions to define relationships among decision model components. In some instances, it may also be necessary to develop a comprehensive list of discipline-specific terms and their definitions to disseminate among stakeholders and technical advisers.

The other important branch of reducible uncertainty, epistemic uncertainty, has several forms; the three we commonly encounter in natural resource management decision making are **statistical, observational,** and **structural uncertainty.** Statistical uncertainty arises because we use sample data to estimate the parameters of decision models. Most often you see statistical uncertainty expressed as confidence limits or standard errors for parameter estimates (Chapter 5). Statistical uncertainty can be reduced through conducting additional studies or collecting data (e.g., monitoring) to improve parameter estimates. Statistical uncertainty is incorporated in decision models using statistical probability distributions (Chapter 5).

Observational uncertainty, also referred to as **partial observability** or **statistical uncertainty,** is due to our inability to accurately assess the state of the system being managed. For example, all of the animals in a population usually cannot be counted during sampling, so biologists typically used sample designs and estimators to obtain an estimate of population size. The estimate then may or may not equal the true value due to sampling error or sample bias. Observational uncertainty can be incorporated into decision making using statistical probability distributions. However, it can be difficult to incorporate into certain instances, particularly sequential dynamic (i.e., Markov) decisions that use dynamic programming to find the optimal decision (Chapter 8).

The last and most often overlooked epistemic uncertainty is **structural uncertainty**, also termed **system uncertainty**. Structural uncertainty arises from an incomplete understanding of system dynamics. In other words, we are uncertain about what factors or ecological mechanisms affect the outcome of a decision. For example, consider the notion that the size of a hypothetical animal population is positively related to the availability of some habitat component. A positive relation between habitat availability and population size can take an infinite number of forms and the assumptions about the true form can have a substantial effect on decision making (Figure 7.1). If the linear model is used, the decision maker would decide to increase habitat as much as the budget will allow because he expects to have a proportional increase in the objective, population size. In contrast, the asymptotic model in Figure 7.1 suggests that the population response decreases after a certain amount of habitat is available and increasing habitat beyond that point would likely waste management resources. Alternatively, the exponential model suggests that managers would be wasting resources if they could not increase habitat beyond a certain point. Structural uncertainty is incorporated into decision models by including alternative models of system dynamics. The use of multiple models is a relatively recent phenomenon in natural resource decision making, though several of the case studies in Chapter 9 incorporate alternative models. However, the most widespread and consequential use of alternative models is by the meteorologists predicting the weather or likely paths of destructive storms, such as hurricanes.

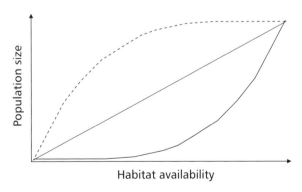

Figure 7.1 Three positive relations between habitat availability and the size of a hypothetical population: additive linear relation (thin solid line), asymptotic relation (heavy broken line), and exponential relation (thick solid line).

Disagreements about the factors affecting the system being managed can be an important source of conflict among stakeholders, because alternative models can provide very different estimates of the effectiveness of alternative actions. The decision then often becomes one of deciding which model is right. We and others believe that such an approach is unnecessary and only leads to gridlock. Rather, we advocate the use of multiple models to incorporate structural uncertainty and eliminate "your model versus my model conflicts." In dynamic decision-making situations, monitoring data can be used to resolve the uncertainty about system dynamics and improve decision making and knowledge through time. This is the basis of adaptive resource management, using multiple models and monitoring to improve our understanding of the system being managed; thereby improve decisions (more later).

The alternative models that are used in natural resource decision making require model weights to create a composite or **model-averaged estimate** of the outcomes of management actions. The alternative models should be collectively exhaustive and the weights must sum to one. For example, consider two models that estimate a change in population size in response to management actions (Figure 7.2). The first model assumes that there is no effect of habitat availability on the population and the second assumes that habitat affects population size. The weights for the two models should sum to one, so assume that "no effect" has a weight on 0.75 and "habitat effect" has a weight of 0.25. Each model also predicts a different population size resulting from each management action. The no effect model predicts a population size of 25 and the habitat effect model

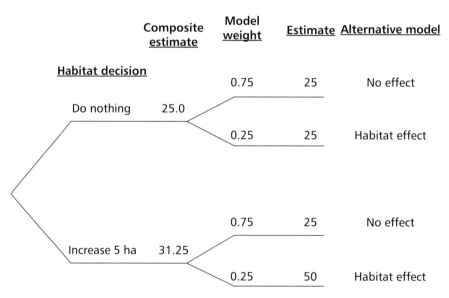

Figure 7.2 A decision tree of habitat decision with two alternative models of population response to changes in habitat availability: no effect and habitat effect. The optimal decision is increase habitat 5 ha with an expected value of 31.25.

predicts a population size of 50 if habitat in increased by 5 ha. To create a model-averaged estimate, the values predicted by each model are multiplied by their corresponding weights and added together. For example, the weighted "no effect" estimate of 25×0.75 is added to the "habitat effect" estimate 50×0.25, which equals 31.25. Thus, the estimated population size for an increase in 5 ha of habitat is 31.25 (Figure 7.2).

Model weights can be obtained in a number of ways. If data are available, Bayesian model posterior probabilities or Akaike model weights can be used as the basis for the model weights (Chapter 5). Alternatively, they could be based on the vote or consensus of stakeholders or the judgment of experts (Chapter 6). Of these, estimating model weights with data is usually the most defensible approach. However, when there is potential for conflict among stakeholders about the models, it is often best to let the stakeholders agree upon the approach even if it means not using available data. Keep in mind that the **sensitivity** of the decision to the model weighting should be evaluated during sensitivity analysis (more on that later). In sequential dynamic decision-making situations, these weights also are initial weights that change through time as monitoring data are used to evaluate which model is better at predicting the outcome of management decisions. We consider sensitivity in detail in the section below.

Effects of uncertainty on decision making

Approaches for incorporating uncertainty in decision making

Once we've identified our sources of uncertainty, and hopefully expressed them in terms of probability distributions (Chapter 5) and alternative models, we need to incorporate this information into decision making. The most common approach is, in fact, the one we advocated in Chapter 6, where we selected the decision than provides the maximum **expected value**. That is, given a utility value for each possible consequence, and the conditional probability (Chapter 5) of that outcome given each decision, we select the decision that maximizes the weighted average of the objective over the statistical distribution, with weights provided by the conditional distribution. For simple problems of the type considered in Chapter 6 the optimal decision can often be obtained by inspection; more complex (e.g., dynamic) problems can require more sophisticated, computationally intensive approaches (Chapter 8).

Another approach that is sometimes advocated is termed minimax, and refers to the selection of decisions that minimize the loss for the worst case or maximum loss outcome. Taken literally, the method used only the utilities for the outcomes and not their probabilities, and works by first establishing the minimum utility for each decision, and then selecting the decision that maximized these values (Lindley, 1985). The minimax method has been criticized on a number of grounds, including the claim by Lindley that it leads to incoherent decisions (that is, decisions that are essentially guaranteed not to meet our objectives). Therefore, we will not cover the minimax approach in detail.

The final approach we consider, and describe here in detail, is called **risk profiling**, which is used to evaluate the tradeoffs between risk and return to the decision maker. The terms risk and uncertainty are often used synonymously, albeit incorrectly, and this is often a source of confusion when working with stakeholder groups. Uncertainty, as we defined earlier is anything that falls short of absolute certainty. It is quantified using probability, alternative models and model weights. **Risk** then is a state of uncertainty for which there exists an outcome that is *undesirable*, such as the extinction of a species. The probability of the undesirable outcome can be quantified using a probability distribution, but the measurement of risk generally varies among decision makers and decision situations. For example, the extinction of a threatened species is certainly an undesirable outcome and it can be estimated using probability (i.e., the probability of extinction). An extinction probability will never equal zero (e.g., there is always a chance a comet could destroy the earth), so determining what extinction probability is deemed acceptable is subjective and depends largely on the decision makers.

Up until this point, with the exception of minimax, we have based decisions on the expected value of the decision. The expected value approach, however, can lead to decisions that may not appeal to decision makers. Consider for example a problem where the decision maker has two decisions to make: plant forage for the animal population being managed or do a controlled burn (Figure 7.3). If forage is planted, there are two outcomes with equal probability, the population increases by 10 or is unchanged. If there is a controlled burn, there are two outcomes with equal probability, the population increases by 1000 or decreases by 500, perhaps due to the burn getting out of control. The expected value of burning is 50 times greater than planting forage (Figure 7.3).

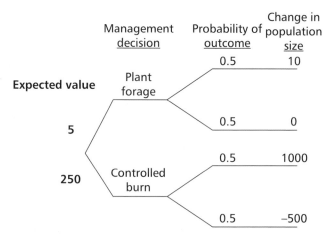

Figure 7.3 A decision tree depicting a management decision with two alternatives: plant forage and controlled burn. The optimal decision is controlled burn with an expected value of 250.

However, many of us would perceive a controlled burn as much riskier because it could lead to a large decrease in population size (note the level of subjectivity there – we will get into that later). The decision to burn would probably be considered even riskier if the population was a threatened or endangered species. Note here too that there is no completely objective value to risk. It is relative to the decision problem and is therefore, a subjective measure. Remember that the expected value of a decision is based on the long-run frequency. That is, it is the expected increase in the population (in this example) after making several decisions. For example, if we make a one-time decision to plant forage, the worst we could do is no (zero) increase in the population and the best is to increase the population by 10. If we make that decision 100 times, on average, the worst we will do is increase the population by 3 and the best we increase it by 7. As we increase the number of times that a decision is made, the difference between worst and best narrows and will eventually meet at the expected values.

The expected value of the decision does not take into account **risk attitudes**. Risk attitude is the chosen response of decision makers to uncertainty that matters and is a subjective judgment that reflects the values and perceptions of decision makers. Variability in risk attitudes among stakeholders can have a substantial effect on decision making in multi-stakeholder and multi-objective decision situations. To illustrate risk attitudes, consider a situation where a person is given the choice between two management decisions, one with a guaranteed positive benefit to a population being managed and one without. In the guaranteed decision, the population increases by 25 with *absolute certainty*. In the uncertain decision, there is a 50% probability that the population will increase by 50 and a 50% probability that the population will not increase (zero growth). The expected value for both decisions is 25, meaning that a decision maker that was insensitive to risk would not care (i.e., would be indifferent) whether they made the guaranteed decision or uncertain decision. However, individuals and groups may have different risk attitudes.

There are three basic types of risk attitudes. A decision maker is considered risk-averse if they would accept a positive outcome with absolute certainty, rather than take a chance and receiving a less valued outcome. Risk-neutral decision makers are indifferent between certain and uncertain management actions provided that the expected value is the same for both. Risk-prone or (risk-seeking) decision makers would rather take the chance with the uncertain management if the potential result of the decision is sufficiently large. Returning to the above example, a guaranteed increase of 25 individuals is not sufficient to make a risk-prone decision maker act or choose that decision if there is a chance to increase the population by 50. As we discussed earlier, risk attitudes and the perception of risk can vary among stakeholders and with the decision situation, so much so that stakeholder groups often consist of members with all three risk attitudes. The variation in risk attitudes can be a source of disagreement and conflict among stakeholders. Thus, it is often necessary to evaluate the risk attitudes of stakeholders.

The size or magnitude of the outcome that the decision maker would accept instead under absolute certainty is defined as the **certainty equivalent**, and the difference between the expected value under the decision with uncertainly and the certainty equivalent is called the **risk premium**. The risk premium can be thought of as the "price" paid (for, in a sense, foregoing the lost opportunity to obtain higher gains of the occasional "better" outcome) to avoid the risk of a less valuable or bad outcome. For risk-averse individuals, the risk premium is positive, for risk-neutral persons it is zero, for risk-prone individuals the risk premium is negative. **Risk tolerance** is the ability of the decision maker to accept or absorb risk. To assess the risk tolerance of individuals and groups, we can use a utility curve to develop a risk profile of the decision maker. A utility curve displays the relations between the objective value, such as population size, number of animals harvested, or user-group satisfaction into utility value or utility units that represents how much the decision maker is satisfied with the outcome (Figure 7.4). Thus, it is a way to translate various valued objectives (numbers of animals, cost, and persistence probability) into a common scale-utility that is usually bounded by zero and 1. The risk profile represents the relation between the objective value and the utility of the decision-outcome, providing a means of specifying a **utility function** (Chapter 3). We demonstrate how to develop a risk profile in Box 7.1.

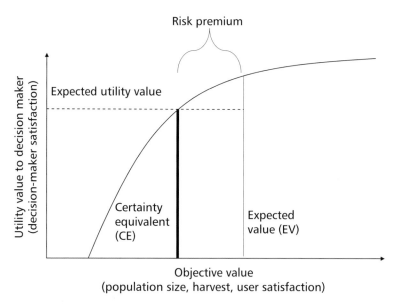

Figure 7.4 A graphical illustration of the relationship between the certainty equivalent, the expected value of a decision, and the risk premium. For a risk-adverse individual, the expected utility value line intersects the utility curve before it intersects with the expected value of the management objective. Note that for a risk-prone individual, the shape of the curve would be convex and the expected utility line would intersect the expected value line before the certainty equivalent so the risk premium would be negative.

Box 7.1   Creating a risk profile using a utility curve

The risk profile for decision makers can be created using certainty equivalents. Here, we illustrate using a species-conservation decision where the fundamental objective is to maximize the number of years out of the next 100 years that a population persists. The process begins by rescaling the best and worst outcome from a management decision at zero and 1. This will constrain the utility values to that scale. The fundamental objective is on the unit scale 0 to 100 years with zero being the worst outcome. Using these values, we now have two points for constructing the curve.

To find other points in the curve, we identify certainty equivalents by asking the stakeholders to provide a value that would make them indifferent to decision A and B. For example, in this decision tree, the stakeholders would be asked what guaranteed level of persistence would they accept rather than taking a chance that the expected value is 50 years and a 50% chance of a possible outcome of zero years.

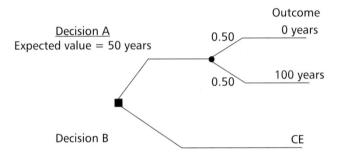

Let's assume that the decision makers decided that 25 years was the certainty equivalent (CE). This means that the decision maker was indifferent between 25 years and an uncertain outcome with an expected value of 50 years. We rescaled the utilities to match the minimum and maximum number of years (our fundamental objective) so we know that the utilities on 0 and 100 years. Thus, the utility associated with a CE of 25 years is estimated as:

$$U(CE) = 0.5 \times U(100) + 0.5 \times U(0)$$
$$U(25) = 0.5 \times U(100) + 0.5 \times U(0)$$
$$U(25) = 0.5(1) + 0.5(0)$$
$$= 0.5$$

To find other points in the curve, we identify certainty equivalents for different values between known utilities. Starting with the reference point we identified earlier, i.e., 25 years with a utility value of 0.5, we ask the decision makers what guaranteed number of years persistence would they accept rather than taking a chance with the risky decision that has an expected value is 62.5 years and a 50% chance that the outcome is 25 years.

(Continued)

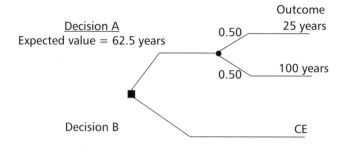

Let's assume that the decision makers decided that 50 years was the CE. From the previous question, we know that $U(100) = 1.0$ and $U(25) = 0.5$ and using the same formula we calculate an expected utility as:

$$U(50) = 0.5 \times U(100) + 0.5 \times U(25)$$
$$U(50) = 0.5 \times 1 + 0.5 \times 0.5$$
$$= 0.75$$

The process is then repeated using the smaller two reference points. Here, we ask the decision makers for the CE associated with a risky decision with 12.5 as the expected number of persistence years and a 50% chance of a possible outcome of 0 years.

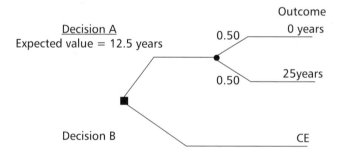

Again we assume that the decision-makers' choice, 10 years, was the CE. Using our formula we calculate an expected utility as:

$$U(10) = 0.5 \times U(0) + 0.5 \times U(25)$$
$$U(10) = 0.5 \times (0) + 0.5 \times 0.5$$
$$= 0.25$$

We now have five pairs of persistence years and utility values for constructing the risk profile curve.

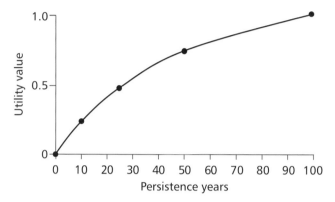

Persistence years	Utility
0	0.00
10	0.25
25	0.50
50	0.75
100	1.00

This risk profile would be incorporated into the decision by replacing the objective value, years persisting, with the corresponding utility values from the curve.

Sensitivity analysis

All decision models should be evaluated using sensitivity analysis prior to choosing and implementing the optimal or satisficing decision. In this chapter, we use the term sensitivity analysis to refer to the systematic perturbation of model inputs or parameters to see the influence on decision making. This usage of sensitivity, which is common in the decision-making literature (e.g., Clemen and Reilly 2001), differs somewhat from the more formal, mathematical definition of sensitivity involving marginal changes evaluated with differential calculus (Chapter 8, Appendix E, Caswell 2001, Williams et al. 2002). Sensitivity analysis is used to identify the components that have the greatest effect on the expected value of the decision and more importantly, the components that have the greatest effect on what decision alternative is estimated to be the best. Remember, a decision model is a simplification of a real-world system and several assumptions were made regarding relations among components, how they were parameterized, and how the utility as calculated. Sensitivity analysis is used to evaluate the behavior of the decision model to make sure it is performing as expected. Equally important, sensitivity analysis provides stakeholders with a means to identify those components that largely drive the decision and those that could be improved, perhaps through an adaptive management process.

The basic idea behind sensitivity analysis (as explored in this chapter) is to vary the values of each parameter or model component and examine how it affects the expected value or the optimal decision. There are a large number of approaches to sensitivity analysis and no single type of analysis is best. Here, we discuss four types of sensitivity analysis that are commonly used in natural resource decision modeling, one-way and two-way sensitivity analysis, response profile sensitivity analysis, and indifference curves. In one-way sensitivity analysis, the value of an individual model component is varied from its minimum to its maximum to evaluate how it changes the expected value of the optimal decision. In a two-way sensitivity analysis, two model components are varied to evaluate how sensitive the expected value is to the interaction between two components. A response profile sensitivity analysis is used to evaluate how the optimal decision changes when a single component is varied. It is used to answer the question–does uncertainty in this component affect what decision I would make? Indifference curves are used to evaluate the sensitivity of the decision to the relative weighting of multiple objectives. It is used to evaluate tradeoffs and determine if the objectives were properly weighted. Because each type of sensitivity analysis provides a different look into the behavior of the model, all decision models should be examined with more than one technique. We illustrate sensitivity analysis using the amphibian conservation decision influence diagram from Chapter 6 (Box 6.3) in Boxes 7.2–7.5.

During one-way sensitivity analysis, the values of each model component (e.g., influence diagram node) except the decision and utility are systematically varied from minimum to maximum (Box 7.2). The expected value and optimal decision are then recorded for each value or state of the model component. The results of a one-way sensitivity analysis are graphically displayed in a tornado diagram that is created using the values obtained during the systematic perturbation. A tornado diagram is a two-dimensional plot with the expected values for the optimal decision plotted along the x-axis and the model components on the y-axis (Figure 7.5). The sensitivity of the expected value to changes in each model component is represented by a horizontal bar that corresponds to the range of expected values observed during the sensitivity analysis. The model components in a tornado diagram are sorted from the most influential component (i.e., widest bar) at the top to the least influential (i.e., narrowest bar), which then resembles a funnel or tornado (Figure 7.5).

In a two-way sensitivity analysis, two model components are varied simultaneously and the expected value of the optimal decision is recorded. The process is then repeated for all possible combinations of model components 2 at a time (Box 7.3). Two-way sensitivity analysis is useful for evaluating potential interactions and nonlinear effects of model components on the expected value of the decision. The results of two-way sensitivity analysis can be displayed in a contour plot of the expected value of the optimal decision for the range of values of two model components.

Response profile sensitivity analysis is used to evaluate the sensitivity of *decisions* to the model components. Similar to one-way and two-way sensitivity

Box 7.2 An illustration of one-way sensitivity analysis

To illustrate one-way sensitivity analysis, we use the amphibian conservation decision influence diagram from Chapter 6 (6.3). The optimal decision is to "restore wetlands" with an expected value of 8.02, shown below in Netica software format.

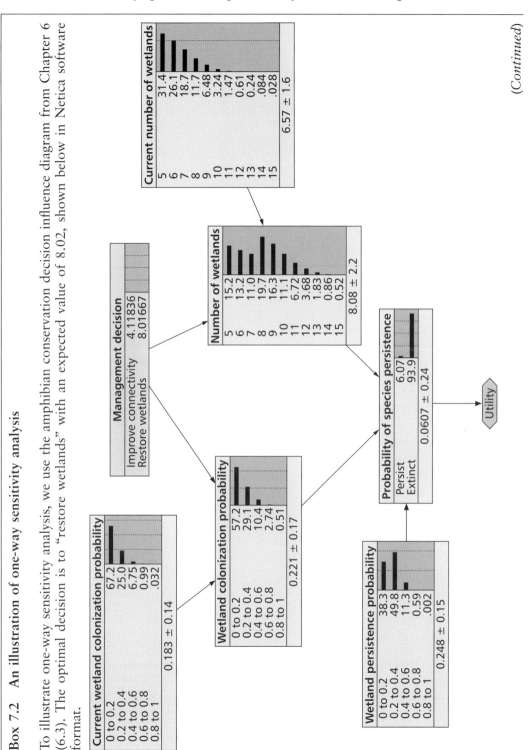

(Continued)

We begin one-way sensitivity analysis by varying values of each model component except the decision and utility are systematically varied from minimum to maximum. When the current number of wetlands is 5 (i.e., known with absolute certainty), the optimal decision is "restore wetlands" with an expected value of 3.8.

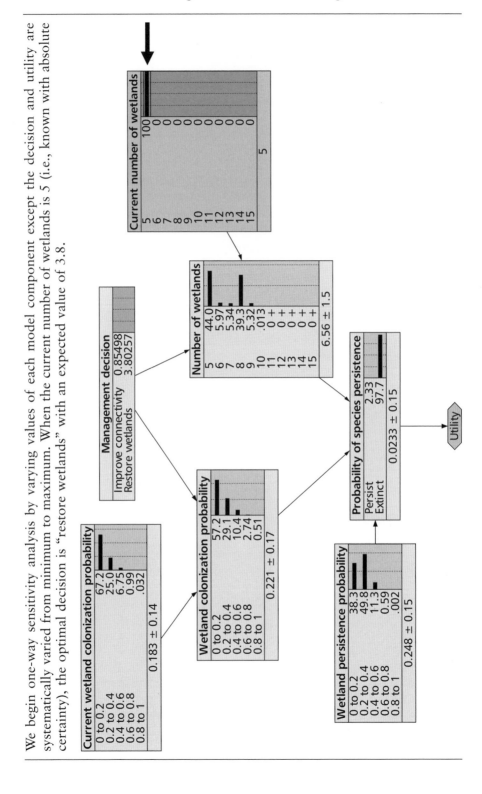

When the current number of wetlands is 15, the optimal decision is "improve connectivity" with an expected value of 31.6.

Current number of wetlands

5	0
6	0
7	0
8	0
9	0
10	0
11	0
12	0
13	0
14	0
15	100

15

Management decision

Improve connectivity	31.5763
Restore wetlands	23.1023

Number of wetlands

5	0
6	0
7	0
8	0+
9	0+
10	0+
11	0+
12	0+
13	.030
14	11.9
15	88.1

14.88 ± 0.33

Current wetland colonization probability

0 to 0.2	67.2
0.2 to 0.4	25.0
0.4 to 0.6	6.75
0.6 to 0.8	0.99
0.8 to 1	.032

0.183 ± 0.14

Wetland colonization probability

0 to 0.2	57.2
0.2 to 0.4	29.1
0.4 to 0.6	10.4
0.6 to 0.8	2.74
0.8 to 1	0.51

0.221 ± 0.17

Wetland persistence probability

0 to 0.2	38.3
0.2 to 0.4	49.8
0.4 to 0.6	11.3
0.6 to 0.8	0.59
0.8 to 1	.002

0.248 ± 0.15

Probability of species persistence

Persist	27.3
Extinct	72.7

0.273 ± 0.45

Utility

(Continued)

The process is repeated for each model component (except the decision and utility nodes). Below is a summary of the optimal decision and associated expected values when each model component is varied from minimum to maximum.

Model component	Minimum		Maximum	
	Optimal decision	Expected value	Optimal decision	Expected value
Current number of wetlands	restore wetlands	3.8	improve connectivity	31.6
Current wetland colonization probability	restore wetlands	1.4	restore wetlands	88.6
Wetland colonization probability	restore wetlands	0.2	restore wetlands	90.6
Number of wetlands	improve connectivity	0.7	improve connectivity	31.8
Wetland persistence probability	restore wetlands	3.8	restore wetlands	81.2

The results of a one-way sensitivity analysis are graphically displayed in a tornado diagram that is created using the values obtained during the sensitivity analysis. A tornado diagram can be created in Excel using the horizontal bar chart option. Alternatively, the same plot can be created in R using the program below.

```
#barplot function requires the following libraries
library(gplots)
library(Hmisc)
## data read from "one.way.sensitivity.csv"
one.way<-read.table("one.way.sensitivity.csv ",header=T, sep=" ,")
## data must be in matrix format for barplot function
plot.data <- as.matrix(one.way)
par(oma=c(0,0,0),mar=c(5,10,2,2))
```

```
barplot2(plot.data,beside=FALSE,las=1,horiz=TRUE, cex.names= 0.5,  col=c("white","black",
"black"),border=c("white","black","black","black"),xlab="Probability of persistence",ylab="",
xlim=c(0,100))
box()
```

After running the program you should get the following:

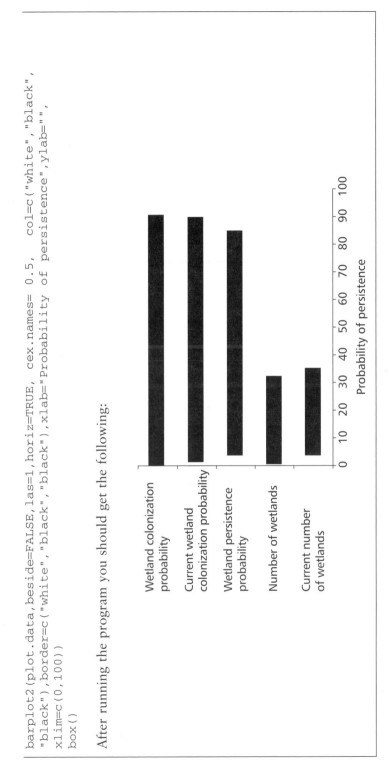

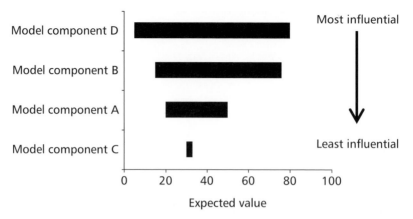

Figure 7.5 Tornado diagram for one-way sensitivity analysis with components sorted from most influential at the top to the least influential at the bottom.

Box 7.3   An illustration of two-way sensitivity analysis

In a two-way sensitivity analysis, two model components are varied simultaneously. The process is then repeated for all possible combinations of model components 2 at a time. Here, we illustrate using the amphibian conservation decision influence diagram from Chapter 6 (6.3). The table below shows the optimal decision and the associated expected value for each component when current wetland colonization probability is at the minimum (zero) and the maximum (1.0). In practice, a two-way sensitivity analysis would span the full range of values for each component (not shown).

	Minimum		Maximum	
Model component	Optimal decision	Expected value	Optimal decision	Expected value
Current wetland colonization probability = 0.0				
Current number of wetlands	restore wetlands	0.4	improve connectivity	12
Wetland colonization probability	restore wetlands	0.2	restore wetlands	90.6
Number of wetlands	improve connectivity	0.1	improve connectivity	12.2

Model component	Minimum		Maximum	
	Optimal decision	*Expected value*	*Optimal decision*	*Expected value*
Wetland persistence probability	restore wetlands	0.3	restore wetlands	73.4
Current wetland colonization probability = 1.0				
Current number of wetlands	restore wetlands	79.4	improve connectivity	99.8
Wetland colonization probability	restore wetlands	0.2	restore wetlands	90.6
Number of wetlands	improve connectivity	24.7	improve connectivity	99.8
Wetland persistence probability	restore wetlands	81.9	restore wetlands	99.9

The minimum and maximum values for the pairs of components can be added to the tornado diagram to evaluate the relative sensitivity of the model to the interaction of two components. Alternatively, a two-way sensitivity analysis can be displayed in a contour plot of the expected value of the optimal decision for each combination of the value of two model components. The contour plot can be created in R using the program below.

```
## data read from "two.way.sensitivity.data.csv"
>two.way<-read.table("two.way.sensitivity.data.
csv",header=T,sep=",")
>attach(two.way)
>names(two.way)
>#create vector of midpoint values of current colonization
probability
>Current.colonization.prob<-c(0.1, 0.3,0.5,0.7,0.9)
>#create vector of values of current number of wetlands
>Current.number.wetlands<-c(5:15)
#create matrix of expected values of optimal decision
z<-matrix(Expected, ncol = 5)
>contour(Current.number.wetlands,Current.colonization.
prob,z,xlab="Current no. wetlands", ylab = "Current colon.
pro12b")
```

Here, we need to create a matrix of the expected values and two vectors of "current wetland colonization probability" and "current number of wetlands" in ascending orders. Note that we are using the midpoint value of each range in the "current wetland colonization probability" node. After reading in the data and running the program you should get the following.

(Continued)

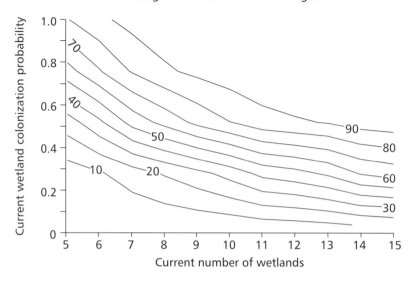

Here, it is looks like current wetland colonization probability has a greater influence on the expected value than the current number of wetlands because the expected value changes more across the range of values. It also appears that there are no unusual or strong interactions between these two components as the contours are nearly parallel.

analysis, one or two model components are varied to evaluate the change in the expected value (Box 7.4). Unlike one-way sensitivity analysis, the expected value for *each decision*, not just the optimal decision, is recorded. These values are then plotted to examine how optimal decisions change over the range of values for the model component(s). When a single component is examined, the expected values of each decision are plotted across the range of values of the component. The optimal decision is the decision with the greatest expected value, so the point(s) where two or more lines cross represent the value of the decision model component where the optimal decision changes (Figure 7.6). In some circumstances, the expected value of one decision will be greatest across the range of values of the component. That is, the optimal decision is the same regardless of the value of the model component (Figure 7.6). Thus, the value of the component has no effect on the decision because the decision maker would always make the same decision. This is known as stochastic dominance and it is an important concept in decision modeling.

The sensitivity analysis of the amphibian conservation decision (Boxes 7.2–7.4) is useful for highlighting important aspects of sensitivity analysis. The one-way sensitivity analysis indicated that the expected value of the optimal decision was most sensitive to (future) wetland colonization probability, but that the "restore wetlands" decision was stochastically dominant (i.e., it was always

Box 7.4 An illustration of response profile sensitivity analysis

To create a response profile for current number of wetlands in the amphibian conservation decision influence diagram (Box 6.3), the value of the component is varied from 5 to15 (i.e., the range) and the expected value is recorded for *each decision* and plotted. After completing the task, the response profile plot is created by plotting the expected values for each decision as using the scatter plot option in Excel or using the following R code.

```
########## Profile sensitivity
## data read from "profile.sensitivity.data.csv"
>profile<-read.table("profile.sensitivity.data.
csv",header=T,sep=",")
>attach(profile)
>names(profile)
>plot(Current.no.wetlands,Improve.
connectivity,lty=1,type = "l",lwd=1,ylab="Probability of
persistence",xlab="Current number wetlands")
>lines(Current.no.wetlands,Restore.
wetlands,col="red",lty=1,lwd=3)
>legend(10,08,c("Improve.connectivity","Restore.wetlands"),
lty=c(1,1),col=c(1,2))
```

After running the R code, you should get the following:

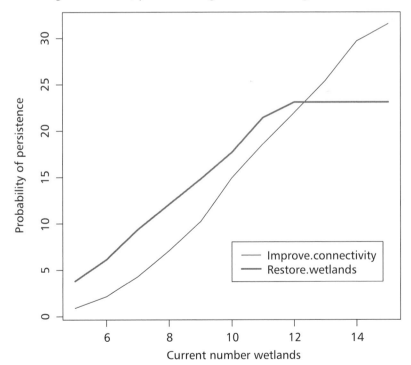

Notice that the optimal decision (i.e., the one with greatest utility) changes from "restore wetlands" to "improve connectivity" when the current number of wetlands is 12 or more. This suggests that current number of wetlands is a key uncertainty.

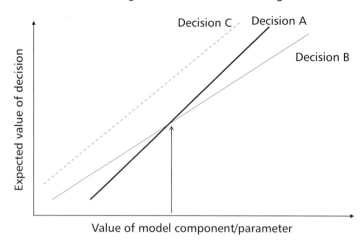

Figure 7.6 Response profile plot of the expected value of three decisions across the range of values of a single model component. The optimal decision is the decision with the greatest expected value. Here, decision C (broken line) stochastically dominates the decision set. The arrow represents the value of the model component where the expected value of decision A (heavy solid line) is greater than decision B. If decision C was not included, this would be the point where the optimal decision changes from B to A.

optimal). In contrast, the expected value of the optimal decision was much less sensitive to current number of wetlands. However, the profile sensitivity analysis indicated that the optimal decision was sensitive to current number of wetlands and depended on its value. This illustrates a couple of important points. First, the expected value can be very sensitive to a model component, but if it does not change the identity of the optimal decision, it may not be a key uncertainty. Secondly, a model component that has a large influence on the expected value but a small influence on what decision is optimal suggests that a decision alternative may have been overlooked. For example, a management action designed to increase species persistence in a wetland might be more effective at maximizing amphibian persistence than the two alternatives that were considered. The insights from a sensitivity analysis can cause decision makers to rethink the decision as well as the model structure.

In Chapter 3, we discussed various methods for combining the outcomes for multiple objectives in a single value. The weighting and scoring of multiple objectives can be potentially problematic, especially if the objectives are on very different scales (e.g., dollars vs. number of fish). **Indifference curves** are used to evaluate tradeoffs when considering multiple objectives. The basic idea is to evaluate alternative weighting schemes to find the point at which the values of the decision alternatives are equal and the decision maker is indifferent to the decision (i.e., no single decision is optimal). To create indifference curves, alternative sets of combined utilities are created by systematically increasing the weight of a single objective and recording the expected value for each decision (Box 7.5). These values are plotted to identify the weight at which the decision maker is

Box 7.5 An illustration of indifference curves

We illustrate indifference curves using the temporary wetland management decision from Chapter 6 (Box 6.2). The table below contains the set of values of the management decision expressed as scores with "do nothing" scored as 3 because it is the least expensive and risky decision and "burn" scored as 1 because it is the costly and riskiest action. Similarly, the dabbling ducking density is valued (scored) from 1 at the lowest density to 6 at the greatest density.

Management actions		Duckling density		Combined value
Decision	Value (M)	no. / ha	Value (D)	(M + D)
Do nothing	3	0–5	1	4
Do nothing	3	5–10	2	5
Do nothing	3	10–15	3	6
Do nothing	3	15–20	4	7
Do nothing	3	20–25	5	8
Do nothing	3	25–30	6	9
Mow	2	0–5	1	3
Mow	2	5–10	2	4
Mow	2	10–15	3	5
Mow	2	15–20	4	6
Mow	2	20–25	5	7
Mow	2	25–30	6	8
Burn	1	0–5	1	2
Burn	1	5–10	2	3
Burn	1	10–15	3	4
Burn	1	15–20	4	5
Burn	1	20–25	5	6
Burn	1	25–30	6	7

The combined objective value was an additive function of each objective value (score). They were originally given equal weight (importance) and were simply summed for the combined utility as shown above. To evaluate the effect of the alternative weighting schemes, we create an alternative combined utility by weighting either the value of one objective while holding the value of the other objective(s) constant and evaluating how the new weighting scheme affects decision making. For example, under an equal weighting scheme, the combined value for the decision "do nothing" and the outcome "0–5 ducklings/ha" is 3 + 1 = 4. If the duckling density value is given twice the weight of the decision value, the combined value of that same decision and outcome combination is $3 + 2 \times 1 = 5$.

To create indifference curves, alternative weighting schemes are created by systematically increasing the weight of a single objective (e.g., the scores in the table above) and recording the expected value for *each decision*. The expected values vs. objective weight are plotted with a line corresponding to each decision alternative to identify the weight at which the decision maker

(*Continued*)

is indifferent to the optimal decisions. This is similar to response profile plots (Box 7.4) and graphically shown as the point where two or more decision alternative lines cross. The indifference curves can be created in Excel using the scatterplot option or in R using the code below.

```
## data read from "indifference.curve.data.csv"
>ind.cur<-read.table("indifference.curve.data.
csv",header=T,sep=",")
>attach(ind.cur)
>names(ind.cur)
>plot(Weight,Do.nothing,lty=1,type = "l",lwd=1,ylab="Expected
value",xlab="Dabbling duck weight")
>lines(Weight,Burn,col="red",lty=1,lwd=3)
>lines(Weight,Mow,col="blue",lty=1,lwd=3)
>legend(6,10,c("Do nothing","Burn","Mow"),lty=c(1,1,1),col=
c(1,2,4),lwd=c(1,3,3))
```

If you run the code, you should get the following plot:

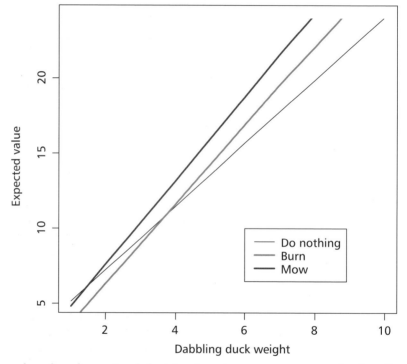

Remember that the optimal decision has the greatest expected value. For the temporary wetland management decision, the expected value is greatest for the "do nothing" alternative when the weight of the dabbling duck density is equal to the value of the management decision. The point where the decision maker would change the decision from "do nothing" to "mow" corresponds to a weight of approximately 1.6. This suggests that to decide to mow requires the decision maker to value the dabbling duck density 1.6 times more than the value (i.e., cost and risk) of mowing.

indifferent to the optimal decisions. Similar to a responsive profile plot, this is graphically depicted as the point where the lines for two (or more) decisions cross.

Value of information

The most influential model components are considered critical to the decision and hence, are given higher priority as potential components for additional study. If collecting additional data could reduce the uncertainty in these components (e.g., variance), monitoring efforts would be most productive by focusing on these variables. However, the value (e.g., relative cost) and usefulness (i.e., reduction of uncertainty) of collecting such data should be evaluated prior to making the decision.

The expected value of perfect information (EVPI) is the increase in the expected value of a decision if the "true" value of a model component(s) or the relationship among components is known with certainty. Thus, it can be used, in part, to identify and rank potential variables for monitoring or additional data collection efforts. Graphically, EVPI is represented as an arc connecting an uncertainty node(s) to a decision node (Figure 6.2a). Recall that these arcs represent timing or sequence and indicate that the uncertainty will be resolved (i.e., the information will be known) before the decision is made (Chapter 6). The expected value of perfect information is calculated assuming no uncertainty in value of the model component of interest. That is, the value is known with absolute uncertainty. Similar to sensitivity analysis, the expected value of the optimal decision is recorded for each state of the model components, assuming it is known with certainty. The values then are weighted by their state-specific probabilities and summed. The result is an estimate of the EVPI in the currency that is valued by the decision makers. For example, if the objective value is estimated in terms of population size or number of animals harvested, EVPI would let decision makers know how many more animals would be in the population or how many more could be harvested if the information were known. That is a very powerful way to communicate the value of additional study or managing in an adaptive resource management framework (more below).

To illustrate the calculation of EVPI, consider a simple conservation decision model where the decision to conduct a management action affects the degree of environmental disturbance that in turn affects the future status of an animal population. The future population status is also affected by the current status of the population and takes two states, "present" and "absent". The value of the decision depends on the management action and the future population status. The optimal decision and value when the probability of current species presence is 50% is "low intensity" with an expected value of 9.65. If we assume that the species is present with a 100% probability, the optimal decision is "none" with a value of 18.25 (Figure 7.7). If we assume that the species is absent with a 100% probability, the optimal decision is "high intensity" with a value of 5.45. The probabilities of current species presence and absence in the model of 50–50% are used to calculate the expected value as $18.25 \times 0.5 + 5.45 \times 0.5 = 11.85$. This is the expected value of the decision if the population status was known with certainty. EVPI is

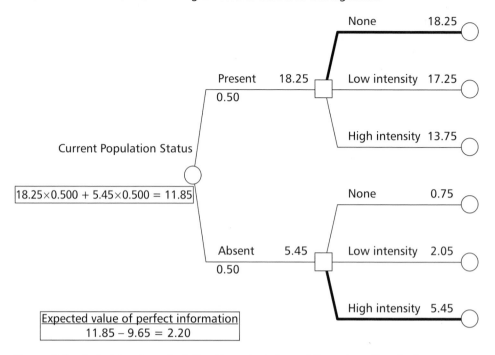

Figure 7.7 A decision tree depicting the calculation of the value of perfect information for the conservation decision with three alternatives: none, low-intensity, and high-intensity actions. Heavy lines signify the optimal decisions when the current population status is known with certainty.

calculated as the difference between the expected value with and without knowing the current population status, $11.85 - 9.65 = 2.20$ (Figure 7.7). If the cost of getting information on the current status of the population was less than the EVPI, it would increase the expected net value of the optimal decision, and hence would be beneficial to obtain. Note too that the decision was very sensitive to current population status and the optimal decision changed three times over the range of probabilities for current population status. This should motivate the decision maker to consider additional study to obtain the information.

Imperfect information

Although the EVPI can be useful as a first approximation, it is usually not realistic to expect information to be perfect. Even the most carefully controlled experiments – or carefully made measurements – will have some uncertainty (e.g., variance) associated with them, which can affect their value. To account for the error in measurements or models requires the estimation of the expected value of imperfect information (EVII). EVII is considerably more complicated to estimate than EVPI. It requires an estimate of the expected outcome and the use of probability and (e.g., Bayes theorem; Chapter 5, Appendix A) to calculate probabilities and expected values.

We illustrate calculating EVII using the previous conservation decision model. Here, the decision maker is interested in evaluating the value of conducting sample surveys in the potential management area to detect the species. One source of imperfection in sample data that would affect the value of conducting studies is incomplete detection. Not all species can be detected during sampling so to incorporate this source of imperfect information, we first estimate the probability of detecting the species given it was present. In other words, we calculate the probability of obtaining sampling results, species detected or species not detected. Assume that 10 samples are needed to detect the species with an 80% probability, given the species is present in the area. The probability of detecting the species if no individuals are present is, of course, zero. Assuming that there is a 50% chance that the species is present, the probability of detecting the species with 10 samples is $0.8 \times 0.5 = 0.4$ and the probability of not detecting them is $1.0 - 0.4 = 0.6$. This new component is added to the model and is graphically displayed in an influence diagram with current species presence pointing into a sampling result node (Figure 6.2b).

If the species is not detected, the species may be absent or the species may be present but was missed during sampling. To incorporate the possibility of falsely concluding the species was absent, we need to use Bayes' theorem to estimate the probability the species is present, given it was not detected, known as the posterior probability of presence. The posterior probability of presence requires four estimates:

The probability of species presence and absence: $p(\text{present}) = 0.5$, $p(\text{absent}) = 0.5$.

The probability of not detecting the species, given it was present (one minus detection probability above:

$$p(\text{not detect} \mid \text{present}) = 1 - 0.8 = 0.2.$$

The probability of not detecting the species, given it was absent:

$$p(\text{not detect} \mid \text{absent}) = 1.0.$$

Using these values and Bayes' rule we obtain:

$$p(\text{present} \mid \text{not detect}) = 0.20 \times 0.50/(0.2 \times 0.5 + 1.0 \times 0.5) = 0.167.$$
$$p(\text{absent} \mid \text{not detect}) = 1 - 0.167 = 0.833.$$

The values are then entered into the model as the conditional probabilities of species presence, given the sampling result not detected. The next step is to estimate the value of the optimal decision for the each sampling result using the new model. For the sampling result, "detected," the optimal decision in "none" with a value of 18.25 that is the same value as presence known with certainty (100%). This is because we are assuming that if it is detected, we are certain that the species occurs in the management area. This would not be the case if the biologists in charge of sampling misidentified the species. When the sampling result is "not detected", the optimal decision is "high intensity" management action with a value of 6.83. Notice that this is different from the expected value of perfect information example, where we assumed absence was 100%. This is because there is a 16.7% chance that the species occurs given it was not detected (i.e., the calculation above), which changes the value of the optimal decision.

To estimate the EVII, we weight the utility value of the optimal decision for each sampling result by the probability of obtaining that sampling result and sum the values as: $18.45 \times 0.40 + 6.83 \times 0.6 = 11.40$. This is the expected value of the decision if sampling is conducted. We subtract that value from the expected value if the species was not sampled and we get $11.40 - 9.65 = 1.75$ as the EVII. Notice that the EVII 1.75 is lower than the EVPI 2.20. This should always be the case; the value of imperfect information should always be smaller than that of perfect information. Of course, this value would be discounted by the actual cost of collecting samples. Assuming that the probability of detecting a species is related to the number of samples, the decision maker could examine how the EVII varies with sample size to determine the optimal level of sampling effort.

Reducing uncertainty

Traditional approaches to reducing epistemic uncertainty

In the previous section, we illustrated that uncertainty affects decision making and that reducing epistemic (mainly structural) uncertainty (i.e., learning), can lead to better decision making and greater returns to the decision maker. There are three basic approaches to learning in the natural resource sciences that can be used to reduce uncertainty about the dynamics of systems being managed. The most familiar and best approach to learning is experimentation that involves replication, randomization, and treatments. Conducting experiments, however, is labor intensive, which precludes application at the large spatial scales and long timeframes that are often necessary to understand the dynamics most managed systems. The second approach is retrospective data analysis where statistical models are used to discern patterns in data. In contrast to experimental approaches, retrospective analyses can be conducted using data collected across broad spatial or temporal extents. Retrospective studies often provide the basis for constructing decision models. However, the relations derived from observational data often are confounded with other factors. In addition to these problems, both of these approaches involve additional study and require additional resources (e.g., funds) and more time before a decision is made. This can cause conflict between researchers and managers due to perceived competition for limited management resources. More importantly, decisions often cannot wait for the studies to be completed. A decision to conduct additional study often means that the current management policy (i.e., the status quo) continues to be implemented. This means that decision makers will incur a loss in terms of the objectives for each decision opportunity unless that policy is optimal or satisficing that potentially translates into greater risk to threatened and endangered species when they are part of the decision.

Reducing uncertainty through ARM

The third approach to learning in the natural resource sciences is called **adaptive resource management (ARM)**. Adaptive resource management is an outgrowth of the field of scientific management originated by Taylor in the early 1900s (Haber 1964) and further advanced by Holling (1978) and Walters (1986) for natural resource management decision-making (see also Williams et al. 2002, Williams et al. 2009). In essence, ARM is about learning while managing. Management decisions are made and feedback, in the form of monitoring data, is used to reduce uncertainty about system dynamics. ARM, however, is not about experimentation of managed systems, haphazardly or randomly implementing management actions to see what happens, or trial and error. Rather, ARM is a special case of structured decision making in which decisions are made to maximize the objective value (or utility) over a long time horizon. Learning then occurs through a formal process of comparing predicted outcomes for two or more models to outcomes observed during monitoring. In contrast, experimental, haphazard, and trial and error approaches to learning-while-managing are primarily focused on learning rather than the management objective and can result in suboptimal decision making, potentially wasting management resources and incurring greater risk to sensitive systems.

ARM requires three basic elements: recurrent (dynamic) decision making in time and or space, two or more alternative models of system dynamics (structural uncertainty), and monitoring. Figure 7.8 depicts the adaptive resource management for a dynamic decision through time. The influence diagram differs somewhat from previous versions of dynamic decision through time (Chapter 6) in that it contains new components representing the information state at three points in time. These represent the model weights or relative belief in alternative models. The diagram also contains two alternative models of system dynamics that could represent alternative scientific hypotheses about how the managed system works and a calculated node that is used to compare predicted outcomes with observed outcomes to update the model weights. At time t, the state of the system is observed (probably imperfectly) and the optimal management action is taken based on the system state and information state (i.e., model weights). The management action results in a change in system state (or not) in the next time period. The observed system state in $t + 1$ is compared with the predicted system state under each alternative model and the model weights are updated using Bayes' rule. The process is then repeated through time (Box 7.6).

In ARM, monitoring serves two purposes. First, monitoring provides an estimate of the current state of the system before a decision is made. Remember that decisions are state dependent. After a management alternative is implemented, monitoring provides information on what changes, if any, occurred to the system. More importantly, this should also provide information on the system dynamics that should reduce uncertainty and improve future decision making and returns (objective value). Monitoring and models are tightly linked in adaptive resource management. The model weights are updated by comparing predictions to observed outcomes. Thus, predicted and measured responses should be on the same unit scale and in the same units. For example, if models

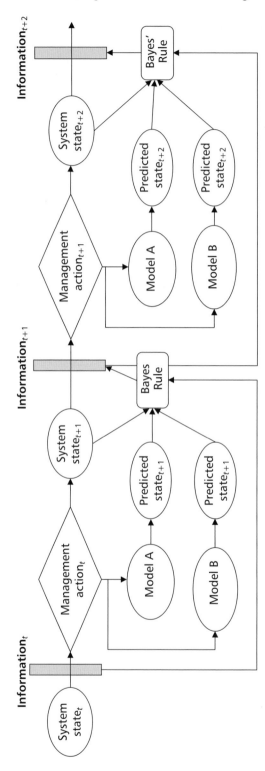

Figure 7.8 Influence diagram of adaptive resource management decision model with two alternative models of system dynamics.

Box 7.6   An illustration of adaptive resource management process

To illustrate the ARM process, consider the robust redhorse (*Moxostoma robustum*) that was believed extinct but was rediscovered in the Oconee River Georgia USA in 1991 by fishery biologists with the Georgia Department of Natural Resources (Warren et al. 2000). Initially, there was scientific disagreement about the factors responsible for depressing robust redhorse populations. Some scientists believed that flathead catfish (*Pylodictis olivaris*), large non-native piscivore, were depressing redhorse populations through predation. Other scientists believed that redhorse were rare because upstream hydropower generation increased streamflow variability, negatively affecting the population. It follows then that the optimal decision for increasing redhorse populations differed based on what mechanism was responsible for the relative rarity of redhorse. If predation was responsible, then the best management alternative might be to control the non-native catfish populations, whereas the best alternative would be to decrease power generation if the mechanism was flow variability. Here, we have a large amount of uncertainty about the factors affecting redhorse populations and that uncertainty likely has a substantial effect on the optimal management decision. If decision making was dynamic and sequential in time or space, this would be a good candidate for ARM.

Let's assume that biologists developed two simple models for predicting redhorse abundance in response to two management actions: decrease power generation and control flathead catfish. The first model assumes that flow variability primarily controls redhorse populations. The model predicts that there will be 30 redhorse if power generation is decreased and 15 redhorse if flathead catfish are controlled. The second model assumes that redhorse populations are primarily controlled by flathead catfish. The model predicts that there will be 25 redhorse if flathead catfish are controlled and 15 redhorse if power generation is decreased.

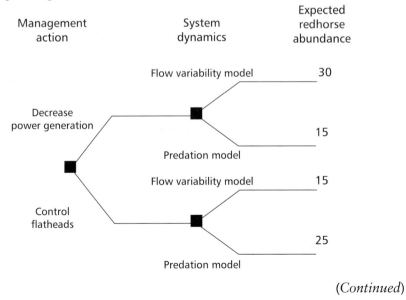

(*Continued*)

Notice that the expected number of redhorse is greatest when the correct decision is matched with the corresponding system dynamics. This means that if decision makers knew that flow variability was the mechanism, they would always choose to decrease power generation (i.e., the optimal decision, 30 redhorse) or if they knew that predation by flatheads was the mechanism, they would always choose to control flatheads (i.e., the optimal decision, 25 redhorse). However, decision makers were unsure and the decision could not wait. Therefore, this structural uncertainty was incorporated using two models each with equal weight (0.5/0.5). The optimal decision then is identified by calculating the uncertainty weighted outcomes. For example, the expected number of redhorse for decreasing power generation is the top half of this decision tree:

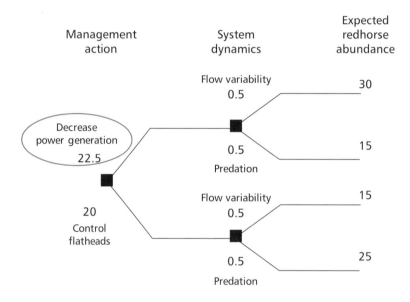

$0.5 \times 30 + 0.5 \times 15 = 22.5$ and the controlling catfish is $0.5 \times 15 + 0.5 \times 25 = 20$. The optimal decision alternative then is to decrease power generation. After implementing this alternative, 21 redhorse are counted during annual monitoring. Notice that this value is closer to the estimated result of power generation using the predation model (15 redhorse). This suggests that there is greater evidence that predation is controlling the redhorse population. We use this evidence to update the model weights.

The initial weights represented the probability or degree of belief that a particular system dynamic controlled redhorse populations. To incorporate the new evidence (i.e., monitoring data), we need to calculate the probability of the each system dynamic (i.e., model), given the observed outcome using Bayes' rule (Chapter 5). First, we need the probability of obtaining the outcome, given the outcome predicted under the decision that was made. This is defined as the conditional likelihood. Assuming a Poisson error distribution (remember these are redhorse counts) and using the Poisson probability distribution function (Chapter 5), the probability of obtaining 21 redhorse if 30 are predicted is 0.0192 or 1.92% and the probability of obtaining 21 redhorse if 15 are predicted is 0.0299. The new model weights are then calculated using the prior weights (0.5/0.5) and the conditional likelihoods and Bayes' rule as:

The probability of flow variability dynamics: p(flow model) = 0.5.
The probability of predation dynamics p(predation model) = 0.5.
The probability of obtaining 21 redhorse, given 30 are predicted under flow model:

$$p(\text{observed } 21\,|\,30 \text{ predicted using flow model}$$
$$\text{under decreased power generation}) = 0.0192.$$

The probability of obtaining 21 redhorse, given 15 are predicted under predation model:

$$p(\text{observed } 21\,|\,30 \text{ predicted using predation model}$$
$$\text{under decreased power generation}) = 0.0299.$$

Using these values and Bayes' rule we obtain:

$$p(\text{flow model}\,|\,\text{observed } 21) = 0.0192 \times 0.50/(0.0192 \times 0.5 + 0.0299 \times 0.5)$$
$$= 0.391$$
$$p(\text{predation model}\,|\,\text{observed } 21) = 1 - 0.391 = 0.609.$$

There calculations are easy to implement in R code, for example

```
> p_flow<-0.5
> p_pred<-1-p_flow
>
> p_flow<-dpois(21,30)*p_flow/
(dpois(21,30)*p_flow+dpois(21,15)*p_pred)
> p_pred<-1-p_flow
> p_flow
[1] 0.3908099
> p_pred
[1] 0.6091901
```

The resulting new flow variability weight (termed posterior model weights) is 0.391 and the posterior predation model weight is 0.609. These model weights are used to make the decision at the next time step.

The posterior model weights calculated during the previous time step are used to make the decision at the following time step. The optimal decision is then identified by calculating the uncertainty weighted outcomes.

(Continued)

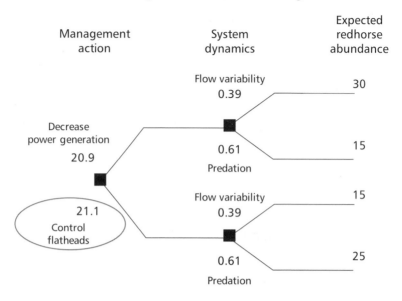

Here, the expected number of redhorse for decreasing power generation is $0.39 \times 30 + 0.61 \times 15 = 20.9$ and the controlling catfish is $0.39 \times 15 + 0.61 \times 25 = 21.1$. The optimal decision alternative is then to control flathead catfish. After implementing this alternative, 18 redhorse are counted during annual monitoring. This value is closer to the outcome estimated using the flow variability model and suggests that there is greater evidence that flow variability is controlling the redhorse population. As before, we use this evidence to update the model weights using Bayes' rule using prior model weights of 0.39 and 0.61 for the flow variability and predation models, respectively and the conditional likelihoods of obtaining 18 redhorse after controlling flathead catfish. The calculations are as before but now using the current posterior weights as priors, and using the conditional likelihoods under the "control flatheads" decision

The probability of flow variability dynamics: p(flow model) = 0.39.
The probability of predation dynamics p(predation model) = 0.61.
The probability obtaining 18 redhorse, given 15 are predicted under flow model:

$$p(\text{observed } 18 \mid 15 \text{ predicted using flow model}$$
$$\text{under flathead control}) = 0.0706.$$

The probability obtaining 18 redhorse, given 25 are predicted under predation model:

$$p(\text{observed } 18 \mid 25 \text{ predicted using predation model}$$
$$\text{under flathead control}) = 0.0316.$$

Using these values and Bayes' rule we obtain:

$$p(\text{flow model} \mid \text{observed } 21) = 0.0706 \times 0.39/(0.0706 \times 0.39 + 0.0316 \times 0.61)$$
$$= 0.59.$$

$$p(\text{predation model} \mid \text{observed } 21) = 1 - 0.59 = 0.41.$$

The R code to perform this (assuming that the updated values of 0.39 and 0.61 are still in memory is;

```
> p_flow<-dpois(18,15)*p_flow/
(dpois(18,15)*p_flow+dpois(18,25)*p_pred)
> p_pred<-1-p_flow
> p_flow
[1]  0.589338
> p_pred
[1]  0.410662
```

The posterior model weights that were calculated using Bayes' rule are 0.59 and 0.41 for the flow variability and predation models, respectively. As before, these are used to weight the estimated redhorse abundance for each combination of decision alternative and model of system dynamics. Here, the expected return for decreasing power **generation** alternative is greater than the control flathead alternative and would be the optimal decision. The process then repeats.

estimate population size, then monitoring should estimate the number of animals in the population rather than estimating some index of population size, such as relative abundance. Decision makers should also take great care to avoid systematically biased measures, such as raw counts that are unadjusted for incomplete detection. Biased measures can provide misleading information that can lead to bad management decisions. In fact, misleading information can have negative value (e.g., see EVII above).

Choosing what to monitor as part of an adaptive resource management plan will vary from plan to plan and depends largely on the resources available to the decision maker (e.g., personnel, equipment, funds) and the sources and levels of uncertainty in the decision model. Of course, the most important components to monitor are the outcomes that are used to calculate the utility (e.g., population size, distribution, harvest). These must be monitored. Monitoring other components of the decision model should focus on the key drivers of the outcomes that are identified during sensitivity analysis (e.g., habitat availability, annual temperature and precipitation, etc.). These data are required to estimate the expected outcome after the decision for each alternative model and can also be useful for explaining unanticipated outcomes.

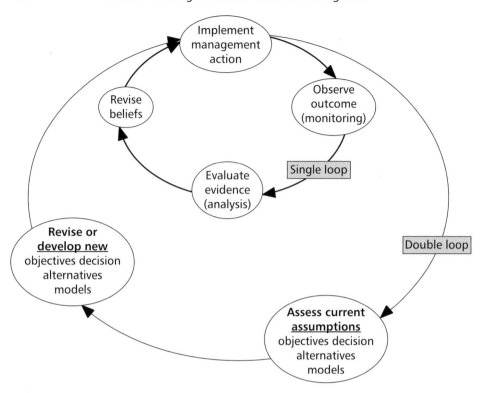

Figure 7.9   A graphical representation of single- and double-loop learning under adaptive resource management.

So far, we have described ARM as a special case of structured decision making that involves dynamic decision making, multiple models representing alternative notions of system dynamics, and monitoring. Monitoring is used in an iterative process that provides feedback to reduce uncertainty about the dynamics of the system being modeled. This iterative process has been defined as single-loop learning (Figure 7.9). Single-loop learning in the context of ARM begins after the initial structured decision-making processes is completed. That is, objectives and decision alternatives have been identified and the alternative decision models built. Learning within the single loop occurs with respect to the given (fixed) set of objectives, alternatives, and models and learning occurs relatively frequently (e.g., annually). Through time, decision makers may find that their current set of models is inadequate or that management objectives or decision alternatives are insufficient and need to be changed. In double-loop learning, the management objectives, decision alternatives, and models are reassessed and potentially revised to reflect changes in scientific knowledge and management objectives and alternatives (Figure 7.9). Learning in the outer loop occurs at a much slower rate and with lower frequency (e.g., decadally) compared to single-loop learning.

There is no general rule when to initiate the reassessment as it will vary from program to program and largely depends on the decision makers, stakeholders, and technical experts. For example, the US Fish and Wildlife Service conduct endangered-species status assessments approximately every 5 years, so the reassessment of objectives, alternatives, and models (i.e., the outer loop) may coincide with planned status assessments. However, decision makers should try to minimize the frequency of the reassessments to allow for sufficient amount of time to accumulate information.

ARM is fairly flexible and can be modified to fit the decision situation. For instance, decisions, monitoring, and feedback (i.e., updating weights) can occur at different intervals. For example, the decision to use controlled burns to manage vegetative structure may be revisited every 3 to 5 years but monitoring the vegetative structure can occur annually. Some decisions may also not be revisited as frequently as others due to slow system response or as a result of legislative mandates. For example, the US Federal Energy Regulatory Commission licenses private, municipal, and state hydroelectric projects for periods that range from 20–50 years. The licenses include conditions for which licensees must comply, such as dam operation restrictions designed to minimize impacts on fish and wildlife. Thus, dam operation decisions are revisited over long time intervals. In these instances, it may be more efficient to incorporate spatial feedback where the decision is made on one project (e.g., a hydropower dam), monitoring is conducted and used to resolve uncertainty about system dynamics, and the updated beliefs used to identify the optimal decision at another location (e.g., another hydropower dam). Sequential dynamic decision making in space, however, requires that the same set of decision alternatives are available to decision makers at each managed system. In addition, the systems should be sufficiently similar so that the same model set can be used to predict the outcomes of management actions.

There are two basic forms of adaptive resource management, passive and active. The ARM process is the same for both forms. Alternative models are used to identify the optimal decision and monitoring data are used to improve the understanding of system dynamics. In **passive ARM**, decisions are chosen as if the current uncertainty about the system dynamics will not change. That is, the decision is chosen based on long-term gain in management objectives (i.e., objective value) assuming the model weights will not change. Information gained through monitoring for passive ARM is incorporated, but not in a planned way. In contrast, **active ARM** takes into account how reducing uncertainty can affect the long-term gain. For example, a particular decision at a point in time may resolve key uncertainties quicker or more efficiently than the other decisions and resolution of the uncertainty will result in better decision making and greater long-term gain in management objectives. Active ARM generally does not *necessarily* involve experimentation or probing of the system. Probing can, in fact, reduce the long-term gain. Probing is only valued when uncertainty is very high and management loss is expected to be high if the uncertainty is resolved. Taking into account learning to improve management is termed **dual control**, and works by considering our learning about the system to be another system state. Because management influences future system states, it also influences what we will learn about the system, which in turn improves management in a positive feedback loop. This fully integrated form of management learning should always provide long-term resource gain that is superior to passive or non-adaptive approaches, although the gains may be relatively small for some systems (Williams et al. 2002).

ARM also can be useful for resolving potential conflicts among stakeholders, provided the disagreement is about science. If stakeholders agree on objectives but have different ideas about how the system works, the differing ideas can be incorporated into the decision model as alternative models. Monitoring then can be used to resolve the uncertainty. However, not all apparent disagreements on science are truly about the science or "your model versus my model." Sometimes, scientific disagreement is used to mask disagreements over stakeholder objectives. In these instances, ARM is not the appropriate tool, but the ARM process can be used to reveal these potential problems.

There are some common misconceptions about ARM; chief among these is that it is research. Adaptive resource management is first and foremost, management. All decisions are made to maximize management objectives. Learning occurs as a byproduct of management and it improves decision making, thereby increasing the long-term gain to managers. Another misconception is that ARM is too risky. All natural resource management decision making involves risk in one form or another (e.g., overharvesting a population vs. aggravating user groups) and uncertainty increase risk. ARM reduces uncertainty and risk. ARM also does not distract from management goals because learning is focused at improving management. Most natural resource management agencies already incorporate most, if not all, of the components of ARM. Managers have a set of decisions and management objectives and a decision is chosen based on the belief that it will achieve a more desirable outcome. The belief can be based on a model or may be based on the opinion of the decision makers and their advisers. Monitoring is also generally conducted to evaluate the status and trends in populations of interest. Given these components, all that is needed for ARM is to explicitly link decisions to objectives with an explicit model and ensue that the monitoring matches the model outputs so predictions can be matched with outcomes.

References

Caswell, H. (2001) *Matrix Population Models: Construction, Analysis, and Interpretation.* 2nd edn. Sinauer Assoc., Sunderland, Mass.

Clemen, R.T. and T. Reilly, (2001) *Making Hard Decisions.* South-Western, Mason, Ohio.

Haber, S. (1964) *Efficiency and Uplift: Scientific Management in the Progressive Era, 1890–1920.* University of Chicago Press. Chicago, IL.

Holling, C.S. (1978) *Adaptive Environmental Assessment and Management.* John Wiley and Sons, New York, NY.

Lindley, D.V. (1985) *Making Decisions.* Wiley. New York.

Walters, C.J. (1986) *Adaptive Management of Renewable Resources.* Blackburn Press, Caldwell, NJ.

Warren, M.L., B. M. Burr, S. J. Walsh, H. L. Bart, R. C. Cashner, D. A. Etnier, B. J. Freeman, B. R. Kuhajda, R. L. Mayden, H. W. Robison, S. T. Ross, and W. C. Starnes. (2000) Diversity, Distribution, and Conservation Status of the Native Freshwater Fishes of the Southern United States. *Fisheries* 25(10), 7–31.

Williams, B.K, J.D. Nichols, and M.J Conroy. (2002) *Analysis and Management of Animal Populations*. Elsevier-Academic. New York.

Williams, B.K, R. C. Szaro, and C. D. Shapiro. (2009) Adaptive Management: The US Department of Interior Technical Guide. [Online] URL: http://www.doi.gov/archive/initiatives/AdaptiveManagement/TechGuide.pdf

Additional reading

Regan, H.M., Colyvan, M. and Burgman, M.A. (2002) A taxonomy and treatment of uncertainty for ecology and conservation biology. *Ecological Applications* **12**, 618–628.

8

Methods for Obtaining Optimal Decisions

In Chapter 6, we showed how to create models for predicting the outcome of management actions and for identifying the optimal or satisficing decision using graphical models (e.g., decision trees and influence diagrams) and computer simulation. While these approaches are widely used in natural resources, they are by no means the only approaches to identifying optimal decisions. This chapter provides a relatively complete description and outline of the different approaches used to obtain optimal decisions. We classify these approaches based on the type of decision to be made (static or dynamic/sequential), whether the resource system is treated as deterministic or stochastic, and the complexity of objectives and constraints. We start by presenting deterministic, static decisions that can be solved using constrained or unconstrained optimization methods, including classical, linear, and nonlinear programming. Building on this, we discuss sequential decision making and adaptive resource management wherein decisions are revisited through time and/or space, and where the resource system contains stochastic influences that affect decision making. We discuss the advantages (e.g., realism) and disadvantages (e.g., computational limitations) of these approaches and provide guidelines and suggestions to implement each. Our focus is on application and implementation of the above approaches in natural resource conservation/management scenarios when complexity is high. Our primary goal is to provide an overview of the wealth of approaches used in natural resource management rather than a comprehensive treatment of each. Readers that desire more detail are encouraged to read Appendix E and Williams et al. (2002).

Decision Making in Natural Resource Management: A Structured, Adaptive Approach,
First Edition. Michael J. Conroy and James T. Peterson.
© 2013 John Wiley & Sons, Ltd. Published 2013 by John Wiley & Sons, Ltd.

Overview of optimization

By **optimization** we mean the seeking of values of a management control or other input that result in a desired value (usually a maximum or a minimum) of a specified **objective function**. Optimization frequently takes into account **constraints** on values that the controls or values of the objective can take. Examples of optimization problems in conservation include:

- Finding the population level that results in maximum yield from a population at equilibrium.
- Deciding on a schedule of releases or reintroductions over time to restore a population to viability.
- Allocating finite resources to the maintenance of 2 species with conflicting habitat needs.

In every case, the optimization problem requires specification of 3 elements, all of which we have already considered:

- a list or range of alternative actions;
- an objective function that captures the fundamental objectives of the decision maker;
- a model that functionally connects the decisions and other relevant variables to the outcomes that are used to evaluate the objective.

A simple, verbal statement of the optimization problem is then: "Given a range of possible decisions that can be chosen, chose the combination of decisions that provides the best result in terms of the objective." However, to address optimization in a rigorous and quantitative way, we need a more formal statement of the optimization problem. First, we will define a set of 1 or more control or **decision variables**.

$$\underline{d}' = (d_1, d_2, \ldots, d_n)$$

For example, suppose that we are trying to define the optimal set of harvest regulations to achieve an objective (such as sustainable harvest) for a waterfowl population. We then might have $\underline{d}' = (d_1, d_2)$, where d_1 is the daily bag limit (e.g., 0–6 ducks) and d_2 is the maximum season length (e.g., 0–75 days).

Next, frequently **constraints** exist or are imposed by the decision maker on the range of decisions (or **decision space**). Constraints include costs, logistics, or physical or logical limits on which decisions can co-exist. For the duck harvest problem, we implicitly imposed constraints by not considering decision outside the range $0 \leq d_1 \leq 6$, and $0 \leq d_2 \leq 75$; other combinations (e.g., $d_1 = 0$ with $d_2 > 0$ or the reverse) are eliminated as nonsensical.

Costs or other resource limitations frequently act as constraints. For example, we may be able achieve a desired population objective by a combination of controlled fire (d_1) and trapping of exotic predators (d_2). Given per-unit costs for each of these **actions** of c_1, c_2 and a total cost limit of C, we may impose the decision constraint

$$c_1 d_1 + c_2 d_2 \leq C$$

Constraints of this nature can themselves be the products of an economic or other model that operates in parallel with biological process model. As we will see later, **system dynamics** – that is, how the system we are managing evolves through time in response to our decisions and other factors – can also be viewed as a constraint on decision. Collectively, we can refer to the permissible decisions (after constraints) as an **opportunity set** $\underline{D}$; thus any decisions that are selected will be a subset.

Finally, we need to specify an **objective**, F ($\underline{d}$), that we are seeking (again, ordinarily to maximize or minimize) as a function of the decision variables. For example:

- Maximize harvest yield from the population over the next 10 years.
- Minimize risk of extinction over 100 years.
- Maximize reproductive success next year.

Our formal statement of the optimization problem then is: "Select $\underline{d}$ to maximize $F(\underline{d})$ subject to $\underline{d} \in \underline{D}$.

Note that although the problem is stated as "maximize" we can easily convert it to a problem where we are minimizing $F(\underline{d})$ (for instance, if $F(\underline{d})$ is cost or probability of extinction).

Finally, by stating the objective F as a function of the decision variables $\underline{d}$ we have left out many intermediate details that we have discussed in earlier chapters, including how we expressed competing or conflicting resource objectives (Chapters 3 and 6), and the specific influence of decisions on the system state outcomes (Chapters 6 and 7) that are incorporated in the objective. However, these details only affect the specific tools available for optimization to the extent that they affect the specific mathematical form of the objective function (e.g., whether the function is continuous and differentiable; whether dynamic decisions are involved; or whether stochastic effects are involved).

The exact methods used to obtain optimal decisions will depend both on the complexity of the system and the type of controls that are involved, and range from very simple approaches requiring little mathematical sophistication, to complex approaches requiring high-speed computers. We have relegated the details of the various computational approaches to an appendix (Appendix E) and in this chapter simply apply these approaches to problems having a range of complexity.

Factors affecting optimization

As noted above, the methods used for optimization depend on system complexity and the types of controls and their interactions. These can be described by a series of dichotomies that can operate in combination (Figure 8.1).

	DETERMINISTIC		STOCHASTIC
STATIC DECISIONS	UNCONSTRAINED	Unconstrained optimization	Decision trees Influence diagrams Simulation Simulation-optimization
	CONSTRAINED	Classical Programming Linear programming	
DYNAMIC DECISIONS		Nonlinear programming Calculus of variations Dynamic programming	Simulation-optimization Stochastic dynamic programming Markov decision processes Heuristics

Figure 8.1   Methods for obtaining optimal decisions. See Williams et al. (2002: Chapters 22 and 23).

Single vs. multiple decision controls

The simplest problems involve a single control variable d that is used to obtain an objective result $F(d)$. Often, fairly simple approaches such as graphical methods, sorting in lists, or even visual inspection will suffice to solve such problems (Appendix E). More generally, decision problems will involve 2 or more decision variables. Note that these could be either decisions involving multiple controls at a single point in time; a single type of control carried out over time; or a combination of these. Depending on the dimensionality of such problems they may be solved with simpler methods, but usually will require more sophisticated methods based on multivariate calculus or other approaches, particularly if constraints are involved.

Unconstrained vs. constrained optimization

In addition to the number of controls, the imposition of constraints adds complexity to the optimization problem that may require specialized approaches such as classical, nonlinear, or linear programming. By contrast, unconstrained problems tend to be simpler to solve, e.g., by simple function maximization.

Static or equilibrium optimization vs. dynamic optimization

Most natural resource systems we are interested in managing evolve through time, i.e., have **dynamics**. As a simple example, a population growing at a constant rate r will evolve through time according to the logistic growth model

$$N(t+1) = N(t) + r_{max} N(t)[1 - N(t)/K]$$

where r_{max} is the maximum growth potential and K is the carrying capacity. For specified r_{max} (e.g., 0.1), K (e.g., 500) and initial abundance (e.g., 100) we can predict future abundance, e.g.,

$$N(5) = 145$$

When the system is dynamic, we can consider 2 types of decisions. In the first, the decision is **static**, that is, does not depend on the evolution of the system at time t. This type of decision is appropriate in at least 2 types of circumstances. The first is when the decision is applied once (say at the beginning of a period of interest) but its effects are felt through time. In the above example, a decision that results in the one-time increase of K could then be evaluated in terms of its impact on population growth over time, in a static fashion. The second situation occurs when the population is managed so as to maintain **equilibrium** (system state unchanging in time). In such cases time can effectively be ignored and the decision treated in a static way; certain harvest optimization problems such as **maximum sustainable yield (MSY)** can be solved in this way.

A more complicated (and more common) situation arises when the decisions recur through time. Now, the decision that is made is influenced by the current conditions (state) of the system. In addition, decisions that are made earlier in the decision timeframe will influence the system in ways that affect future decision options; thus, dynamics effectively imposes another constraint. We will address this issue more completely below.

Deterministic vs. stochastic system response

Another complication that occurs in real decision problems is the fact that real systems usually do not respond in a simple, deterministic fashion to management decisions, but instead are subject to random environmental, demographic, and other forms of stochasticity (Chapter 6). That is, there is no longer a deterministic relationship between the decisions $\underline{d}$ and the objective outcome $F(\underline{d})$. For example, assume a population is growing according to a simple growth model

$$N(t+1) = (1+r)N(t)$$

We can model the impacts of a per-capita harvest decision d as

$$N(t+1) = (1+r-d)N(t)$$

and deterministically evaluate the impacts of different values of d ($0 < d < r$) on population growth rate, cumulative harvest, or other objective over a specified timeframe, and then select the value of d to maximize our objective. If instead r varies randomly through time (say because of environmental stochasticity), we no longer have a deterministic relationship between management actions and the objective. This means that some of the mathematical optimization methods will only approximately (if at all) give us an optimal solution. We will return to this issue in more detail later in the chapter.

Again, all of the above factors can and do exist in various combinations in real problems. For instance, a problem can involve static decisions with multiple decision variables and deterministic response, etc. Figure 8.1 summarizes how the various factors can interact and provides some guidance as to appropriate optimization methods for each situation, discussed further below and in Appendix E.

Static, single-objective, deterministic decisions

We start with the simplest optimization problem, in which (1) the decision is static (not dependent on time) and (2) there is a single objective, i.e., we are not having to deal with objective components that involve multiple values or trade-offs, (Chapter 3) and (3) outcomes are **deterministic** (vs. **stochastic**). We will also assume that there is a single control or decision variable, d, although having multiple controls only slightly complicates the problem. Solving this problem involves finding the value (or sometimes, range of values) of d that maximizes or minimizes $F(d)$. This can be accomplished in a number of ways, including direct ("brute force") searching and the calculus of unconstrained optimization (Appendix E).

We can illustrate unconstrained, single-control optimization with the classic problem in harvest management of determining **maximum sustainable yield** (**MSY**). This is solvable as a static optimization problem given a few assumptions:

- The population is growing deterministically according to the logistic growth model.
- We can remove a specified harvest from the population (perfect controllability) that perfectly balances population growth (by definition this is **sustainable yield**).
- Our management control consists of specifying the desired equilibrium population level (N^*) at which to achieve MSY.

The logistic growth assumption results (in discrete form) in population growth as

$$N(t+1) = N(t) + r_{max}N(t)[1 - N(t)/K]$$

where r_{max} is the maximum per capita growth rate and K is the carrying capacity (an upper limit to $N(t)$).

If we harvest this population each year we have

$$N(t+1) = N(t) + r_{max}N(t)[1 - N(t)/K] - H(t)$$

We can exactly balance population growth by taking

$$H(t) = r_{max}N(t)[1 - N(t)/K]$$

at which point we have $N(t+1) = N(t) = N$. Now the population is at equilibrium, and we can rewrite H as a function of N, so that

$$H(N) = r_{max}N[1 - N/K]$$

Our optimization problem is to find the value of N that maximizes this function. As shown in Appendix E, this problem is amenable to an exact solution based on calculus, but could just as legitimately have been solved (at least approximately) by trial and error or by plotting values of $H(N)$ versus candidate values of N for specified values of the parameters (K and r_{max}). Box 8.1 illustrates how MSY can be found for a specific case.

Box 8.1 Unconstrained single-control optimization: MSY

We illustrate single-control optimization with an example of **maximum sustainable yield** (MSY). Although MSY is useful as a conceptual basis for incorporating ideas of population self-regulation into harvest management, its use as a management tool for setting fish and wildlife harvest levels has largely lost favor because of simplistic assumptions about population growth (Larkin 1977, Williams et al. 2002). Nonetheless, MSY can be a useful starting point for an analysis of harvest decisions. The usual assumption of MSY is that the population is growing according to the logistic model of density dependence. Under this model, equilibrium is maintained by the relationship

$$H(N) = r_{max}N[1 - N/K]$$

We show in Appendix E that this function has a maximum at $N^* = K/2$, found by taking the first derivative with respect to N, setting to zero, and solving for N; by substitution into the above equation this produces an optimum yield (MSY) of $H(N^*) = r_{max}K/4$.

To take a specific case, suppose the population is growing at $r_{max} = 0.2$ with $K = 5500$. We first find $N^* = K/2 = 5500/2 = 2750$. MSY is then $0.2 \times 5500/4 = 275$. We can confirm this result by plotting values of H versus N over the range 0 to K. We provide a small R function to perform the exact calculations for N^* and H^* and generate this graphics for the example parameter values; users may change these value for their specific problem (Appendix G).

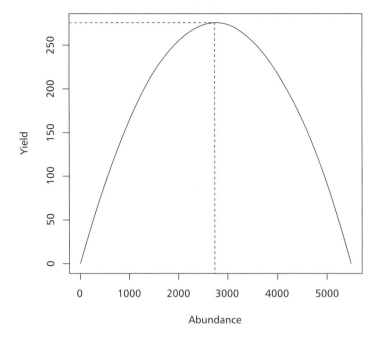

The above results were produced by the "exact" solution using derivatives. An equivalent numerical result can be produced using the optim() procedure in R, which uses a numerical method known as BFGS (Broyden 1970; Appendix E, G).

This approach generalizes fairly easily to the more common situation where we have multiple controls, the only complication being that the optimal decision now requires finding the combination of 2 or more controls that maximizes the objective. We illustrate the multiple control problem for a case involving 2 controls (Box 8.2).

Multiple attribute objectives and constrained optimization

The above development assumes that our objective involves a single output, measured in a common set of units, and that no constraints are involved. In Chapters 3 and 6 we saw that, more commonly, decision problems involve objectives that have multiple attributes or involve constraints. At this point, we will assume that we have been able to successfully cast our decision problem with

- an **objective function** $F(\underline{d})$ that is valued in single units, whether these are units of resource output, monetary value, or **utility** (Chapter 3); and

Box 8.2   Unconstrained optimization with 2 controls: conservation of 2 species

This example was used by Williams et al. (2002: 594) to illustrate constrained optimization with 2 controls; we will start out with an unconstrained problem. The objective is to sustain 2 species that are in partial competition. Williams et al. (2002) expressed the **objective function** as

$$F(\underline{d}) = 224d_1 + 84d_2 + d_1d_2 - 2d_1^2 - d_2^2$$

and stated that $F(\underline{d})$ reflects potential for population sustainability and relative value of the 2 species. We will accept this function at face value for the sake of illustrating optimization approaches, but note that in practice such a function would need to be derived by methods such as we described in Chapter 3 for deriving **utilities** and **utility functions**. The values d_1 and d_2 are the population levels for each species, over which we assume that we have perfect control; i.e., the decision is the level (d_1, d_2) at which to maintain the 2 species. The objective is to maximize $F(\underline{d})$. The gradient (vector first partial derivatives) is set to 0

$$\frac{\partial F(\underline{d})}{\partial \underline{d}} = \begin{pmatrix} \partial F / \partial d_1 \\ \partial F / \partial d_2 \end{pmatrix} = \begin{pmatrix} 224 + d_2 - 4d_1 \\ 84 + d_1 - 2d_2 \end{pmatrix} = \underline{0}$$

and is fairly easily solved for $(d_1^*, d_2^*) = (76, 80)$.

Further analysis (Appendix E) proves that this is a maximum and not a minimum.

Once we obtain the optimal values for (d_1, d_2), we can substitute these into the gradient to verify that the zeroes condition is satisfied, and into the objective function to return the objective value to obtain

$$F(\underline{d}^*) = 11\,872$$

The calculations to produce the objective function and gradient for this problem are easy in R (Appendix G).

As with the single control example, we can produce a numerically equivalent (within rounding error) result by using the optim() function in R, as illustrated in Appendix G.

We will return to this example when we consider constrained optimization next.

• a series of **constraints** $g(\underline{d})$ on the decision space that reflect values that cannot be easily incorporated into the objective function.

Once we have arrived at this point we still have a number of choices about how to obtain the optimal solution to the decision problem. These choices depend on the mathematical forms of the objective function and accompanying constraints, as elaborated below.

The general problem we have is one of **constrained optimization**. For static (and, for the moment, deterministic) objective problems such as the ones considered so far in this chapter, there exist a wide variety of optimization tools (see Appendix E for more details):

- **Classical programming**, which provides optimal solutions for problems meeting certain mathematical conditions, in which the objective function is differentiable (i.e., requires calculus) and the constraints can be written as equalities.
- **Nonlinear programming**, which generalizes classical programming to allow for inequality constraints.
- **Linear programming**, a special case of nonlinear programming, in which both the objective function and the constraints can be written as linear equations.

In classical programming, the constraints are incorporated into the objective function by means of a device known as a **Lagrangian multiplier**. This approach is more fully developed in Appendix E, but is easily illustrated by means of an objective function with 2 controls and a single constraint, so $F(d_1, d_2)$ and constraint $g(d_1, d_2) - a = 0$; F and g can be any differentiable functions. The constraint is incorporated into the objective function as

$$L(d_1, d_2, \lambda) = F(d_1, d_2) + \lambda[a - g(d_1, d_2)]$$

which upon differentiation with respect to $\underline{d}$ and λ (the Lagrangian multiplier) leads to conditions for optimality (Appendix E). The approach generalizes to more complex objective and constraint functions, so long as these are differentiable. We apply the Lagrangian approach to a bivariate decision problem in Box 8.3.

An alternative to the Lagrangian approach is to use **nonlinear programming** (NLP) that permits more general constraints (such as inequalities) and still allows the objective function to have very general mathematical forms (as long as they are differentiable; see Appendix E). Nonlinear programming permits the relationship between the decision and the objective to be fairly complex; for instance, some types of dynamic objectives and decision can be cast as nonlinear programming problems. We apply nonlinear programming to the bivariate species conservation problem in Box 8.4.

Finally, if both the objective function and the constraints can be cast as linear functions, then a special case of nonlinear programming called **linear programming** provides a rapid and efficient means of finding **feasible solutions** (solutions that satisfy the constraining functions) to very large and complex decision problems. For this reason, it can be very tempting to attempt to approximate complex, nonlinear systems via linear models. Sometimes the approximation is reasonable. Furthermore, if the temporal relationships of a dynamic system can be suitably expressed in linear form, then linear programming can be used to solve dynamic problems. We provide an example of the application of linear programming to a simple reserve design problem in Box 8.5.

Box 8.3   Constrained optimization with 2 controls – conservation of 2 species

We can return to the 2 species conservation example to illustrate optimization with constraints.

Retaining the objective function $F(\underline{d})$ from Box 8.1, we now impose a resource constraint of 280 units, and also stipulate that species 1 exerts a 3 times higher demand for resources. These values might come from a mapping of habitats of the system that identify 280 ha of total habitat available, and a resource use study that determines that species 1 requires 3 ha of habitat for every ha required by species 2 (e.g., its home range size is 3 times larger). This results in the constraint:

$$3d_1 + d_2 = 280$$

This problem can be solved by the method of Lagrangian multipliers, in which the constraint is incorporated into the objective function (Appendix E)

$$L(\underline{d}, \lambda) = 224d_1 + 84d_2 + d_1d_2 - 2d_1^2 - d_2^2 + \lambda(280 - 3d_1 - d_2)]$$

The modified objective is optimized by finding the value of d_1, d_2 and λ that satisfy

$$\partial L(\underline{d}, \lambda)/\partial_1 = 224 + d_2 - 4d_1 - 3\lambda = 0$$
$$\partial L(\underline{d}, \lambda)/\partial_2 = 84 + d_1 - 2d_2 - \lambda = 0$$
$$\partial L(\underline{d}, \lambda)/\partial\lambda = 280 - 3d_1 - d_2 = 0$$

These equations can be solved for $(d_1^*, d_2^*, \lambda^*) = (69, 73, 7)$, and we provide the R code illustrating this example in Appendix G. If we substitute the optimal decision and Lagrangian values into the objective we obtain the value 11 774, which is lower than the value of 11 872 that was obtained under unconstrained optimization. Constraints will always result in an objective return no greater than would be obtained by releasing the problem from such constraints.

The optimal values of the Lagrangian multiplier are useful in that they provide expressions of the **sensitivity** of decision problems to constraints. In Appendix E we show for this example that the **elasticity** (proportional sensitivity) of the constraint is provided by

$$e(\underline{g}) = \lambda^* \frac{g}{F^*} = 7\left(\frac{280}{11774}\right) = 0.166$$

providing a measure of the proportional marginal gain we could expect in our objective, for a proportional relaxation in the resource constraint (e.g., through habitat improvement).

Box 8.4   Constrained optimization with 2 controls – conservation of 2 species: inequality constraints

Again, return to the 2 species conservation example with the objective function defined in Box 8.1, but now with inequality constraints. Suppose that instead of dictating a fixed limit of 280 ha, all of which would be allocated to the 2 species, we permit outcomes where potentially less than this amount of resources is used, potentially leaving some resources available for other uses. The objective is to maximize $F(\underline{d})$ subject to the constraint

$$3d_1 + d_2 \leq 280$$

Implementation in Rsolnp() requires specification of objective and constraint function that are in turn called by the optimization program.

In this example, specifying an upper limit of 280 provides the same optimal decision and value as the equality constraint.

$$(d_1^*, d_2^*) = (69, 73)$$
$$F(\underline{d}^*) = 11774$$

By contrast, if we relax the constraint a little by increasing the resource limitation to 290 the optimal decision changes.

$$(d_1^*, d_2^*) = (71.5, 75.5)$$
$$F(\underline{d}^*) = 11831.5$$

Finally, removal of the constraint returns us to the original, unconstrained solution.

$$(d_1^*, d_2^*) = (76, 80)$$
$$F(\underline{d}^*) = 11872$$

Again, note that as we relax the constraints the objective return increases; in general, the objective return under relaxed constraints will be greater than or equal to that with the constraints imposed. It also illustrates that unconstrained optimization can be treated as a special case of NLP.

The R code to compute the optimal NLP solution and to illustrate the above constraint sets is provided in Appendix G.

Box 8.5   Linear objective and constraints: optimal reserve design

Suppose we have 3 biological reserves, and we need to decide how much habitat to conserve on each. Let d_1, d_2, d_3 represent the areas of habitat conserved on each reserve. Suppose further that we have fit an occupancy model to field data and estimated the relationship between the probability of persistence (1- extinction) and the area of each reserve and obtained the model:

$$\text{logit(persistence)} = Y = 5 + 0.001d_1 + 0.0012d_2 + 0.0009d_3$$

The implementation costs per ha differ among reserves at \$5, \$7, and \$6.50 per ha, respectively. We also assume that we only have \$5k to spend, so the total cost constraint is \$5000 and that the maximum conservation area is dictated by the existing reserve size of 1000, 2500, and 2000 ha, respectively (i.e., another set of constraints). Finally, for "political" reasons we have to conserve at least 250 h in each reserve. We can express the objective function as the linear objective

$$F(d) = Y - 5 = 0.001d_1 + 0.0012d_2 + 0.0009d_3$$

(note that the intercept 5 has no bearing since it is not a function of any of the decisions), which we wish to maximize subject to the constraints

$$5d_1 + 7d_2 + 6.5d_3 \leq 5000$$

$$d_1 \leq 1000$$
$$d_2 \leq 2500$$
$$d_3 \leq 2000$$

$$d_1 \geq 250$$
$$d_2 \geq 250$$
$$d_3 \geq 250$$

We can solve this problem using linear programming (LP). The LP solution to maximize the above objective subject to the constraints is

$$(d_1^*, d_2^*, d_3^*) = (325, 250, 250)$$

with objective value

$$F(\underline{d}^*) = 0.85$$

The objective value return can be converted to probability of persistence by the inverse logit transform of 0.9971284. Thus, to fulfill the objective of maximizing species persistence subject to the stated constraints, we should allocate conservation to each reserve as 325 ha, 250 ha, and 250 ha, respectively.

As a side note, we no doubt could have achieved a higher probability of persistence by tossing out the cost constraints, or allowing expansion of the reserves, or bucking politics. For example, if we eliminate the 250 ha minimum constraint we achieve a different result:

$$(d_1^*, d_2^*, d_3^*) = (1000, 0, 0)$$

with objective value

$$F(\underline{d}^*) = 1.00$$

and probability of persistence

$$P(\underline{d}^*) = 0.9975274$$

In this case, we have increased the objective return slightly (from 0.9971284 to 0.9975274) by allocating all effort to the first reserve. In general, release of constraints from an objective problem will result in an objective value at least as high as under the constraint. If we also remove the cost constraint we get:

$$(d_1^*, d_2^*, d_3^*) = (1000, 2500, 2000)$$

with objective value

$$F(\underline{d}^*) = 5.8$$

and probability of persistence

$$P(\underline{d}^*) = 0.99998$$

In this case, the optimal solution lies on the boundary of the area constraint: we can conserve no more habitat than exists. Fairly obviously, if we remove this constraint the problem has no bounded solution: given a linear objective function maximizing the function means maximizing (or minimizing) the controls without bound.

Our example is *not* a support of any of these (or other) constraints, but rather illustrates that intelligent decisions *can* be arrived at even in the face of multiple, seemingly conflicting mandates.

The R code to provide a solution using the lp() procedure is provided in Appendix G.

Dynamic decisions

Up to this point, we have either dealt with static systems (such as the reserve design problem) or, if the system was dynamic, we assumed that any decisions would be static (Chapter 6 examples). So for example, we might be interested in the impact of a reserve design on future population status, but the decision was a one-time one. Alternatively, as in the case of MSY harvest management, the optimal decision was aimed at maintaining a population at equilibrium. In either case, system dynamics were either ignored or incorporated via relatively simple constraints (e.g., the assumptions of population equilibrium and deterministic population growth in MSY harvest management are examples of constraints).

More general conservation decision problems require an explicit consideration of how decisions and system states interact through time. There are at least 3 reasons for this. First, system controls (decisions) generally are not fixed through time, but change based both on where the system now is and where it is desired to go. Secondly, decisions made earlier in a timeframe have the potential to affect the system, and thereby opportunities for decisions made later. Thirdly, natural resource objectives almost always incorporate future resource values. For example, in animal harvest decisions, the number of animals that can be harvested in a given year (the decision) depends, in part, on the number of animals in the population (current system state) and the number of animals harvested affects the number of animals in the population in the future. The interplay between system controls (management actions) and system states in the objective is captured in the **optimal control problem (OCP)**. The OCP is laid out more formally in Appendix E, but can be stated verbally as

- select the trajectory of decisions through time, $\underline{d}(t)$ that maximizes resource utility over the entire timeframe

subject to

- the dynamics over the timeframe, including the impacts on the change in system states $\underline{x}(t)$ wrought by the decisions $\underline{d}(t)$ and other factors.

The 2 major, traditional approaches for solving this problem are the **calculus of variations** and **dynamic programming**. The second of these has seen much more use in natural resource optimization, and so will emphasized here, but both are described in more detail in Appendix E. Dynamic programming (DP) is an algorithm, based on **Bellman's optimality principle** (Appendix E) that permits solution to OCP via backwards induction from a terminal solution. It is readily extendible to stochastic systems but suffers from the "curse of dimensionality" in that it is only practicable for relatively simple, low-dimensioned systems (more later).

Other approaches exist and are described in Appendix E, including those based on **heuristic approaches** or **optimization-simulation** that are not guaranteed to find optimal solutions to OCP. We focus on DP because a) it is known to provide OCP solutions provided the system can be captured in DP form (i.e., has sufficiently low dimension), b) allows for stochasticity and other forms of realism, and c) has seen important applications in North American conservation.

A very general statement for OCP in discrete systems (see Williams et al. 2002: 607) is of the form:

Maximize

$$J = \sum_{t=t_0}^{t_f} F[\underline{x}(t), \underline{d}(t), \underline{z}(t)] + F_T[\underline{x}(t_f)]$$

where x, d, and z are states, decisions, and random variables, respectively, at points t from t_0 (an initial time) to t_f (the end of the timeframe of interest). This objective is modified by the constraints of system dynamics, i.e.,

subject to

$$\underline{x}(t+1) = \underline{x}(t) + f(\underline{x}(t), \underline{d}(t), \underline{z}(t), t)$$

As seen in Appendix E, this is solved by a backwards inductive approach using DP. The original DP approach was developed for deterministic systems, in which we would eliminate $z(t)$ from the above expressions. The more general form of DP known as **stochastic dynamic programming (SDP)** extends DP to stochastic systems. We apply DP to the optimal harvest problem but now where the harvest decision evolves dynamically (but deterministically) through time (Box 8.6). We will return to this as an SDP program, when we consider decision making under uncertainty, below.

Box 8.6 Dynamic optimization: Harvest under the discrete logistic growth model

We return to the discrete logistic model (Box 8.1) with $r_{max} = 0.3$ and $K = 125$. We wish to select the optimal harvest rate that maximizes cumulative harvest over a specific timeframe. We will therefore maximize cumulative harvest and assign no terminal value.

Maximize

$$J = \sum_{t=t_0}^{t_f} [N(t)h(t)]$$

(*Continued*)

subject to

$$N(t+1) = N(t) + r_{max}N(t)[1 - N(t)/K] - N(t)h(t)$$

In general form the objective function is

$$J = \left\{ \sum_{t=t_0}^{t_f} F[N(t)h(t)] \right\} + F_T[\underline{x}(t_f)]$$

where F is the value of harvest obtained at each time step and F_T is the "terminal" or ending value to the state of the system. Although other values (utilities) could be used, we will take the usual approach (e.g., Williams et al. 2002) of using undiscounted harvest at each time step, and of giving no terminal value to the system. Thus,

$$F[N(t)h(t)] = N(t)h(t)$$

and

$$F_T[\underline{x}(t_f)] = 0$$

Although not necessary, we will also constrain the decision space so that the harvest is $0 \le h(t) \le 0.5$. We have implemented this problem in a deterministic, discrete dynamic programming algorithm in R (Appendix G). The program calls 3 functions within a backwards iterative loop:

- a state return function that computes F;
- a terminal value function that computes F_T;
- a state dynamics function that computes the transition from t to $t+1$.

Although this version is simplified (reflecting our relative lack of programming sophistication) these are exactly the same elements in common with modern programs such as SDP (Lubow 1995), which performs both deterministic and stochastic dynamic programming (discussed below).

If we run this program for a state space defined as integer values over [1,100] and harvest rate decisions [0,0.5] in increments of 0.1, we will observe that the algorithm converges on a stationary policy within 20 iterations.

The above example assumes that we have a deterministic model of state space and decision influence. However, DP is readily generalized to stochastic problems, in which case the solutions are determined by evaluating

$$J^*(x, t) = \max \left(E\{[J(x, t)] + J^*(x, t+1)\} \right)$$

that is, by averaging over the random variables to get the expected value for J. We'll return to this point in the next section.

Optimization under uncertainty

We have already seen that many if not most natural resource decisions involve uncertainty from a variety of sources. In Chapters 6 and 7, we described the major sources of uncertainty, and showed how, although some types of uncertainty can be reduced, generally speaking we must take uncertainty into account as we are seeking to find optimal solutions. We agree with Lindley (1985) that this can result in somewhat artificial distinctions, particularly when considered from a Bayesian perspective, in which our uncertainty measures can be strongly or weakly influenced by data (Chapters 5 and 7). We thus in general advocate describing uncertainty in terms of probability distributions and show how we can consider decision making in terms of value-weighted outcomes (Chapter 6). That is, if we are able to describe the utility of an outcome (Chapter 3) and the conditional probability of an outcome given a decision, it would appear that all we need to do is to find the decision that provides the "better" outcome distribution. As you might expect, the mathematics for finding this answer (assuming one exists) often is more complicated than the above simplification would suggest.

Maximum expected values approach

One approach that we have already seen is to describe the decision outcomes in the form of branching outcomes that have differing utility values and occur with different (but specified) probabilities. For instance, in Chapter 6 we used simple decision trees and influence diagrams to represent the relationship between decisions and outcomes and showed how one approach for optimization is to compute the conditional expectation of each candidate decision, and then select the decision that maximizes (or minimizes) the expected objective value. For simple, static problems with uncomplicated objective responses this can be simply accomplished by direct examination or ranking of the objective values (basically the approach we used in Chapter 6). In other cases, we might need to use formal optimization methods (considered earlier) to maximize the objective. Of course, approaches based on maximizing expected values lead to decisions that, by definition, should be good "on average" but may stray very far from the mark in any one instance, leading to consideration of alternative criteria such as minimax (Appendix E) that seek to avoid very bad outcomes (but sacrifice average performance) and the use of risk profiles to represent utilities (Chapter 7).

As decision problems increase in complexity, because of constraints, dynamics, or both, approaches such as LP or NLP for finding optimal decisions for stochastic systems will no longer work. For example, suppose we have a problem like the one on Box 8.5 but now some of the coefficients describing the objective value or constraints are no longer fixed, but instead follow random variable distributions. We might be tempted to solve this problem via LP by replacing the random variables with their expectations. However, this would likely result in decisions that were infeasible (i.e., did not fulfill the constraints) over the random coefficient distributions.

Simulation-optimization

For some LP and NLP problems it is possible to incorporate stochasticity via constraints, but these approaches tend to be difficult to implement particularly for larger problems. An alternative approach called simulation-optimization combines features of formal optimization with simulation. Basically, we set the problem up as if it was a deterministic optimization problem (so, a deterministic objective and constraints). We can then introduce uncertainty by adding random factors to the deterministic model, and then performing optimization based on the stochastic results (an alternative is to optimize based on deterministic inputs and then add a stochastic "noise" to the result). The simulation-optimization is then repeated a large number of times and used to construct a distribution of optimal decisions and values.

Simulation-optimization approaches can provide a good approximation for many problems and can be relatively simple to implement, so they can be a reasonable choice. They do not, however, provide solutions that are guaranteed to be optimal, so if a problem can be cast using a more exact approach (such as stochastic dynamic programming, see below) that will often be preferable. We give an example of simulation-optimization applied to the earlier species conservation example in Box 8.7. The results in this example suggest that we could probably do an adequate job of management by optimizing under the deterministic model. However, it is always a good idea to investigate the behavior of our optimal decision under stochastic simulation, even if we do not invoke stochastic effects in the optimization (see the exercises below).

One issue that will sometimes arise when we do stochastic simulation-optimization with equality constraints is that optimal decisions will force decision variables into inadmissible (e.g., negative) ranges. In such cases it will usually be necessary to use NLP or other procedures that allow for inequality constraints (e.g., forcing all the decision variables to be zero or above).

Stochastic dynamic programming

Stochastic dynamic programming (SDP) extends DP to allow stochastic uncertainty in the objective and dynamics. The objective is now evaluated by

$$J^*(\underline{x}(t), t) = \max\left(E\{[J(\underline{x}(t), t)] + J^*(\underline{x}(t+1), t+1)\}\right)$$

The expectation is calculated over the distribution of the random variables $\underline{z}(t)$, typically discretized, with probability p_i for each discrete value z_i:

$$J^*(\underline{x}, t) = \max\left(\sum_i^k J(\underline{x}(t), \underline{z}_i(t), t)p_i + J^*(\underline{x}(t+1), \underline{z}_i(t+1), t+1)p_i)\right)$$

Box 8.7   Simulation-optimization: Conservation of 2 species

We will return to the 2-species conservation example solved earlier using NLP to illustrate optimization with inequality constraints where we now have uncertainty in the relationship between the decision and the objective. Again, the objective is to sustain 2 species that are in partial competition. The values x_1 and x_2 are population levels for each species, over which we assume that we have control (i.e., the decision is the level (x_1, x_2) at which to maintain the 2 species. Now we are going to assume that we have stochastic uncertainty in the resource constraint. That is, instead of a fixed constraint of 280, realized resource availability follows a normal distribution with mean 280 and sd of 5. However, we will also assume for simplicity that once realized resource availability is known that we have perfect control, that is, the ability to make an optimal decision based on this value. We then use simulation-optimization to visualize the resulting variation in decisions and objective values

Recall that the original problem was to maximize $F(\underline{x})$ where

$$F(\underline{x}) = 224x_1 + 84x_2 + x_1x_2 - 2x_1^2 - x_2^2$$

subject to the constraint

$$3x_1 + x_2 = 280$$

We solved this problem earlier using Lagrangian multipliers, in which case we obtained a single solution based on the objective and constraint coefficients in the above 2 equations. Recall that the optimal decision as $\underline{x} = (69,73,7)$ providing an objective value of 11 774. We incorporate stochastic uncertainty in the resource constraint drawing random values of the constraint from the above normal distribution, incorporating these values into the Lagrangian while keeping the other coefficients fixed, solving for the optimum, and repeating this process a large number (say 10,000) times. A typical result for this example provides a range of optimal decisions of $x_1 = (63–74)$, $x_2 = (67–78)$, and objective value ranging from 11 570–11 865.

The R code to implement the simulation-optimization example is provided in Appendix G.

Stochastic effects also are typically part of the state transition function, so that

$$E\{\underline{x}(t+1)\} = \underline{x}(t) + \sum_{i=1}^{k} p_i f(\underline{x}(t), \underline{d}(t), \underline{z}_i(t), t)$$

We apply SDP to the dynamic harvest problem considered earlier (Box 8.8). For simplicity we considered only stochasticity in the harvest control; more generally

---

Box 8.8　SDP: Partial controllability of harvest

Here, we extend the previous DP harvest example to allow for a simple variety of stochasticity, due to partial controllability (Chapter 7). The objective is stated as

Maximize

$$E(J) = E\left(\sum_{t=t_0}^{t_f} [N(t)h(t)] \right)$$

subject to

$$N(t+1) = N(t) + r_{max} N(t)[1 - N(t)/K] - E\{h(t)N(t)\}$$

where partial controllability is represented by a discrete random variable distribution $m = (0.75, 1, 1.25)$ indicating that realized harvest rate is 75%, 100%, or 125% of the intended harvest rate; we specified the probabilities associated with each value of the multiplier m as $p = (0.25, 0.5, 0.25)$. Optimization proceeds as before but now with averaging over this discrete distribution for both the stage return and state dynamics function. Appendix G implements SDP for the harvest example. Notice that the optimal decision is similar, but that convergence to stationarity takes longer. Also notice that (at the equivalent stage) the cumulative objective return is lower for SDP than for DP. This is because stochasticity effectively acts as a constraint, whose removal (if possible) would inevitably result in better performance of the optimal decision.

　　Finally, finding the optimal decision – regardless of how it is arrived at – in stochastic systems should never be the last step. It is always a good idea to substitute the optimal (usually state-specific) decision back into the dynamic model and simulate the populations change through time (in fact this is a good idea even if we are pretending the system changes deterministically).

---

we would also include environmental stochasticity (e.g., due to rainfall variation).

　　SDP allows proper incorporation of stochastic effects into the optimal decision-making problem. It is also straightforward to extend SDP to allow for **structural uncertainty** (Chapter 6), usually represented by alternative models about how the system is hypothesized to respond to management and other factors. The model weights are then treated as another system state that is at least partially under control of management and is known as **adaptive stochastic dynamic programming** (ASDP; Williams et al. 2002). We provide a numerical illustration of ASDP in Appendix E.

Analysis of the decision problem

Sensitivity analysis

Sensitivity refers to the change that occurs in model output, given a small change in some model parameter or other input (like a decision variable). Sensitivity can be very useful, both for exploring the relative importance of various model parameters and inputs, as well as for focusing effort on reducing uncertainty in these quantities (Chapter 7). Here, we define sensitivity in terms of the partial derivatives of some output $f(x)$ with respect to x (the parameter or other input):

$$\Gamma(x) = \frac{\partial F(x)}{\partial x}$$

In some cases, sensitivity can be directly computed from analysis using calculus or by other method (e.g., eigenanalysis for matrix models of age structure). More commonly, we need to use simulation to approximate $\Gamma(x)$. An effective approach is based on numerical approximations to the partial derivatives and was used by Conroy et al. (2002) in a simulation model for American black ducks (*Anas rubripes*). In this approach, a particular parameter value x is selected for sensitivity analysis. The parameter is perturbed a small amount (Δx) and model output $F(\underline{x})$ is obtained with x and Δx as inputs. The approximation is calculated as

$$\Gamma(x) \approx \frac{F(x + \Delta x; \underline{x}_0) - F(x; \underline{x}_0)}{\Delta x}$$

where $\underline{x}_0$ are the remaining (unperturbed) parameters. Since $\Gamma(x)$ generally varies over the range of x and sometimes over $\underline{x}_0$, it is usually necessary to evaluate it over a range of parameter values.

A closely related quantity is called **elasticity**, which measures proportional sensitivity in relationship to proportional values of the input. Elasticity is computed as

$$e(x) = \frac{\partial F(x) / F(x)}{\partial x / x} = \frac{\partial F(x)}{\partial x} \frac{x}{F(x)} = \Gamma(x) \frac{x}{F(x)}$$

with again $\Gamma(x)$ is typically approximated as above. Elasticity has the advantage of being unitless, so that comparisons can be easily made of the relative sensitivities of parameters on vastly different scale.

We illustrate sensitivity analysis in Box 8.9 with the 2 species conservation problem previously considered in Box 8.2.

Simulation of optimal decision impacts

This chapter has emphasized the finding of optimal decisions for problems involving both deterministic and stochastic decision outcomes. Although simulation approaches can also find optimal solutions under some conditions, optimization methods will often be more efficient, and in some cases (e.g., dynamic

Box 8.9 Sensitivity analysis: conservation of 2 species

We return to the previous example for conservation of 2 species that are in partial competition (Box 8.2). Again, the objective function reflects potential for population sustainability and relative value of the 2 species, and the values (that we seek to optimize) are d_1 and d_2, the population levels for each species. The original objective was stated as maximize $F(\underline{d})$

$$F(\underline{d}) = 224d_1 + 84d_2 + d_1d_2 - 2d_1^2 - d_2^2$$

and we determined the maximum by solving the gradient

$$\frac{\partial F(\underline{d})}{\partial \underline{d}} = \begin{pmatrix} \partial F / \partial d_1 \\ \partial F / \partial d_1 \end{pmatrix} = \begin{pmatrix} 224 + d_2 - 4d_1 \\ 84 + d_1 - 2d_2 \end{pmatrix} = \underline{0}$$

to obtain

$$(d_1^*, d_2^*) = (76, 80)$$
$$F(\underline{d}^*) = 11872$$

Here, we consider sensitivity to values of the objective coefficients (perhaps because of uncertainty in what these exact values are). Thus, we will replace the constants in $F(d)$ with coefficient variables, which can be perturbed one at a time in a sensitivity analysis.

$$F(\underline{d}) = c_1d_1 + c_2d_2 + c_3d_1d_2 + c_4d_1^2 + c_5d_2^2$$

The gradient is now

$$\frac{\partial F(\underline{d})}{\partial \underline{d}} = \begin{pmatrix} \partial F / \partial d_1 \\ \partial F / \partial d_2 \end{pmatrix} = \begin{pmatrix} c_1 + c_3d_2 + 2c_4d_1 \\ c_2 + c_3d_1 + 2c_5d_2 \end{pmatrix} = \underline{0}$$

We now perturb the coefficients one at a time and produce numerical approximations to sensitivity and elasticity values for the resulting optimal decisions and objective values. For example, if we perform perturb c_1 from its initial value of 224 but keep the other coefficients fixed (at 84, 1, −2, and −1) we obtain approximate elasticity values of

$$e(d_1^*; c_1) = 0.84$$
$$e(d_2^*; c_1) = 0.4$$
$$e[F(d_1^*, d_2^*; c_1)] = 1.43$$

Perturbing c_2 produces

$$e(d_1^*; c_2) = 0.16$$
$$e(d_2^*; c_2) = 0.6$$
$$e[F(d_1^*, d_2^*; c_2)] = 0.57$$

We can proceed in like manner through the remaining coefficients and find that whereas the objective value is proportionally most sensitive to c_1 (1.43 exceeds in absolute value the other elasticities) the decisions appear to be most sensitive to c_4 and c_5 (−1.14 exceeds in absolute value the other elasticities)

$$e(d_1^*; c_3) = 0.43 \qquad e(d_1^*; c_4) = -1.14 \qquad e(d_1^*; c_5) = -0.3$$
$$e(d_2^*; c_3) = 0.686 \qquad e(d_2^*; c_4) = -0.54 \qquad e(d_2^*; c_5) = -1.14$$
$$e[F(d_1^*, d_2^*; c_3)] = 0.512, \quad e[F(d_1^*, d_2^*; c_4)] = -0.97, \quad e[F(d_1^*, d_2^*; c_5)] = -0.53$$

The R code to perform these calculations is provided in Appendix G.

decisions of stochastic systems) are uniquely capable of guaranteeing optimality. However, this will still be valuable, because once we have found the apparently optimal decision it will usually be of interest to evaluate its impact via simulation, particularly for dynamic or stochastic systems where the implications of the selected decisions may not be immediately obvious. Essentially all that is involved in simulating the decision is to take the optimal decision as a given, and then substitute it back into the dynamic equation that describes population response. In Box 8.10, we return to the stochastic dynamic harvest decision made earlier (Box 8.8) and apply the optimal state-specific decision to the population via simulation.

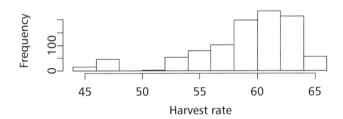

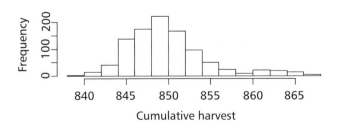

Simulating the optimal decision: Partial controllability of harvest.

Box 8.10   Simulating the optimal decision: Partial controllability of harvest

We return to the SDP harvest example from Box 8.8. In that example we found the optimal state-specific harvest rate decision under the dynamics and partial controllability distribution described earlier. That is, our optimal harvest decision is strictly a function of population size at any time t, $h(t)^* = f[N(t)]$. The SDP program produces the list of optimal decisions associated with each population level. We then start with an arbitrary initial population state (e.g., $N(0) = 10$) and simulate forward. The simulation involves the steps

- Taking the current value of $N(t)$ and obtaining an optimal harvest decision h^* as described above.
- Drawing a random controllability multiplier $m(t)$ from the discrete distribution (same as the one used to perform the SDP).
- Obtaining the realized harvest rate as $h(t) = h^* m(t)$.
- Simulating next year's population size by the dynamic equation

$$N(t+1) = N(t) + r_{max}N(t)[1 - N(t)/K] - h(t)N(t)$$

- Continuing the simulation for the specified number of years (e.g., 100).
- Returning to the initial conditions (starting N) and repeating the simulation the specified number of times.

We conducted such a simulation (1000 replicates for 100 years each) of the optimal harvest decision, keeping track of year-100 abundance and cumulative harvest (histograms below). The results provide a picture of what could be expected if the optimal decision were implemented, given the specified sources of uncertainty. For instance, it may be of concern that a non-trivial fraction (about 10%) of the outcomes in 1000 trials resulted in 100-year abundance of < 50; this could motivate reconsideration of the decision problem, for instance placing an objective penalty on such undesirable outcomes. Of course, such a penalty would constitute a constraint on the harvest objective, and we should expect to receive a lower average harvest as a result.

The R code for this example is presented in Appendix G.

Suboptimal decisions and "satisficing"

The preceding coverage of optimization has been premised on the idea that we are seeking an optimal solution to the decision problem, that is, one that maximizes or minimizes our objective function subject to our defined constraints. However, it can frequently be the case that several decision alternatives (i.e., combinations of choices for our decision controls) provide nearly the same objective result. It such cases it may not be deemed worth the trouble to select the optimal decision, and the stakeholders may be satisfied with an outcome that is

---

Box 8.11   Satisficing: sub-MSY management

We illustrate suboptimal decision making with a simple harvest example. Under the logistic model of growth, we know that the optimal population value for MSY is $N^* = K/2$, which results in a yield value of $H^* = r_{max}K/4$. For instance, if $K = 150$ and $r_{max} = 0.1$, MSY is achieved at $N^* = 75$, with resulting yield of 3.75.

Supposed we are content with 99% of MSY. We can examine the yield equation for values of N^* that fulfill this requirement. It will turn out that for the above values, levels of N between 68 and 82 will "satisfice". If we are content with 95% of MSY then levels of N between 59 and 91 will "satisfice". Of course, we may be unequally pleased with lower versus higher abundance values, perhaps due to population conservation concerns. This can lead to a more risk adverse harvest strategy than MSY as harvest is deliberately sacrificed in order to maintain higher population levels. This type of harvest strategy is sometimes referred to as a **right-shoulder strategy,** referring to the tendency to select harvest strategies to the right of the maximum on the yield curve.

Finally, we note that the labeling of this type of decision as "suboptimal" is a bit of a misnomer. What we have effectively done is to incorporate a population **utility** as part of the **objective function**. For example, we have already seen an alternative approach based on **utility multipliers** applied to a very similar problem in Chapter 3. See other examples for satisficing in Chapter 6, and discussion of alternative approaches based on **risk profiles** in Chapter 7.

The R code for this example is presented in Appendix G.

---

technically suboptimal. This notion relates to the idea of "**satisficing**" (Byron 1998) introduced in Chapter 6, which can be pursued in an informal way, as when decision makers compare the performance of competing but suboptimal decisions, or using more formal measures of satisfaction or acceptance (**utility**). We gave examples of more formal analysis of satisfaction in Chapter 7, when we introduced the idea of **risk profiles**. We provide a simple example of satisficing related to harvest management in Box 8.11.

Other problems

We have provided a wide array of approaches for obtaining optimal decisions, and hopefully readers will find some or even several methods that suit their particular needs. It will hopefully be clear that no single method will necessarily suffice, and we encourage readers to familiarize themselves with a broad range of optimization and simulation tools.

Some problems will prove especially challenging because of their size, complexity, or other features, and readers may find no one method that suffices. For example, many complex, dynamic systems involve a large number of **state variables** (variables that measure an attribute of a system, such as abundance or habitat condition, that may vary over time and space) and decisions, and therefore are not really suitable for solving by DP or SDP methods that would otherwise be preferred. Sometimes reasonable solutions can be found by simplifying such systems and applying methods such as DP or NLP, then simulating the results using more realistic dynamic models. In other cases it might be preferred to use methods such as simulation-optimization or heuristics (see Appendix E) that, while not guaranteed to be optimal, may provide "satisficing" solutions.

Finally, we recognize that there remain other, more fundamental and perhaps intractable obstacles to optimal decision making. We deal elsewhere with political and institutional obstacles. In some cases we are faced with systems that may be undergoing fundamental shifts, for which it is nearly impossible to forecast the impacts of decisions beyond short timeframes. A possible example of this involves societal (and natural resource managers') response to anthropogenic climate change (Conroy et al. 2011, McDonald-Madden et al. 2011). While such issues present profound challenges, we remain convinced that they are best dealt with via an objective-driven and science-based approach, as we have outlined here.

Summary

In this chapter, we covered the major approaches used to obtain optimal decisions. These range from simple approaches based on single or non-conflicting objectives and static, deterministic systems, to complicated decisions involving multiple attributes, tradeoffs, and dynamic, stochastic systems. Readers are encouraged to become at least basically familiar with the major approaches, and to apply them to problems in their specific resource areas. The concepts and tools here are of course fundamentally connected to those in other chapters, where we developed objectives, related them functionally to decisions, and considered the roles of uncertainty in decision making.

References

Broyden, C.G. (1970) The convergence of a class of double-rank minimization algorithms, *Journal of the Institute of Mathematics and Its Applications* **6**, 76–90.

Byron, M. (1998) Satisficing and optimality. *Ethics* **109**, 69–73.

Conroy, M.J., M.W. Miller, and J.E. Hines. (2002) Review of population factors and synthetic population model for American black ducks. *Wildlife Monographs.* 150.

Conroy, M.J., M.C. Runge, J.D. Nichols, K.W. Stodola, and R. J. Cooper. (2011) Conservation in the face of climate change: The roles of alternative models, monitoring, and adaptation in confronting and reducing uncertainty. *Biological Conservation.* **144**, 1204–1213.

Larkin, P.A. (1977) An epitaph for the concept of maximum sustained yield. *Trans. Am. Fish. Soc.* **106**, 1–11.

Lindley, D.V. (1985) *Making Decisions*. Wiley. New York.

Lubow, B.C. (1995) SDP: Generalized software for solving stochastic dynamic optimization problems. *Wildlife Society Bulletin* **23**, 738–742.

McDonald-Madden, E, M.C. Runge, H.P. Possingham, and T.G. Martin. (2011) Optimal timing for managed relocation of species faced with climate change. *Nature Climate Change* **1**, 261–265.

Williams, B.K., J.D. Nichols and M.J. Conroy. (2002) *Analysis and Management of Animal Populations*. Academic Press, London.

PART III. APPLICATIONS

9

Case Studies

We have participated in SDM and ARM projects that ranged from local (statewide scope) to international, and devote this chapter with examples of these as case studies. Technical aspects of these studies have been published in the journal articles and reports cited. We encourage you to review these for in-depth detail on the modeling and optimization approaches. Here, our objectives are to provide a behind-the-scenes perspective of the SDM process applied to real problems. The first of our case studies involves adaptive harvest management for American black ducks in North America. The second case study involves the management of water resources for human use and endangered aquatic fauna in the southeast US. The third deals with the setting of sport fishing regulations for the State of Georgia. All of these examples will detail the step-by-step process that cumulated in the actual management decisions and the monitoring follow-up where applicable.

Case study 1 Adaptive Harvest Management of American Black Ducks

Background

General background for adaptive harvest management on North American waterfowl

In North America (specifically in the United States and Canada), authority for harvest regulations for migratory game birds including waterfowl (ducks, geese, and swans, family *Anatidae*), coots and rails (*Rallidae*), doves (*Columbidae*),

Decision Making in Natural Resource Management: A Structured, Adaptive Approach,
First Edition. Michael J. Conroy and James T. Peterson.
© 2013 John Wiley & Sons, Ltd. Published 2013 by John Wiley & Sons, Ltd.

woodcock and snipe (*Scolopacidae*), and crows (*Corvidae*), resides with the Federal governments because of international treaties between the United States, Great Britain (acting for Canada in 1916), Mexico, Japan, and Russia. In the US, states select seasons, bag limits, and other options within the framework established at the Federal level, and coordinate with the Federal government by means of Flyway Councils that roughly parallel the major migratory corridors described by Lincoln (1935), from west to east the Pacific, Central, Mississippi, and Atlantic Flyways (Martin and Carney 1977, Rogers et al. 1979, Williams et al. 2002).

By the 1980s, many of the components of what we would now recognize as **adaptive harvest management (AHM;** ARM applied to harvest decision making) were in place. Thus, the decision-making process had both structure (objectives, decision alternatives, predictive models of influence) as well as means of providing adaptive feedback (sequential decision making, alternative structural hypotheses, and monitoring). However, if the process was adaptive, it was passively so, in that harvest decisions did not explicitly incorporate the value of learning in seeking to achieve objectives. Frustration with continued uncertainty about biological processes hampered efforts to achieve consensus on harvest strategies. In addition, political interference in the US regulatory process in 1994 threatened the integrity of the process, creating an opportunity to implement elements of AHM. In addition to the decision-making and monitoring elements noted earlier, existing optimization approaches (Bellman 1957, Anderson 1975, Lubow 1995, 1997) were now incorporated into the process, with structural uncertainty now represented via alternative predictive models (Anderson 1975, Johnson et al. 1993, Nichols et al. 2006, Johnson 2010). The initial focus of AHM was the mid-continent population of Mallards (Pospahala et al. 1974), but more recently efforts have been underway to implement AHM for other stocks of Mallards, other dabbling ducks, and geese. Because of the first author's extensive involvement with American black ducks throughout his career, we focus on the incipient development of AHM for this species (henceforth, "black ducks").

Historical background for the black duck issue

Black ducks are principally confined to eastern North America, and historically have occupied breeding habitats in the boreal forest region at low densities. The combination of habitat affinity and low densities creates challenges for surveys of black ducks (in contrast to prairie breeding species such as Mallards). Thus, historically much reliance was placed on winter surveys of black ducks, particularly in coastal habitats of the Atlantic Flyway where birds occur at high concentrations and are easier to count (Conroy et al. 1988). Declines in counts of black ducks on the mid-winter waterfowl surveys suggested that populations were both declining in overall abundance and contracting in range, with general loss of black ducks from much of the western range. Reasons for the decline were vigorously debated, with some suggestions of excessive harvest as the culprit, but strong competing arguments implicating completion and hybridization with Mallards or habitat loss as more important factors (Rusch et al. 1989).

Calls for a moratorium on black duck harvest begin appearing as early as the 1960s, and reached a fever pitch by the 1980s, culminating in lawsuits against the U.S. Fish and Wildlife Service under MBTA and the National Environmental Policy Act of 1969 (NEPA; Feierabend 1984). Although the judge eventually ruled against the plaintiffs in this and a subsequent lawsuit, the legal attention and related discussion among managers resulted in redoubled efforts at research and monitoring directed at elucidating factors responsible for limiting black duck populations. In addition to observation evidence, suggestions were made for addressing the issue of harvest impacts via experimental manipulation of regulations (e.g., Anderson et al. 1987). Conroy et al. (2002) summarized the technical issues that surround black duck population dynamics, developed and parameterized a synthetic model capturing key population processes, and proposed elements of an AHM process for black ducks. In the meantime, a consensus was growing toward use of AHM in principle to manage black duck populations, culminating in the formation of a Black Duck AHM working group, described below in more detail, which met for the first time in Athens, Georgia in 2000. Technical details of this approach can be found in Conroy et al. (2002), Conroy (2004, 2010); here we emphasize how the SDM process developed to support harvest management decisions for black ducks.

Decision problem

Superficially, the decision problem for black ducks is the same as that for Mallards or other migratory waterfowl. At the Federal level in the US this is translates in the selection of annual Flyway frameworks for season lengths and daily bag limits. However, a notable difference between efforts to implement AHM for black ducks, and those directed at mid-continent Mallards, is that AHM for Mallards was (and remains) dominated by the US. Most of the harvest of Mallards occurs in the US, and while Canada is included as a participant in the process (e.g., Canadian Federal and Provincial representatives participate in Flyway Council meetings), neither the Federal nor provincial governments have yet based harvest regulations on an AHM process. By contrast, most of the black duck breeding range is in Canada; furthermore, historically about one-half of the continental harvest of black ducks occurs in Canada, albeit much of this by US residents. Thus, an AHM process for black ducks must include both countries, and this issue becomes highly relevant at several steps of the process, as discussed further below. This need is complicated, in turn, by several unavoidable issues. First, in contrast to the majority view in the US, there has been no general agreement in Canada that harvest is the key or even a dominating factor in limiting black duck populations. Many populations of black ducks in Canada are stable or increasing, particularly in the eastern core of the range (eastern and northern Quebec and the Maritimes). Secondly, hunter numbers in Canada are relatively low compared to the US and participation in hunting is declining. From a Canadian perspective, concern relates as much to the issue of numerical parity in harvest (or the appearance thereof), as it does in absolute numbers of harvested ducks, an issue that we will return to later. Thirdly, in contrast to harvest regulations in the US, which are set annually based on current survey information, harvest regulations in Canada are set the previous year or earlier. Although there are provisions for emergency changes to regulations, the asynchrony in regulation setting creates challenges for an adaptive approach. Fourth, the Provincial–Federal relationship in Canada is notably different from that in the US; this and the complex relationship of Quebec to the rest of Canada further complicates regulation setting in that nation. Finally, to date most of the technical expertise in AHM in North America has come from the US side, which can result in a certain amount of defensiveness and resistance among Canadians. Each of the above issues must be addressed, for black duck AHM to succeed on a meaningful scale, since continental AHM for black ducks cannot occur without the partnership of Canada.

Stakeholders, decision makers, and development of an SDM/ARM approach to black duck harvest management

The stakeholders for harvest management of black ducks include sport hunters and the commercial interests that serve them, but also birdwatchers and other non-consumptive users of the resources in both countries. More broadly, because these are trust resources under Federal laws and international treaties, all citizens of both nations have a fundamental stake in the outcome of harvest deliberations. The public's interest is represented by proxy by the agencies involved in the regulatory process, with oversight via the public comment period for proposed regulations, and, occasionally, by litigation (notably the lawsuits mentioned earlier).

We have already noted the formation of a Black Duck AHM Working Group (BDAHMWG) around 2000. Early on it was recognized that there was distinction that needed to be made as to the various roles of the participants of the group that lead to the formation of the BDAHMWG. This group included (and includes) Federal, State, and Provincial managers who were responsible directly to policy makers for making decisions about harvest regulations under the MBTA frameworks. These individuals were voting members of the Working Group.

Membership is fluid depending on current job assignments of the staff involved, with for example the Chiefs of Migratory Bird Management frequently delegating authority to subordinate staff. In addition to the voting members, a number of additional members serve in an ad hoc or advisory, but non-voting role. These included technical staff of the USFWS, CWS, and States, US Geological Survey (USGS) scientists, university scientists, and private consultants. In addition to the policy and technical roles, a very important role is that of meeting facilitation. Ideally this is someone with training in SDM and meeting facilitation. At various times, the first author has served in this role, along with a role as a USGS / University scientist on technical issues. At other times, a private consultant with expertise in SDM and meeting facilitation has served under contract in this capacity.

The public has not been routinely invited to BDAHMWG meetings; however, the group regularly reports before meetings of the Atlantic or Mississippi Flyway

Councils, which include attendees from sportsmen's groups, conservation NGOs, and other stakeholders.

Management objectives and decision alternatives

At the very earliest meetings of the BDAHMWG, it became obvious that managers in both the US and Canada had objectives that were more complex than simply maximizing sustainable harvest (or harvest opportunity) for black ducks. Managers have the general perceptions that black duck populations are currently well below desired (or recent historical) levels. Numerical, albeit arbitrary, goals have been set for continental black duck populations (386 000 on the Mid-Winter Waterfowl Surveys) and recently re-expressed in terms of target number on the breeding grounds. Thus, it was deemed desirable to maintain populations above these levels, even if it involved sacrificing some harvest opportunity. Likewise, simply optimizing harvest for maximum sustainability can result in spatially unequal allocation of harvest, while still achieving an overall numerical optimum. For socio-political reasons such outcomes would be deemed less desirable than outcomes that achieved relative parity (roughly defined as $50 \pm 10\%$ distribution of the harvest in either country), even at the cost of some loss to overall harvest opportunity.

Although implicit in the BDAHM deliberations, a formal objective network (Chapter 3) was not developed. However, we constructed an objective network using MJC's understanding of the objectives, as captured during his tenure with the BDAHM Working Group during 2000–2010 (Figure 9.1). We derived a quantitative objective function that incorporated harvest, population, and parity fundamental objectives, by modification of the objective function of maximum long-term harvest. First, it was deemed desirable to maintain populations at or above the North American Waterfowl Management Plan (U.S. Dept. of the Interior 1994) goal of approximately 376K wintering black ducks. After group discussion this goal was converted to a breeding population (BPOP) equivalent of 830K. Using this goal, we constructed a functional relationship between BPOP and harvest utility (a multiplier ≤1 that discounts the harvest objective when the population goal is not met (Conroy 2010; Figure 9.2), resulting in our optimizing algorithm (described below) tending to avoid strategies that violate the population goal but otherwise would satisfy the MSY objective. As an alternative to the above penalizing function, we also considered modifying the objective to deliberately strive for an equilibrium abundance greater than that which would be achieved under an MSY strategy; for deterministic, logistic growth this produces what can be described as a "right-shoulder strategy" (Anderson et al. 2007), graphically described in Figure 9.3. In practice, we found little difference between these approaches and thus here emphasize the first.

Because harvest rate decisions are considered independently in the Canada and the US, it is in theory possible to achieve harvest and population goals with harvest allocated very differently between the 2 countries. For instance, a harvest strategy could fulfill population-discounted harvest yield yet do so by allocating 100% of the harvest in 1 country. To avoid such outcomes, however unlikely, we constructed a parity utility function similar to the population constraint,

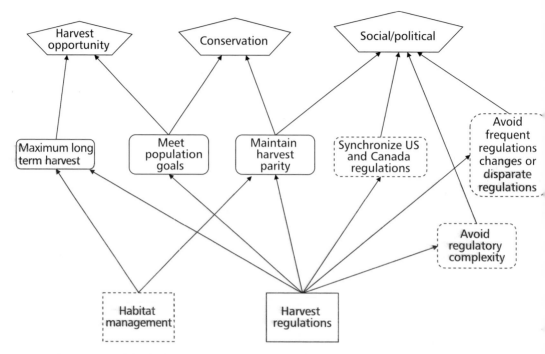

Figure 9.1 Objectives network for black duck adaptive harvest management with fundamental objectives (pentagons), means objectives (rounded rectangles), and decision alternatives (rectangles). Objectives and decisions with dashed borders not currently part of AHM framework for black duck adaptive harvest management.

imposing a penalty if one nation's proportion of the harvest is predicted to fall outside of an acceptable range, currently considered to be 0.4–0.6 of the continental harvest (Conroy 2010; Figure 9.4). Finally, harvest, population, and parity goals were combined by multiplying the utility components for each. This process provided a quantitative objective function that could be used in subsequent optimization, but does not fully resolve other important management objectives.

Although the decision alternatives have been stated as the selection of season lengths and bag limits, to date black duck AHM has proceeded under the assumption that regulatory packages would translate exactly into specified harvest rates for Canada and for the US, in the latter case the harvest rates being defined as harvest as a proportion of the population remaining after the take of harvest in Canada. Obviously, this ignores the issue of **partial controllability** (Chapter 7); we will address this issue further at the end of the section. Additionally, as previously noted, that harvest regulations in Canada and the US are set according to different schedules, with US regulations set "just in time" (after current BPOP survey and other data are available and just before hunting seasons), whereas Canadian regulations are set well in advance (usually the previous year, or earlier). This effectively formed a constraint on the decision space,

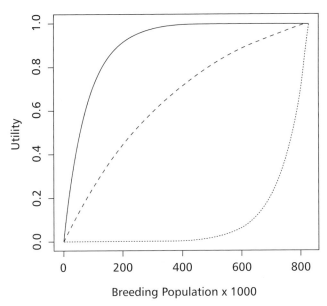

Figure 9.2 Penalty functions used to incorporate population goals into objective function. Penalty applied when projected spring abundance falls below desired level $N^* = 830\,000$. Solid line penalty applied gradually as $N < N^*$, dashed line penalty applied in nearly a linear fashion, steep penalty applied immediately below N^*. See text for details.

whereby potentially US regulations would change in a given year but Canadian ones would not. We ignored this issue and instead assumed that harvest rates in each country can be perfectly controlled (see below) but this discrepancy needs to be resolved for future implementation of international AHM. A second issue related to the possibility, deemed politically undesirable, that regulations could move in greatly different directions internationally, with for example very restrictive regulations in the US and very liberal ones in Canada. Fortunately, such instances rarely or never occur when a parity constraint is in place, but could be eliminated altogether by forcing linkage between the US and Canadian regulations, e.g., by dictating proportionality between regulator levels. Finally, it was desirable (for sociopolitical reasons) to avoid, if possible, situations where hunting seasons are closed, or where regulations shift dramatically between years. Such situations can be eliminated by imposing further constraints on the decision space, but recognizing that these (like all constraints) will reduce the long-term harvest return that otherwise would be achieved.

Model development

General approaches to modeling and estimation were similar to that outlined in Conroy et al. (2002), but with major differences reflecting changes in survey methodology, and application of Bayesian approaches (Conroy 2010). The

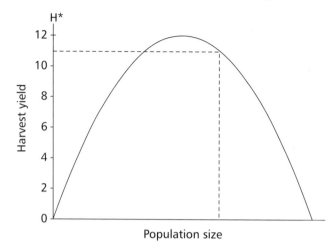

Figure 9.3   Right-shoulder strategy approach for incorporating population objective into harvest objective for black duck adaptive harvest management. See text for details.

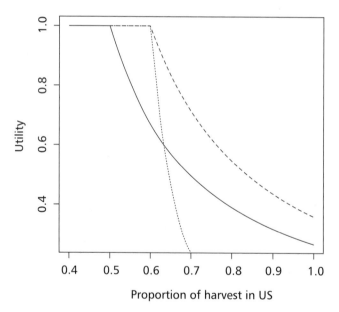

Figure 9.4   Penalty functions used to incorporate parity goals into objective function for black duck adaptive harvest management. Utility penalty applies when one country's proportion of total harvest p exceeds a desired upper proportion p_0. Solid and dashed line, moderate penalty (shallowly decreasing utility) applied to $p_0 = 0.5$ and $p_0 = 0.6$, dotted line severe penalty applied to $p_0 = 0.6$. Illustrated for harvest in US. See text for details.

system state is annual abundance of black ducks and Mallards based on spring aerial surveys. Two empirically based submodels were included: a recruitment model that predicts a fall abundance of juveniles as a function of black duck (density dependence) and Mallard abundance (competition), and a winter survival is used to predict post-harvest survival of black ducks, incorporating density-dependent (compensatory) responses if any. Winter survivors were assumed to become part of the next year's breeding population. Model parameters are currently estimates using a state-space approach (Conroy 2010), informed by data on from the annual breeding population surveys, band recovery data, and harvest survey data.

Incorporation of uncertainty

The parameters of the model were derived under specific assumptions about density and Mallard influences on recruitment (under 2 alternative models: presence or absence of Mallard competition; Figure 9.5) and harvest influence on survival (under additive or compensatory assumptions), resulting in four alternative structural models. Uncertainty was represented by AIC weight under these alternative models (Conroy et al. 2002, Conroy 2010).

As an alternative to the "alternative models" approach above, we re-expressed the model uncertainty in terms of parametric uncertainty in two key parameters: c_2, representing the influence of Mallard completion on reproduction rates, and a_1, representing the additive effect of harvest mortality on annual survival. First, we estimated these and the other model parameters under the global model where both of these parameters were present (competition + additivity model). We then represented parametric uncertainty via the discretized posterior distributions for each of these parameters, taking as cut-points the 0.025, 0.25, 0.5, 0.75, and 0.975 quantiles.

Evaluation of decision alternatives

We used the predictions under each model (previous section) weighted by AIC model weights, or the parametric uncertainty distributions describe above, in conjunction with stochastic dynamic programming (Lubow 1995, 1997; Chapter 8) to obtain optimal, state-specific harvest strategies. For each weighted or model-specific prediction, optimization was performed taking into account environmental stochasticity (the discrete random effect distributions for black ducks and Mallards described above) and for each of the 9 scenarios incorporating combinations of population and parity constraints; 9 additional scenarios were also considered that forbade closed seasons, restricting harvest rates to the range 0.05–0.3 (Table 9.1). Once optimal strategies were obtained for the model-averaged and model-specific predictions, the strategies were simulated for 100 replications using the discrete environmental stochasticity distributions to obtain trajectories of abundance, harvest rate, parity (proportion of harvest in Canada), and total harvest per year over 100 years. Optimal stationary harvest strategies for combination of black duck and Mallard abundance indicated broad zones of state combinations where similar regulations would presumably apply in order to achieve the target harvest rates (Figure 9.6). However, the analysis revealed a number of cases where large jumps in regulations would be associated with small differences in system state (number of black ducks and Mallards). This is of concern given known errors and biases in the estimation of these states from survey data, and may be cause for further constraining the degree or frequency at which regulations are permitted to change. Nevertheless, the algorithm resulted in strategies that, when simulated, seemed likely to achieve the fundamental objectives of population conservation (albeit at levels somewhat below the target of 830 000 in the breeding population), harvest, and parity (Figure 9.7).

Sensitivity analysis

We conducted two types of sensitivity analysis on the decision model. First, the basic model structure and parameterization follows Conroy et al. (2002), who conducted extensive sensitivity analysis by use of derivatives (Chapter 8, Appendix E). These analyses were focused on the relationships among the model parameters, in particular the parameters controlling the effects of Mallard competition and harvest, whose sensitivities varied in response to Mallard abundance and harvest rates, respectively (Conroy et al. 2002). We updated these analysis using the state-space approach described in Conroy (2010) and used the median values for all parameters except c_2 and a_1 describing, respectively, the relationship between Mallard abundance and black duck recruitment, and black duck abundance and non-harvest mortality in the state-space model to simulate population growth. Discretized posterior distributions (quantiles) were used for the latter 2 parameters under a "no harvest" scenario to assess the general behavior of the model for these 2 state variables. Parameter-averaged projections for black duck populations rapidly (<10 years) equilibrated at approximately 1400K (Figure 9.5). Once we obtained optimal harvest strategies (below) we simulated these, with particular attention to the sensitivity of the strategy and the simulated outcomes to the assumed model structure, as well as to the specification of the objectives and any constraints on decision alternatives (e.g., forbidding closed seasons).

Current status of black duck AHM

In the analyses to date, several issues have been glossed over that will eventually need to be dealt with. As noted earlier, we have assumed complete controllability of harvest rates, so that harvest rate is considered to be the "decision." However, it is well known that harvest is only partially controllable via regulations (Williams et al. 2002), and thus it will be necessary to incorporate the statistical relationship between regulation and realized harvest rates into optimization (e.g., Conroy et al. 2005). Secondly, the current model and optimization approach is based on a single stock of black ducks, and ignores important differences in habitat and population status. MJC has done some work on a multi-stock modeling and optimization approach for black ducks (see Conroy 2004), but this

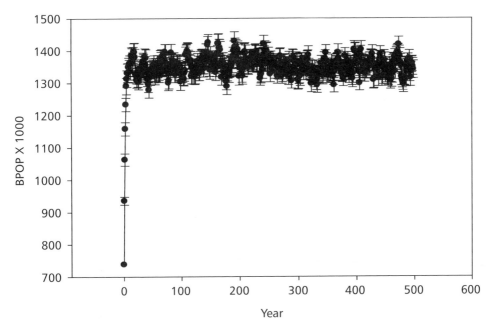

Figure 9.5 Simulated 100-year black duck population trajectories in the absence of harvest averaged over parameter distributions for Mallard competition (c_2) and harvest additivity (a_1) coefficients.

Table 9.1 Scenarios considered in stochastic dynamic optimization using ASDP.

	Harvest rate		Population		Parity	
Scenario	Canada	US	Goal	Slope	Goal	Slope
n000[1]	0	0	–	–	–	–
n001	0–0.3, 0.05	0–0.3, 0.05	830	2	0.6	−2
n002	0–0.3, 0.05	0–0.3, 0.05	830	10	0.6	−2
n003	0–0.3, 0.05	0–0.3, 0.05	830	−10	0.6	−2
n004	0–0.3, 0.05	0–0.3, 0.05	830	2	0.6	−10
n005	0–0.3, 0.05	0–0.3, 0.05	830	10	0.6	−10
n006	0–0.3, 0.05	0–0.3, 0.05	830	−10	0.6	−10
n007	0–0.3, 0.05	0–0.3, 0.05	830	2	0.5	−2
n008	0–0.3, 0.05	0–0.3, 0.05	830	10	0.5	−2
n009	0–0.3, 0.05	0–0.3, 0.05	830	−10	0.5	−2
r001-r009	0.05–0.3, 0.05	0.05–0.3, 0.05	(see above)	(see above)	(see above)	(see above)

[1]No harvest scenario, used to simulate model behavior.

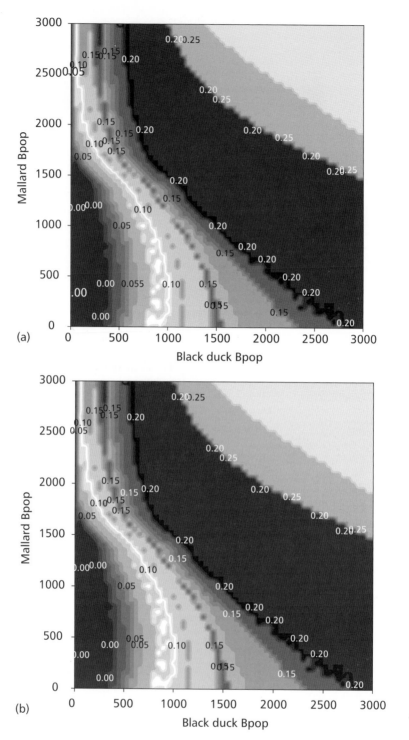

Figure. 9.6 Optimal black duck harvest strategies for (a) Canada and (b) US under parametric uncertainty (averaged over parameter distributions for Mallard competition (c_2) and harvest additivity (a_1) coefficients). Updated BPOP estimates and model corrections (see text). Horizontal axis is BPOP ×1000 for black ducks, vertical axis is BPOP ×1000 for Mallards. Entries are contours of harvest rates of adult males in increments of 0.05. Scenario n002 (Table 9.1).

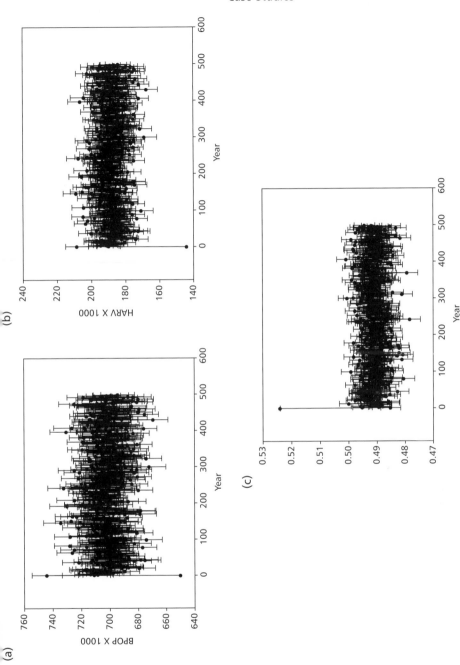

Figure 9.7 Simulation of optimal black duck harvest strategies for Scenario 2 (Table 9.1, Figure 9.6) under parametric uncertainty averaged over parameter distributions for Mallard competition (c_2) and harvest additivity (a_1) coefficients. (a) Annual breeding populations of black ducks, (b) Annual harvest of black ducks, (c) Proportion of annual harvest in Canada.

approach is fundamentally limited by data. In addition, multi-stock optimization can be computationally burdensome or even impracticable, and the advantages with respect to management are not necessarily large (Johnson 2010). Thirdly, important challenges remain for a coordinated harvest management strategy because of differences in the regulatory approaches in the US and Canada, and these challenges must be overcome for black duck AHM to succeed. Finally, many other species of waterfowl lack an adaptive harvest strategy, and AHM for waterfowl has so far only been indirectly linked to habitat management. Some progress is being made in both of these areas, but much more work is needed.

As of this writing, the black duck AHM process had been vetted at annual working group meetings and Flyway meetings for over a decade. The latter are attended by sportsman and other advocacy groups as well as members of the general public and the press. For the immediate future, AHM for black ducks is currently being used to guide the harvest regulations process for the US, but is not the sole basis for decision making. To date, Canada has not agreed to use AHM for black duck harvest management, but has agreed to continue coordination of surveys and research efforts with the option of eventual AHM implementation.

Case study 2 Management of Water Resources in the Southeastern US

Background

Human population growth and development are among the most important issues facing natural resource managers and planners throughout the developed world. This is evident in the southeastern US where populations grew 37% over the last 20 years. The rapidly growing human population has led to increased land conversion from forested to agriculture and urban uses and increased water demands from agriculture, industry, and municipalities. These changes have had, presumably, unintended negative impacts on aquatic resources. Stream ecosystems in the southeastern US contain the richest aquatic fauna in North America, north of Mexico, and among the highest proportion of imperiled aquatic species (Warren et al. 2000; Williams et al. 2008). Approximately 28% of the known native fish species in the Southeastern US are extinct, endangered, or threatened (Warren et al. 2000). Similarly, the streams in the Southeast are home to approximately 95% of the imperiled freshwater mussel species in North America (Strayer 2008; Williams et al. 2008). The decline of these fauna has been attributed to several factors including habitat alteration resulting from intensive agriculture and urban development, river regulation and impoundment, and water use (Richter et al. 1997; Warren et al. 2000; Williams et al. 2008). Thus, natural resource managers in the southeast US are faced with difficult decisions on how to satisfy the socioeconomic needs of a growing human population, while maintaining or restoring properly functioning aquatic systems.

Traditionally, water-resource management in the southeastern US (and other parts of the developed world) focus on the use of streamflow standards for protecting aquatic biota (Annear et al. 2004). Many of these traditional methods emphasize the protection or maintenance minimum flows needed to support the survival of aquatic organisms. In recent years, these traditional approaches have increasingly come under criticism due to the recognition that multiple components of the hydrologic regime (e.g., high *and* low flows) affect physical and chemical processes that, in turn, affect the ecological dynamics of stream-ecosystems (Tharme 2003). In addition, traditional approaches don't allow managers to incorporate multiple objectives directly in water management decisions and none explicitly incorporate the effects of other human (e.g., land) and natural (e.g., climate change) influences on management objectives.

In 2005, the USGS sponsored a workshop, "Linking hydrological change and ecological response in streams and rivers of the eastern United States," that included more than 150 scientists and managers from state and Federal agencies. The participants identified the need to develop a science-based approach to water-resource decision making that allowed managers to predict how stream ecosystems responded to alternative flow-management decisions. Motivated by the needs identified during the workshop, the USGS initiated a national priority science thrust initiative, "Water availability for ecological needs" The goal of the project was to develop a system for evaluating the effect of water management decisions on management objectives in an explicit landscape context. Technical details of the project can be found in Freeman et al. (2012). In what follows, we describe how the SDM process was used to develop the decision support system for managing water resources.

Decision problem

The Flint River basin, located in southwest Georgia was selected as the focal management area to develop a decision model framework for managing water resources in the southeast US. The Basin contains five federal-listed aquatic species and seven additional species that are considered at risk by Georgia Department of Natural Resources (GADNR). The upper portion of the Basin is experiencing substantial urban development and increased water use due to the expansion of the Atlanta metropolitan region. The lower portion of the Basin is one of the most highly productive agricultural regions in the Southeast with a substantial portion of the agricultural lands, 563 million acres, irrigated from aquifers and surface waters. The water use within the Basin has affected stream-flows over the last few decades (Rugel et al. 2012) and there was evidence that the decreased flows negatively affected the aquatic fauna (McCargo and Peterson 2010; Peterson et al. 2011). The mainstream Flint River was also a valued recreation resource, providing numerous boating and fishing opportunities and spectacular views to residents and visitors. Thus, water-resource managers in the Flint Basin needed to conserve and recover sensitive aquatic species, while meeting the socioeconomic needs of the public.

Stakeholders and decision makers

Water-management decisions had the potential to affect multiple stakeholders throughout the Flint River basin. A small group consisting of scientists and biologists from the USGS, USFWS, and GADNR Wildlife Resources Division (WRD) was established to identify the potential stakeholders in the Flint River basin. These stakeholders included three State agencies, the GADNR WRD and Environmental Protection Division (EPD) and the Georgia Farm Bureau; two non-governmental agencies: the Nature Conservancy and the Instream Flow Council; and two Federal agencies, the USFWS and the US Department of Agriculture (USDA). Of these, three stakeholders also were decision makers. The EPD was responsible for permitting water withdrawal in the Basin and the WRD for regulating the use of wildlife in Georgia. The USFWS had the authority to restrict or modify proposed actions that potentially affected Federally listed species.

Representatives of each stakeholder agency were contacted and invited to participate in the SDM process. Of these, representatives of the Georgia Farm Bureau and USDA did not participate for unstated reasons. Representatives of the EPD participated in the initial stakeholder meetings (discussed below) but did not appear at subsequent meetings although they were invited to all meetings. Although the reasons were unknown, the withdrawal of EPD from the process significantly influenced the outcome of the process, which we discuss later.

During the first two meetings (discussed below), the stakeholders created and modified a prototype decision model. It was clear, however, that technical experts were needed to provide advice on the feasibility of alternative modeling approaches and to build and parameterize the decision model. Thus, the stakeholders identified and recruited members of a technical advisory team that included two to three members each from hydrology, geomorphology, aquatic ecology, and landscape ecology disciplines. We note here that during the initial stages of the process, a few of the original technical advisors left or were asked to leave the team for a variety of reasons. Fortunately, the project had an excellent manager who eventually identified and recruited technical advisors that were not only skilled in their respective disciplines, but also were good team players. The final group of technical advisors was in place by the fourth meeting and these scientists worked closely with the stakeholders until the structure of the final decision model was completed.

The replacement of personnel in this study was by no means unique to this case study. Rather, we highlight this process to illustrate a few of important points. The first is that success of a study can be significantly affected by the make-up of the working group. Good team members need to have technical and people skills, so recruiting the right people should be a high priority. Secondly, the right people may not be part of the team from the start, so decision makers should not be afraid to find replacements. This should, however, be done relatively early to minimize disruption in the process. The third point is that a champion is often crucial to the successful completion and implementation of SDM and ARM. In this case study, the project manager was largely responsible for ensuring that the right people were available and that the project had the resources needed to complete the SDM process. The manager's role was so critical that it is doubtful whether the project could have been completed without him.

Management objectives and decision alternatives

Stakeholder representatives, primarily managers and administrators with the respective organizations, met in a series of two- to three-day workshops to identify the problem, objectives, and decision alternatives. These workshops were not open to the general public and were facilitated by the second author. During the first workshop, the stakeholders identified the problem statement as managing the water resources in the Flint Basin to maximize the satisfaction of citizens. The fundamental objective, derived from this problem statement, was to maximize citizen satisfaction. The stakeholders then identified a set of fundamental objective attributes based on user-groups: non-consumptive users and consumptive users. Six attributes were selected for consideration in the decision: the number of boatable days of the river, the aesthetic value of the river, water quality, water availability of human use, and the status or health of the native aquatic biota with an emphasis on sportfish (Figure 9.8). The stakeholder group also identified the initial management decision alternatives based on agency authority and mandates. These three alternatives included: reservoir siting and construction, water withdrawal and use, and the implementation of land-use regulations.

The identification of objectives and management alternatives is generally an iterative process in SDM, so after a series of meetings the Flint Basin stakeholders removed, added, and revised the initial fundamental objective attributes. For instance, maximizing sportfish populations was combined with all of the native biota into a single objective: maximize the number of native aquatic species.

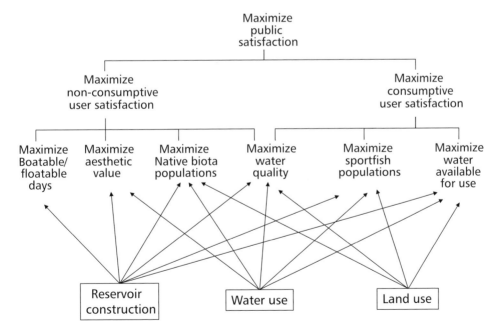

Figure 9.8 Fundamental objectives hierarchy with decision alternatives (rectangles) for water-resource management decisions in the Flint River basin, Georgia.

Boating and aesthetic value attributes were similarly combined into a single objective maximize number of recreational opportunities. Such modifications are common throughout the process as stakeholders encouraged to think hard about potential tradeoffs and the things that they value most. The final fundamental objectives determined by the stakeholders were to maximize the number of native fish and freshwater mussel species and maximize the water available for human use. The final decision alternatives included water use (permitting) and reservoir construction. The stakeholders did not agree on a method for combining the objectives into a single utility and (presumably) preferred to examine the expected values of each in response to management alternatives.

Model development

The decision model went through a series of revisions (approximately 6-8 revisions) starting with the large and unwieldy prototype model created by the stakeholders (Figure 9.9) and ending with the final model parameterized by the technical advisors (Figure 9.10). The final decision model is detailed in Freeman et al. (2012) and is summarized here. The model tracked the presence of fish and mussel species in individual stream segments, which were defined as streams sections bounded by tributary junctions. The model operated on an annual time step and tracked the colonization, extinction, and reproduction of each species in individual segments in response to seasonal streamflow components, water quality, stream channel characteristics, and stream isolation (Figure 9.10). Colonization, extinction, and reproduction transition probabilities were estimated using empirical meta-demographic models parameterized with existing data. Seasonal streamflow components were estimated using a spatially distributed, deterministic, physically based hydrologic model calibrated using existing stream gage data. Stream channel characteristics were estimated using existing 30-m resolution digital elevation models and land-cover data.

Incorporation of uncertainty

The Flint River basin water resource decision model included multiple uncertainties. Environmental uncertainty in climate statistics was incorporated using statistical distributions to represent uncertainty in daily precipitation and temperature. Similarly, statistical uncertainty was incorporated using statistical distributions to represent the prediction error in the stream flow estimates, stream channel geomorphology, and meta-demographic model parameter estimates. Structural uncertainty was incorporated using 8 alternative models to estimate metapopulation dynamics. The weights of the alternative models were based on the Akaike weights estimated during model selection of the empirical meta-demographic models.

Evaluation of decision alternatives

The stakeholders preferred to examine the expected values of the two objectives in response to management alternatives, which necessitated the use of simulation

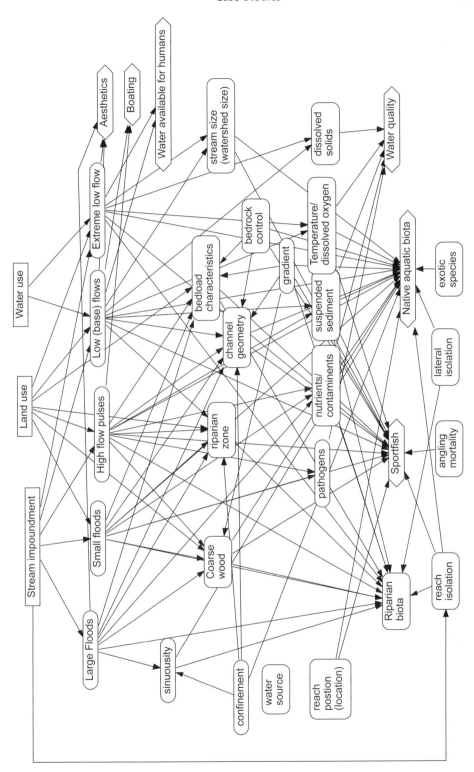

Figure 9.9 Prototype influence diagram for water-resource management decision making in the Flint River basin, Georgia.

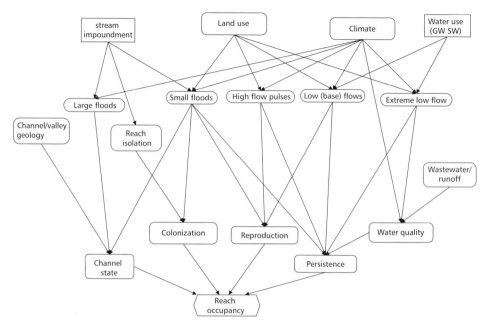

Figure 9.10 Final influence diagram for water-resource management decision making in the Flint River basin, Georgia.

for estimating the effects of decision alternatives. Here, we illustrate the use of the decision model to evaluate a water withdrawal decision on Potato Creek, a midsized tributary of the Flint River described in Freeman et al. (2012). The model estimated the number of species occurring in the stream segments immediately downstream from a water-withdrawal facility in response to four decision alternatives: water use 0.5, 1, 2, and 4 million gallons per day (MGD). For each decision alternative, 500 simulation replicates were run using each alternative model of metapopulation dynamics. Environmental and statistical uncertainties were incorporated by randomly selecting values from statistical distributions. For each simulation run, species richness was calculated as the sum of the number of species present for each year. Model-averaged estimates of species richness were calculated by averaging the model specific estimate but the associated model weight.

The results of the simulations indicated decreases in the number of species with increasing water use (Figure 9.11a). Species losses are expected under water-use alternatives greater than one MGD. Under the one MGD alternative, 2.8% or a little more than 1 species is expected to be lost, whereas the losses are expected to be much greater for the four MGD alternative. The decision makers could consider any one of these decreases in species as acceptable depending on how they weight the two predicted outcomes. Let's look at how the uncertainty could influence the water use decision. In Figure 9.11b, the expected change in the total number of species differed between the alternative

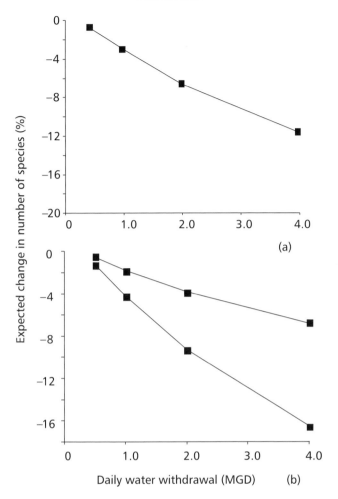

Figure 9.11 The expected decrease in the number of species from current levels for four alternative levels of water use for the model averaged estimates shown (top figure) and for two alternative models of extinction probability: the median summer flow (solid line) and 10-day low flow model (broken line; bottom figure).

metapopulation dynamic models. This structural uncertainty would likely have a substantial effect on water-resource management decision making.

Sensitivity analysis

The decision model was examined using one- and two-way sensitivity analysis. These indicated that the model was most sensitive to assumptions about the metapopulation dynamics (Figure 9.12). However, we were surprised to learn that error in the flow model and stream channel classification had such a strong influence on the expected value of the species richness. The findings from the

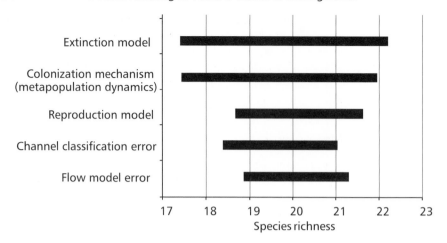

Figure 9.12 Reduced tornado diagram from one-way sensitivity analysis of estimated species richness objective for Flint River basin water-resource management decision model. Diagram only includes five most influential model components.

sensitivity analysis prompted members of the technical advisory team to pursue an additional study to improve the understanding of fish and mussel metapopulation dynamics, hydrology, and stream channel geomorphology. To date, these studies have resulted in significant improvements to the hydrologic models (Viger et al. 2010). The monitoring of fish and mussel communities and evaluation of stream channel geomorphology and evolution are ongoing.

Current status of water resource management

As noted earlier, the GA EPD were the decision makers responsible for water permitting in the state of GA. Representatives of the EPD participated in the initial stakeholder meetings but left before the process was completed. During the SDM process, the state of Georgia also lost a legal battle with Alabama and Florida over water use in the Apalachicola–Chattahoochee–Flint River basin. These events have greatly complicated the decision-making process at a time when the remaining stakeholders are ready to examine alternative water development strategies to meet the needs of humans and conserving and recovering the listed species in the Flint Basin. The SDM process will remain on hold until the negotiations among the various entities are completed. In the meantime, the decision model is currently being used to identify and prioritize aquatic conservation areas in the basin.

Case study 3 Regulation of Largemouth Bass Sport Fishery in Georgia

The two previous case studies involved multiple decision makers, technical advisors, and management actions that occurred over relative broad spatial extents (i.e., they were relatively complicated decision situations). Structured decision making and ARM, however, can be applied to any decision situation regardless of complexity and in fact may be more likely to succeed when used to tackle conventional decisions. Here, we describe how SDM was used to determine the best management action for a popular sport fishery in the southeastern US. Modeling details can be found in Peterson and Evans (2003).

Background and decision problem

West Point Reservoir is located on the Georgia and Alabama border and was once one of the most productive largemouth bass (*Micropterus salmoides*) fisheries in North America. The reservoir was formed by impounding the Chatta-hoochee River downstream of Atlanta. The high productivity of the reservoir was largely due to high nutrient levels from wastewater effluent originating from the Atlanta metropolitan area. The largemouth bass fishery in West Point Reservoir was relatively popular and attracted anglers from throughout North America. Local marinas also hosted numerous bass fishing tournaments year round. Thus, the largemouth bass fishery had a substantial influence on the economy of the surrounding rural community.

Increased water quality concerns in the late 1980s resulted in modifications to the Clean Water Act that required the reduction of nutrient inputs from municipal wastewater facilities. Consequently, total phosphorus concentrations in West Point Reservoir decreased greatly from the late 1980s to the present. The decreased nutrients were also believed to cause a substantial decline in the largemouth bass population (Maceina and Bayne 2001) and concurrent decrease in largemouth bass catches at West Point Reservoir. In response, largemouth bass tournament and recreational angling effort at West Point Reservoir (i.e., the number of visits and tournaments) declined precipitously, negatively affecting the revenue of marina operators and the local economy. Throughout this period, biologists with the GADNR received numerous complaints from anglers and marina operators expressing their dissatisfaction with the largemouth bass fishery. This prompted the GADNR to consider changes to largemouth bass management in the reservoir.

Stakeholders and decision makers

Georgia Department of Natural Resources biologists were responsible for monitoring and managing the fishery resources of West Point Reservoir. However, the actions of managers were generally limited to non-regulatory activities, such as habitat enhancement. Changes to fishery regulations required the approval of the GA Board of Natural Resources, which was comprised of 18 citizens appointed by the Governor and confirmed by the Georgia Senate. Thus, the decision makers included the fishery managers *and* political appointees whose wishes needed to be taken into account during the SDM process. This complicated the process to some extent and affected the identification of decision alternatives.

As discussed earlier, the largemouth bass fishery potentially affected several groups that could be broadly categorized as: anglers, marina operators, and local businesses that catered to anglers. These groups were stakeholders that could have been included in the SDM process. However, the general consensus among the GADNR fishery biologists was that the latter two groups would be content if the largemouth bass anglers were satisfied. Thus, the stakeholders included largemouth bass anglers, which consisted of two user-groups: recreational anglers and tournament anglers.

Management objectives and decision alternatives

During the spring 2001, GADNR fishery biologists met to discuss their objectives and management alternatives for the largemouth bass fishery in West Point Reservoir. The biologists identified the problem statement as managing the largemouth bass fishery resources in West Point Reservoir to maximize largemouth bass angler satisfaction. The fundamental objective was to largemouth bass angler satisfaction for the two user-groups. To identity the factors that satisfy largemouth anglers, we examined the results of a 1997 statewide telephone angler survey (Bason 1997) that indicated that recreational anglers preferred catching large-bodied largemouth bass to more liberal creel limits. For example, 38% of respondents preferred a limit consisting of one largemouth bass larger than 457 mm when asked to choose between creel limits and minimum size limits. These results were consistent with the expectations of GADNR biologists, but the biologists also believed that the values of tournament largemouth bass anglers differed from those of recreational anglers, with the former preferring larger creel limits to larger-sized bass. During the discussions, the biologists also believed that some value should be given to the consistency or stability of bass populations from year-to-year. Thus, we identified the following fundamental objective attributes (means objectives) for each user-group, maximize the number of legal size bass, maximize the number of large largemouth bass (defined as bass > 457 mm), and maximize the stability of the bass population.

Changes in three fundamental objective attributes in response to management actions could be estimated using population dynamic models. However, we needed a means to quantify and weight changes in those attributes in terms of angler satisfaction. To quantify the values of the West Point largemouth bass anglers, we conducted a survey that asked a random selection of anglers to rank the importance of three bass fishery characteristics that corresponded to our three attributes. The survey indicated that tournament and recreational anglers placed the greatest value on population stability, but differed in the other two attributes (Peterson and Evans 2003). Tournament anglers placed greater value (higher rank) on greater numbers of legal-sized fish, whereas recreational anglers preferred greater numbers of large bass. Using the results of the survey, we use the multi-attribute valuation procedure (Chapter 3) to weight each outcome and combine them into a composite angler satisfaction score (i.e., the utility). Details can be found in Box 3.6.

The fishery biologists and managers discussed potential decision alternatives and the feasibility of implementing alternatives that were unusual or unique. The general consensus was that the GA Board of Natural Resources was unlikely to approve novel regulations and actions. Therefore, the potential management options were limited to traditional minimum length limit regulations with four alternatives: no minimum, 305 mm (12 in), 356 mm (14 in), and 406 mm (16 in) minimum total length limits.

Model development

The response of largemouth bass population to decision alternatives was modeled using a stochastic, age-structured demographic model (i.e., a Leslie matrix population model) that consisted of 14 age classes. The model operated on an annual time step and assumed that age-3 and older individuals were mature adults. The number of eggs produced each year was the sum of product of average fecundity of each cohort and the corresponding density of mature females (i.e., half the adult population) surviving the previous time step. Fry density was estimated as a function of the total number of eggs produced and fry survival, which was estimated as a density-dependent function of fry-carrying capacity. Individuals in each age class were promoted to the next age class using age-class-specific annual survival rates (i.e., natural survival minus angling mortality). Angling mortality rates depended upon minimum length limit and was only applied for cohorts that exceeded the size limit. Survival of age-13 fish is assumed to be 0. The average body size (length) of all cohorts was modeled as a function of von Bertalanffy growth parameters.

There were several other factors that potentially affected the bass population in West Point Reservoir. For instance, the GADNR biologists hypothesized that a length-limit regulation change could affect angler behavior and result in changes in catch and release attitudes that would affect angling mortality. The biologists also hypothesized that the future growth of bass could be changed (increased) in response to increased harvest (i.e., due to decreased density) and changes in the trophic state of West Point Reservoir. Similarly, the fry-carrying capacity West Point Reservoir could be changed in response changes in the trophic state. These relationships were incorporated into the final decision model (Figure 9.13).

The decision model was parameterized using a combination of empirical estimates from previous studies and eliciting professional judgment (Chapter 6). Parameter estimates for the population demographic model components were obtained from published studies. The remaining components were estimated using the professional judgment of GADNR fishery managers and scientists. For example, trophic state probabilities were obtain from water quality experts and were based on the belief that proposed wastewater treatment facilities on the Chattahoochee River would lead to relatively higher inputs of nutrients into the watershed.

Incorporation of uncertainty

We incorporated the uncertainty in the population parameter estimates (e.g., survival) and angling mortality by assigning probability distributions (Chapter 5). Structural uncertainty was incorporated using four alternative models that represented the hypothesized effect of largemouth bass density (i.e., effect or no

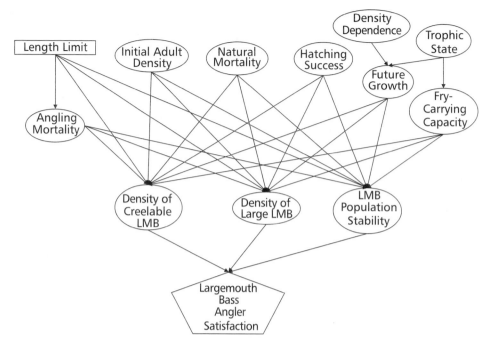

Figure 9.13 Influence diagram for West Point Reservoir largemouth bass fishing decision.

effect) and reservoir trophic state (i.e., effect or no effect) on bass growth. The weights of the alternative models were based on expert judgment (Chapter 6).

Evaluation of decision alternatives and sensitivity analysis

Using the decision model, we estimated that the optimal decision for the West Point Reservoir length limit, given the available information, was 305 mm with an expected angler satisfaction of 61.5%. Angler satisfaction under the existing 406 mm length limit was 49.8%, The highest possible angler satisfaction estimated using was 76%, so angler satisfaction under the optimal decision represented a relatively large increase from the status quo.

Before implementing the optimal decision, we conducted sensitivity analysis (Chapter 7) on the model to identify the most influential components and estimate their effect on the largemouth bass length-limit decision. One way sensitivity analysis indicated the model was most sensitive to angling mortality and fry-carrying capacity and least sensitive to the density dependent growth hypothesis (Figure 9.14). The latter result was surprising to the biologists working on the project because they had initially assumed that changing length limit regulations would lead to increased growth, which would improve fishing. The sensitivity analysis indicated that the assumptions about density-dependent growth had a very minor effect on angler satisfaction.

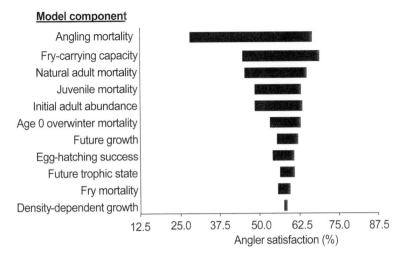

Figure 9.14 Tornado diagram for West Point Reservoir largemouth bass fishing decision model.

Both angling mortality and fry-carrying capacity had a substantial influence on angler satisfaction. However, the response profile plot for each component indicated that the optimal length limit decision changed only once over the range of fry-carrying capacities (i.e., from 305 mm to 356 mm at 410 fry/ha), whereas it changed four times over the range of angling mortalities (Figure 9.15). This disturbed the fishery managers because angling mortality was estimated using the expert judgment of biologists working on similar reservoirs in the region. Clearly in this instance, it would be critical to estimate angling mortality with the greatest accuracy because inaccurate estimates could lead to the choice of the suboptimal (incorrect) length limit. Therefore, we decided to estimate the value of information for this component.

Value of information

The most influential model components identified during sensitivity analysis were considered critical to the decision and hence, may be high priority as potential subjects for future study. For example, the sensitivity analysis of the decision model indicated that angling mortality had a substantial influence on the length limit decision. To improve the accuracy of this estimate, a study could be designed to estimate changes in angling mortality with changing length-limit regulations in an adaptive management approach (Chapter 7). The results of the study then would be used to update (adjust) the LMB decision model to produce improved length-limit decisions in the future.

Angling mortality had a substantial influence on the value of each length-limit decision, but the true value of angling mortality was not known with certainty. Assuming that it could be estimated *without error* during some future study, the expected value of angling satisfaction following the study is estimated as a

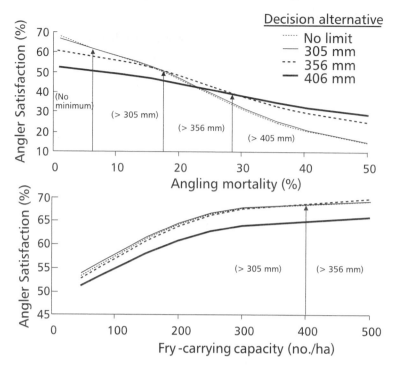

Figure 9.15 Response profile sensitivity analysis for angling mortality and carrying-capacity components of the West Point Reservoir largemouth bass fishing decision model.

probability-weighted value of optimal decision (largest angler satisfaction) under each (known) value of angling mortality. For the LMB decision model, this value was 63.2% and the expected value of perfect information is estimated as 63.2–61.5 = 1.7%. If the cost of the mortality study were minor in comparison, the study would increase the expected net value of the optimal decision and hence would be beneficial to complete.

Current status of largemouth bass management

The optimal decision for managing largemouth bass in West Point Reservoir was a 305-mm length limit. We were concerned that the model was very sensitive to the estimates of angling mortality and believed that the implementation of the 305-mm was risky without additional study of angling mortality. Therefore, the GADNR biologists choose the next best decision, a 356-mm length limit with an angler satisfaction of 60.8%, as their satisficing alternative. The biologists then met with members of the general public (i.e., independent anglers, bass clubs, and sportsman's organizations) and presented the results of the modeling. Using the influence diagram in Netica, they were able to demonstrate important concepts to the general public and eventually obtained support for changing the regulations. The GADNR regional biologists then proposed changing the length limit to 356-mm to the Natural Resources Board who adopted the change in 2002.

Summary

In this chapter, we reviewed three applications of the principles of SDM and ARM to decision making problems familiar to the authors, emphasizing the steps whereby we succeeded in convincing resource managers to consider these principles for problem solving. We believe that these examples have succeeded, to the extent that decision makers remained engaged in an SDM /ARM approach. In two of these cases, the approach has been carried through to the actual implementation of management decisions, such as harvest regulations, while in the third, legal and political events have intervened to stall the process.

In the next and last chapter, we consider some cases where the application of SDM and ARM has proven less than completely successful. We discuss "lessons learned" from these cases, and close with some recommendations to biologists and managers seeking to use a structured approach to management decision making.

References

Anderson, D.R. (1975) Optimal exploitation strategies for an animal population in a Markovian environment: a theory and an example. *Ecology* 56, 1281–1297.

Anderson, D.R., K.P. Burnham, J.D. Nichols, and M.J. Conroy. (1987) The need for experiments to understand population dynamics of American black ducks. *Wildlife Society Bulletin* 15, 282–284.

Anderson, M.G. et al. (2007) Report from the Joint Task Group for clarifying North American Waterfowl Management Plan objectives and their use in harvest management. Unpublished Report.

Annear, T., I. Chisholm, H. Beecher, A. Locke, P. Aarrestad, C. Coomer, C. Estes, J. Hunt, R. Jacobson, G. Jobsis, J. Kauffman, J. Marshall, K. Mayes, G. Smith, C. Stalnaker, and R. Wentworth. (2004) Instream Flows for Riverine Resource Stewardship. Instream Flow Council, Cheyenne, Wyoming.

Bason, J. (1997) Georgia Department of Natural Resources statewide angler survey. Report of University of Georgia Survey Research Center to Wildlife Resources Division-Fisheries Section, Social Circle, Georgia.

Bellman, R. (1957) *Dynamic Programming*. Princeton University. Press, Princeton, NJ. 342 p.

Conroy, M.J., J.R. Goldsberry, J.E. Hines, and D.B. Stotts. (1988) Evaluation of aerial transect surveys for wintering American black ducks. *Journal of Wildlife Management* 52,694–703.

Conroy, M.J., M.W. Miller, and J.E. Hines. (2002) Review of population factors and synthetic population model for American black ducks. *Wildlife Monographs*. 150.

Conroy, M.J. (2004) Development of an integrated, adaptive management protocol for American Black Ducks. Final Report from University of Georgia to USGS. December, 2004. Athens, GA. http://coopunit.forestry.uga.edu/blackduck/final_report/public.

Conroy, M.J., C.J. Fonnesbeck, and N.L. Zimpfer. (2005) Modeling regional waterfowl harvest rates using Markov chain Monte Carlo. *Journal of Wildlife Management* 69, 77–90.

Conroy, M.J. (2010) Technical support for adaptive harvest management for American black ducks. Final Report from University of Georgia to USGS. December, 2010. Athens, GA. http://coopunit. forestry.uga.edu/blackduck/final10.pdf.

Feierabend, J.S. (1984) The black duck: an international resource on trial in the United States. *Wildlife Society Bulletin* 12, 128–134.

Freeman, M.C., G.R. Buell, L.E. Hay, W.B. Hughes, R.B. Jacobson, J.W. Jones, S.A. Jones, J.H. LaFontaine, K.R. Odom, J.T. Peterson, J.W. Riley, J.S. Schindler, C. Shea, J.D. Weaver. (2012) Linking River Management to Species Conservation using Dynamic Landscape-Scale Models. *River Research and Applications* 48.

Johnson, F.A., B.K. Williams, J.D. Hines, W.L. Kendall, G.W. Smith, and D. F. Caithamer. (1993) Developing an adaptive management strategy for harvesting waterfowl in North America. *Trans. N. Amer. Wildl. Nat. Resour. Conf.* 58, 565–583.

Johnson, F.A. (2010) Learning and adaptation in the management of waterfowl harvest. *J. Environ. Manage.* 92, 1385–1394.

Lincoln, F.C. (1935) The waterfowl flyways of North America. US Dept. Agric. Circ. No. 342.

Lubow, B.C. (1995) SDP: Generalized software for solving stochastic dynamic optimization problems. *Wildl. Soc. Bull.* 23, 738–742.

Lubow, B.C. (1997) Adaptive stochastic dynamic programming (ASDP): Supplement to SDP user's guide, version 2.0. Colo. Coop. Fish and Wildlife Res. Unit, Colorado State University, Fort Collins, CO.

Maceina, M.J. and D.R. Bayne. (2001) Changes in the black bass community and fishery with oligotrophication in West Point Reservoir, Georgia. *North American Journal of Fisheries Management* 21, 745–755.

Martin, E.M. and S.M. Carney. (1977) Population ecology of the mallard: IV. A review of duck hunting regulations, activity, and success, with special reference to the mallard. *US Fish and Wildlife Service Resource Publ.* 130.

McCargo, J.W. and J.T. Peterson. (2010) An evaluation of the influence of seasonal base flow and geomorphic stream characteristics on Coastal Plain stream fish assemblages. *Transactions of the American Fisheries Society* 139, 29–48.

Nichols, J.D., M.C. Runge, F.A. Johnson, et al. (2006) Adaptive harvest management of North American waterfowl populations- recent successes and future prospects. *J. Ornithology* 147 (Suppl. 1), 28.

Peterson, J.T. and J.W. Evans. (2003) Decision analysis for sport fisheries management. *Fisheries* 28(1), 10–20.

Peterson, J.T., J.M. Wisniewski, C.P. Shea, and C.R. Jackson. (2011) Estimation of mussel population response to hydrologic alteration in a Southeastern U.S. stream. *Environmental Management* 48, 109–122.

Pospahala, R.S. et al. (1974) Population ecology of the mallard: II. Breeding habitat conditions, size of the breeding populations, and production indices. *US Fish and Wildl. Serv. Resource Publ.* 115.

Richter, B.D., D.P. Braun, M.A. Mendelson, and L.L. Master. (1997) Threats to imperiled freshwater fauna. *Conservation Biology* 11, 1081–1093.

Rugel K., C.R. Jackson, J.J. Romeis, S.W. Golladay, D.W .Hicks, and J.F. Dowd, (2012) Long-term effects of center pivot irrigation on streamflows in a Karst environment: lower Flint River Basin, Georgia, USA. *Hydrological Processes* 26, 523–534.

Rogers, J.P, J.D. Nichols, F.W. Martin, C.F. Kimball, and R.S. Pospahala. (1979) An examination of harvest and survival rates of ducks in relation to hunting. *Trans. North American Wildl. Nat. Resource Conf.* 44, 114–126.

Rusch, D.H., C.D. Ankney, H. Boyd, J.R. Longcore, F. Montalbanso, J.K. Ringelman, and V.D. Stotts. (1989) Population ecology and harvest of the American black duck. *Wildlife Society Bulletin* 17, 379–406.

Strayer, D. L. (2008) *Freshwater Mussel Ecology: A Multifactor Approach to Distribution and Abundance.* University of California Press, Berkeley, California.

Tharme, R. E. (2003) A global perspective on environmental flow assessment: emerging trends in the development and application of environmental flow methodologies for rivers. *River Research and Applications* 19, 397–441.

U.S. Dept. of the Interior (1994) (with Environment Canada and Secretario de Desarrollo Social México). "1994 update of the North American Waterfowl Management Plan". *U.S. Fish and Wildlife Service,* Washington, D.C.

Viger, R.J., L.E. Hay, J.W. Jones, and G.R. Buell. (2010) Effects of including surface depressions in the application of the Precipitation-Runoff Modeling System in the Upper Flint River Basin, Georgia. U.S. *Geological Survey Scientific Investigations Report* 2010–5062.

Warren, M.L., B.M. Brooks, S.J. Walsh, H.L. Bart, R.C. Cashner, D.A. Etnier, B.J. Freeman, B.R. Kuhajda, R.L. Mayden, H.W. Robinson, S.T. Ross, and W.C. Starnes. (2000) Diversity, distribution, and conservation status of the native freshwater fishes of the southern United States. *Fisheries* **25**(10), 7–31.

Williams, B.K., J.D. Nichols, and M. J. Conroy. (2002) *Analysis and Management of Animal Populations: Modeling, Estimation, and Decision Making.* Elsevier-Academic Press. New York.

Williams, J.D., A.C. Bogan, and J.T. Garner. (2008) *Freshwater mussels of Alabama and the Mobile basin in Georgia, Mississippi, and Tennessee.* University of Alabama Press, Tuscaloosa, Alabama.

10

Summary, Lessons Learned, and Recommendations

Summary

We began this book with our thoughts about how a structured decision-making approach could be beneficial to natural resource management. We gave examples where a lack of structure resulted in, at least, inefficient decision making (e.g., the disconnect between management and "surveillance" monitoring programs, Nichols and Williams 2006), but often led to conflict and gridlock (Chapter 1). We then described in some detail the elements of SDM (Chapter 2), with emphasis on the proper definition of objectives (Chapter 3), and working with stakeholders (Chapter 4). In Part II, we developed the tools for SDM and ARM, drawing from the fields of statistical inference (Chapter 5), predictive modeling (Chapter 6), uncertainty analysis (Chapter 7), and optimization (Chapter 8), providing worked examples with accompanying computer code along the way. In Chapter 9, we illustrated the principles of SDM and ARM by means of several case studies involving conservation decision making in North America. Although by no means perfect, these examples were each reasonably successful in applying SDM and ARM to difficult problems.

Lessons learned

In the previous chapter, we illustrated principles of SDM and ARM with 3 case examples from natural resource management in North America. We presented these example because each illustrates (to varying degrees) the steps required to

Decision Making in Natural Resource Management: A Structured, Adaptive Approach, First Edition. Michael J. Conroy and James T. Peterson.
© 2013 John Wiley & Sons, Ltd. Published 2013 by John Wiley & Sons, Ltd.

apply SDM and ARM to a resource decision-making problem, and each has at least partially succeeded, where we define "success" as:

- Resource managers or other decision makers accept SDM and ARM as a tool for making decisions.
- Appropriate technical and policy infrastructure is in place to assure that the process has at least a reasonable prospect of continuity.
- Decisions are current being made or will be made in the near future that are informed by an SDM/ARM process.

We do not claim that these examples, or others where SDM/ARM has been applied to conservation issues, have completely or uniformly succeeded in these 3 areas, nor do we assert that other examples of SDM/ARM applications (e.g., AHM for waterfowl harvest management in North America) have been unmitigated successes, but we would assert that on balance these examples have been successful.

However, we certainly do not claim that SDM/ARM will succeed in all cases, as evidence by our personal experience with instances where these approaches have "failed" for various reasons. In the spirit of learning and ARM, we think we can learn from these failures and from those aspects of our successes that were not completely so. Thus, in this section of our concluding chapter, we briefly review some instances where SDM/ARM has been proposed or attempted, but has been rejected by decision makers or has failed to take hold for other reasons. We summarize these, and the less-than-completely successful aspects of our examples in Chapter 9, into "lessons learned" in the hopes that others will be able to more successfully navigate around these hazards.

Structured decision making for Hector's Dolphin conservation

During a "mini-sabbatical" in 2007, MJC conducted a 2-day workshop on SDM/ARM with colleagues at the University of Otago in Dunedin, New Zealand. Among other NZ issues, we applied SDM and ARM to the problem of conservation of Hector's dolphin (*Cephalorhynchus hectori*), the world's smallest dolphin and an endemic of NZ coastal waters (Baker 1978, Slooten and Dawson 1994). The details of the application of SDM/ARM to this issue are provided by Conroy et al. (2008). Hector's dolphin is classified as endangered under IUCN criteria (Klinowska 1991), and is considered to be vulnerable to a wide variety of mainly anthropogenic impacts, including pollution, disturbance from boating, and entanglement in gill-nets employed in commercial, recreational, and subsistence fishing (Slooten and Dawson 1994, Cameron et al. 1999, Slooten et al. 2000). The precarious status of the population has been affirmed by survival estimates and population projections (Slooten et al. 1992, 2000, Burkhart and Slooten 2003) with focus on apparently low survival rates (Cameron et al. 1999 but see Fletcher et al. 2002).

We believed that the Hector's dolphin conservation problem contains several elements that could make an SDM/ARM approach useful. First, there was a fairly clear conservation goal (avoidance of extinction, de-listing from IUCN endangered status), as well as societal goal (economic and social interesting in maintaining fisheries) that potentially were in conflict. Secondly, there existed profound uncertainty about the causes of the declines of Hector's dolphin populations, and thus uncertainty about the efficacy of actions such as gill-net restrictions intended to remediate population declines. The interaction of scientific uncertainty and value conflict potentially creates a gridlock, where disagreements over science become a rationale for delaying or avoiding decision making (Slooten et al. 2000).

Our workshop environment did not permit a full, realistic incorporation of both science experts and stakeholders – decision makers, but we did have involvement from critical agencies, specifically NZ Dept. of Conservation (DOC), and NZ Ministry of Fisheries (MFish), academics from the University of Otago (Depts. of Zoology and Mathematics and Statistics), and non-governmental organizations. However, there was no direct representation of fisheries interests, other than by proxy from MFish. We initially approached the problem of developing and valuing objectives, and eliciting expert opinions on science matters, via a role-playing exercise in which participants were asked to assume the roles of each of the broad stakeholder groups (conservation interests, fisheries interests). In this process, we fairly quickly identified some fundamental objectives important to each group (i.e., dolphin recovery and maintenance of tourism for "conservationists"; maintenance of the fishery for the "fisheries" group) as well as a series of sub- or means objectives for each (Conroy et al. 2008). Divergence occurred when it came to assigning relative utilities to the fundamental objectives, because "conservation" stakeholders tended to place very high value on recovery and give little penalty to the costs of conservation actions, even when ineffective, whereas "fisheries" interests did the opposite. For the purposes of the workshop, we decided to evaluate decisions under two sets of utilities: a "consensus" set that simply averaged over the stakeholder values, and a "fishery" or "industry" set that gave more weight to "fisheries" values and heavily penalized decision outcomes that involved gear restrictions but were ineffective in dolphin recovery.

As noted, "conservation" and "fisheries" groups tended to disagree on the science of dolphin conservation. Specifically, the former were both more pessimistic about future dolphin status under status quo management (i.e., current gear restrictions) and more optimistic about the efficacy of further restrictions for recovery, while the latter reversed these sentiments. There was little empirical support in either direction, with the single study designed to test the effects of gear restrictions on mortality reduction finding no effect but with low statistical power (Cameron et al. 1999). We therefore used our stakeholder proxy groups as competing expert panels, and derived tables of conditional probabilities for decision influence (Chapter 6) under each panel, subsequently treating these as competing alternative models (Chapter 7).

The ensuing influence diagram allowed us to explore a number of scenarios, including the influence of relative belief in our expert models (i.e., the model weights) and the effect of the different utility sets, on the selection of an optimal decision from among 3 choices: status quo (current restrictions), moderate increases in restrictions, and high increases in restrictions. The results were revealing in that either status quo or moderate increases in restrictions were generally selected, regardless of model weights or utility sets. High increases in restrictions were only favored under the consensus utilities, and then only if model weights were 0.5 or more in favor of the "conservation" model.

We were initially optimistic that the SDM/ARM approach would be considered and even embraced by decision makers in NZ. However, we soon began to get indications of resistance, interestingly much of it from government decision makers who were either advancing a "conservation" agenda, or who perceived that public attitudes would favor a more aggressive conservation strategy – even if such were not effective at actually conserving dolphins. During a subsequent presentation before MFish staff in Wellington, MJC and colleague R. Barker from U. Otago were "gobsmacked" to hear a fairly high level MFish staffer state his view to the effect that it was the *intention* of conservation that mattered, not the outcome and that adverse impacts to other constituencies (like fishery interests) should not affect decision making. In late 2007, the World Wildlife Fund launched a petition drive to introduce "emergency measures" to protect Hector's dolphins, and in 2008 most gill-netting was banned within 4 nautical miles of the South Island's eastern and southern coasts (NZ Ministry of Fisheries 2008). These measures seem largely justified on the basis that mortality from gill-netting was one of the only anthropogenic factors under effective management control. Unfortunately, in the absence of monitoring designed to reduce uncertainty about the efficacy of such controls, it will not be possible to learn from this "trial and error" experience. Furthermore, if, as seems likely, controls have limited success and dolphins continue to decline due to other factors, we anticipate a loss of agency credibility and eventual political repercussions as stakeholders who have effectively been shut out of the process seek redress by other means.

We have learned at least 2 lessons from this experience. First, for an SDM/ARM process to work, the relevant stakeholder groups really need to be part of the process and have "ownership" in the decision making. Because this was fundamentally an exercise – and not a real commitment to the use of SDM/ARM in decision making – that did not happen in this case, and the views of one group were ultimately ignored. Secondly, one set of views about the decision problem should not be allowed to dominate – even if that is the set promoted by the "experts." In this case, a small number of outspoken experts that promoted the view, with little scientific support, that gill-netting is the critical factor limiting Hector's dolphin populations, and (more questionably still) that mitigation of these factor will result in population recovery. The obvious danger of such an approach is that it distracts from other factors (and thus possible mitigating actions) that could be equally or more important. Finally, if Hector's dolphins do become extinct, there will be no way to know if it could have been prevented by other measures; likewise, if they persist it will be unknown if fishing restrictions can be credited.

Landowner incentives for conservation of early successional habitats in Georgia

Conroy and Carroll (2009) described this case study in detail; here we summarize the main points with respect to "lessons learned". As background for this problem, populations of the important game species Northern Bobwhite (aka bobwhite quail; *Colinus virginianus*) along with several other species of birds such as Grasshopper Sparrows (*Ammodramus savanarum*) and Eastern Meadowlark (*Sturnella magna*) have precipitously declined in most of the southeastern United States during the latter decades of the 20th century and continuing to the present. Although other factors are no doubt to blame, a combination of reduction in the total land acreage devoted to agriculture (and thus, increased forestation), together with fire suppression and intensive culture on land remaining in agriculture, has greatly reduced the savannah and fire-dominated forest systems thought to be favorable to bobwhite. However, there exist a number of modifications to standard agricultural practices that could improve habitat for bobwhite and allied species incentives (Thackston and Whitney 2001).

Significantly, most of the remaining lands that are amenable (at least in the short term) to improvement to favor bobwhite and other species are in private ownership. Under Georgia and US laws, owners have broad discretion over their properties, and except under extraordinary circumstances cannot be legally required by the government to modify land-use practices to accommodate wildlife. One tool for voluntary modification of land-use practices is provided by landowner incentives. Starting in 2001, the State of Georgia implemented a landowner incentive program called the Bobwhite Quail Initiative (BQI), funded by specialty automobile tag sales. The program pays landowners to enroll agricultural fields and follow a set of guidelines calculated to enhance habitat quality for bobwhites and other early successional birds.

Both authors were involved in various phases of the technical development of SDM/ARM approach to BQI decision making. We worked with John Carroll and Rick Hamrick (UGA and Mississippi State University, respectively) and graduate student Jay Howell to analyze data from bobwhite monitoring programs. Heretofore, decisions about enrollment of individual parcels were made on an ad hoc basis until program funds were depleted. We suggested, alternatively, an objective- and model-based approach based on our empirical models to prioritize parcel enrollment, with the objective defined in terms of predicted habitat quality and bird abundance, with explicit recognition of structural uncertainty about the response (Conroy and Carroll 2009, Howell et al. 2009). Finally, we recommended continued support of a monitoring program to evaluate management outcomes, provide feedback to reduce structural uncertainty, and improve decision making.

Initially, our recommendations were well received, and we were hopeful that a cooperative decision-making–monitoring program would proceed. However, it soon became apparent that interest in biologically based recommendations was limited to the mid-level State biologists with whom we had mostly interacted, and had little support at higher levels. Furthermore, upper-level administrators had little apparent use for monitoring feedback, and soon directed all funding away from research and monitoring programs. Decisions about BQI enrollment continue to be based on an ad hoc approach, with no apparent connection to an identified resource objective, and no feedback from monitoring.

We have learned from this experience at least 2 lessons: first, we apparently failed to identify properly the **fundamental objectives** of the State agency, which clearly included, in addition to resource outcomes such as habitat quality and bird abundance, political and other considerations; the latter may have in fact been dominant. Secondly, we obviously had not succeeded in conveying notions of **uncertainty** about the consequences of management actions. Thus, to the extent that biological outcomes were even considered, they were assumed to be automatically achieved by the enrollment program. Had we identified both of these conditions up front, we might have (cynically) advised that the program managers save themselves the trouble by simply direct distribution of funds to landowners, without a pretense of obtaining a biological outcome. We at least would have recommended against monitoring, since they clearly already knew what the response would be, and monitoring data to the contrary would have been annoying.

Cahaba shiner

We frequently work with managers of state and Federal agencies developing strategies to deal with a variety of natural resource management problems. For many of these problems, we learned that it is generally best to start with a relatively short workshop (2–3 days) where we complete most of the SDM process; from identifying the problem and objectives to building a prototype decision model (Chapters 2–6). This rapid prototyping process serves several purposes; the most important of these is that it allows us to determine if a SDM approach is appropriate without spending a substantial amount of time and resources on the problem. We use this example on the Cahaba shiner (*Notropis cahabae*) to characterize those instances where we discovered, early on, that structured approach was not suitable.

The Cahaba River, located in central Alabama, is one of the longest (305 km) free-flowing rivers in the southeastern US. The river flows through two Physiographic Provinces, the Valley and Ridge and the Coastal Plain, and is among the most geologically and biologically diverse rivers in North America. The Cahaba Basin historically supported more than 50 species of freshwater mussels and 135 species of freshwater fish. Of these, almost half of the mussel species and one tenth of the fish species have been extirpated and several species are rare and declining, including 11 mussels, 5 fish, and 3 snail species that are listed under the federal Endangered Species Act. To help preserve this valued system, Congress established the Cahaba River National Wildlife Refuge (NWR) in 2002 (USFWS 2007). The refuge includes a 5.6-km section of the Cahaba River that is home to five of the federally listed species. The purpose of the refuge was primarily to preserve a unique stretch of the Cahaba River and protect the habitat for the listed species, including the Cahaba shiner. The refuge was also created to provide consumptive (e.g., hunting, fishing) and non-consumptive (e.g., bird watching, education) uses and to facilitate partnerships among the USFWS, state and local agencies, and the local community. The Cahaba River NWR was unstaffed with a minimal maintenance budget at the time of the workshop. There were no other publicly owned lands on or adjacent to the river. However, the Alabama Department of Conservation and Natural Resources leased a small wildlife management area adjacent to river, but it had little control over habitat management on the area.

The upper portion of the Cahaba River watershed includes the Birmingham Alabama metropolitan area, one of the most rapidly developing urban areas in North America. Urban land cover in this portion of the basin increased fourfold over the past two decades. In addition, there were 31 permitted municipal wastewater discharge points and over 100 industrial discharge points in the upper watershed of the Cahaba River (ADCNR 2005). These discharges contributed to 106 miles of the Cahaba River being designated as threatened or impaired waters by the US Environmental Protection Agency (USEPA 2010) and place additional stresses on the already imperiled aquatic fauna.

We convened a workshop with USFWS personnel in 2007 with the objective of developing a decision making framework for habitat conservation and recovery activities in the Cahaba Basin. The participants selected the Cahaba shiner, a small minnow species that occurs in a 30-km section of the Cahaba River (USFWS 1992), as the focus of the workshop because it is among the most critically endangered species in the basin. During the first half of the workshop, we created a problem statement, identified objectives, and developed a means objective network for the Cahaba shiner (Figure 10.1). Three fundamental objectives were identified as: the recovery of the Cahaba shiner, public support for the NWR, and minimize the costs of management. These were chosen, in part, because they were agency mandates, the second of these being part of the legislative language establishing the NWR.

For most of the remaining workshop, we were unable to progress from the initial objectives network to identifying decision alternatives. The impasse focused on the importance of increasing biological knowledge and involving cooperating state and local agencies and the general public through volunteers and how these objectives were related to the fundamental objectives. After making little progress over the course of a day, we decided to push on and identify potential decision alternatives. This was the breakthrough for the group. The USFWS had very limited authority to manage the Cahaba River system (e.g., those authorized under the endangered species act) and only had the authority to directly manage the resources of the NWR. Thus, the USFWS primarily relied on the state and local agencies to implement management actions that could affect the recovery of the Cahaba shiner. Given this realization, we restructured the objectives network to more faithfully represent the relations among objectives. In this network, building public support and credibility was considered essential to convince collaborating agencies and the private landowners to implement actions that would affect the recovery of Cahaba shiner (Figure 10.2). This view of the problem was decidedly different from our initial assessment and suggested a set of potential decision alternatives that focused more on managing the perceptions of the public than managing natural resources.

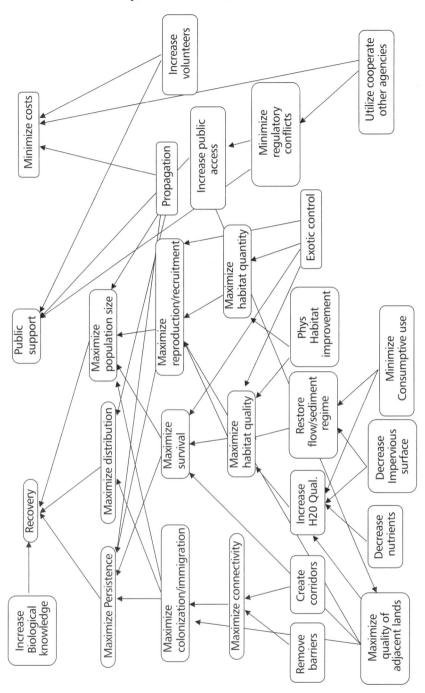

Figure 10.1 Initial objectives network for habitat conservation and recovery of the Cahaba shiner.

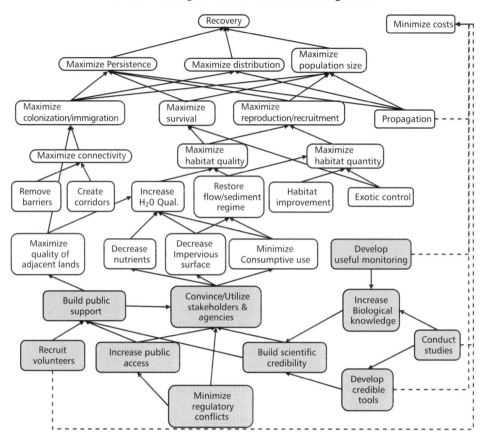

Figure 10.2 Final objectives network for habitat conservation and recovery of the Cahaba shiner. Shaded components represent means objectives that were potentially under the control of USFWS managers.

We have learned several lessons from this experience. The first was that these short workshops were invaluable for determining whether a structured approach was appropriate. In this instance, the problem faced by the USFWS was not a natural resource management problem. Rather, the problem from the perspective of USFWS was one of managing people and their perceptions. Secondly, we also learned that it is important to thoroughly vet the decision problem to ensure that the decision makers are included in the process. If key decision makers do not want to participate (as happened in other workshops), the remaining stakeholders and decision makers may have to address that problem (i.e., getting the entity to participate) rather than the natural resource problem or abandon the process altogether. The third lesson was that if things (e.g., objectives, alternatives) don't seem right, then there is probably a problem. From the start of the workshop, we had a suspicion that the objectives network was inadequate and that there was something missing. We decided to spend more time on the objectives and test the patience of the participants. In doing so, we discovered that the problem was fundamentally different from what we had originally thought. Had we ignored our suspicions, we might have jumped into developing a prototype model that addressed the wrong problem, wasting time and resources.

Other lessons

Through our experiences and those of colleagues, we find that a few key elements are needed to successfully apply SDM and ARM to natural resource problem. Chief among these is the importance of an effective champion to lead the process. Indeed, participants in an adaptive management symposium at the 2011 annual meeting of the American Fisheries Society identified the existence of a champion as the primary reason for success. We define a **champion** as an individual who has the authority, resources, and temerity to keep the process moving forward, ensure that the right people are in the right positions, and that participants have the necessary resources. It does not matter if the champion is a stakeholder or a decision maker (though the latter is generally more effective), only that they are motivated and can motivate others. How do you get a champion? That's a very good question that depends largely on the project and the personnel, so no one-size-fits-all answer is possible. In any event, what we can say is that someone (maybe you) needs to lead the effort to ensure or at least maximize, the chances for success.

Another import lesson that we learned is that successful SDM and ARM applications have the proper balance of biology and objectives. All too often we have witnessed projects overemphasizing the biological and ecological aspects of a problem at the expense of the objectives. This often occurs when more scientists than managers/decision makers are included in the stakeholder group, which can – and usually does – result in very detailed decision models that are of limited use for addressing the real management objectives. The science aspects of most decisions are usually what appeal to most of us, so we tend to gravitate to these areas. Heck, this is the reason why many of us study in natural resources related fields! The overemphasis may also be due to the mistaken impression that better science can resolve disagreements on objectives (see more below). One way to ensure the proper balance is to make sure that managers and decision makers dominate the stakeholder input, at least during the initial stages of the process. This is one of the reasons that we suggested that you limit the participation of scientists in the initial stakeholder meetings (Chapter 4). The managers and decision makers also need to collaborate closely with (i.e., oversee) the technical team activities to ensure that the models are useful for addressing the management objectives.

We have also, in the course of conducting several stakeholder and scientist workshops, learned to be on guard to certain attitudes or behaviors that can easily derail the SDM process. We considered some of these issues when we discussed Identifying Objectives (Chapter 3) and Working with Stakeholders (Chapter 4) but they are worth revisiting. First, you should *be on guard for values masquerading as facts or process*. That is, sometimes a strong statement of fact, or an argument to take the process in a certain direction, is more an indication that the proponent is holding onto a **hidden objective.** Try to recognize this and find a way to clarify that these are values, not facts. Secondly, try to distinguish (both as the facilitator, and for the sake of your group) between *positions* and *objectives.* You will often encounter situations where a participant takes a position on an issue as a tactic to avoid confrontation, thwart an opponent, or otherwise steer the process in a direction they perceive as beneficial to their interests. Find out what their interests – i.e., their **fundamental objectives** – really are, and get them to verbalize these, and not their positions. Also, *be on guard for cherry-picking of the process or the results.* That is, you will sometimes see participants who are content when the process seems to be yielding outcomes that are satisfying to them, and who either want to quit at that point or change the rules of the game when the results turn out contrary to their liking. These participants may need to be reminded that they agreed to a decision-making process and that their fundamental objectives were included, and that they cannot now change that process on the basis of specific outcomes. Ask, instead, "What is it about your objectives, or your understanding of the process that has changed?". Finally, for each of these points, hold yourself to the same scrutiny as your stakeholder-participants. We all have values and desires, and it is easy, without realizing it, to let one's own values permeate the SDM process. Besides the obvious check peer review, a simple test might be: would I, as an SDM facilitator, be satisfied with my response, if I were a participant?

Having laid out as many of the pitfalls and traps as we can think of, we must caution readers that we have no foolproof guidance for developing a structured approach to decision making. We hope that this book will provide useful guidance to those seeking to use this approach. Your best guidance, however, will come from the experiences you gain from applying these principles to real-life conservation problems and from learning from your successes (and, hopefully few, failures). Like a real decision process, there is no guarantee of success, but we believe, with Russo and Shoemaker (2001), that it is better to learn from a bad break than to profit from dumb luck.

References

Alabama Department of Conservation and Natural Resources (ADCNR). (2005) Conserving Alabama's Wildlife: A Comprehensive Strategy. ADCNR, Division of Wildlife and Freshwater Fisheries. Montgomery, Alabama..

Baker, A. (1978) The status of Hector's dolphin *Cephalorhyncus hectori* (van Beneden) in New Zealand. *Reports of the International Whaling Commission* 28, 331–334.

Burkhart, S. M., and E. Slooten. (2003) Population viability analysis for Hector's dolphin (*Cephalorhynchus hectori*): a stochastic population model for local populations. *New Zealand Journal of Marine and Freshwater Research* 37, 553–566.

Cameron, C., R.J. Barker, D. Fletcher, E. Slooten, and S. Dawson. (1999) Modelling survival of Hector's dolphins around Banks Peninsula, New Zealand. *Journal of Agricultural Biological & Environmental Statistics* 4, 126–135.

Conroy. M.J., R.J. Barker, P.W. Dillingham, D. Fletcher, A.M Gormley, and I.M. Westbrooke. (2008) Application of decision theory to conservation management: recovery of Hector's dolphin. *Wildlife Research* 2008, 93–102.

Conroy, M.J. and J.P. Carroll (2009) *Quantitative Conservation of Vertebrates*. Wiley. New York.

Fletcher, D., S. Dawson, and E. Slooten. (2002) Designing a mark–recapture study to allow for local emigration. *Journal of Agricultural Biological & Environmental Statistics* 7, 586–593.

Howell, J.E., C.T. Moore, M.J. Conroy, R.G. Hamrick, R.J. Cooper, R.E. Thackston, and R.J. Cooper. (2009) Conservation of northern bobwhite on private lands in Georgia, USA under uncertainty about landscape-level habitat effects. *Landscape Ecology* 24, 405–418.

Klinowska. M. (1991) Dolphins, *Porpoises and Whales of the World: the IUCN Red Data Book*. IUCN, Gland, Switzerland.

Nichols, J.D. and B.K. Williams. (2006) Monitoring for conservation. *Trends in Ecology and Evolution* **21**, 668–673.

NZ Ministry of Fisheries (2008) (revised 2011). Hector's dolphin. http://www.fish.govt.nz/en-nz/Environmental/Hectors+Dolphins.

Russo, J.E. and P. J. H. Shoemaker, (2001) *Winning Decisions: Getting it Right the First Time*. Currency Doubleday, New York, New York.

Slooten, E. and S.M. Dawson. (1994) Hector's dolphin. pp. 311–333 in S. Ridgeway and R. Harisson (eds), *Handbook of Marine Mammals. Vol. V.: Delphinadae and Phocoenidae*. Academic Press, New York.

Slooten, E., S.M. Dawson, and F. Lad. (1992) Survival rates of photographically identified Hector's dolphins from 1984 to 1988. *Marine Mammal Science* **8**, 327–343.

Slooten, E., D. Fletcher, and B. L. Taylor. (2000) Accounting for uncertainty in risk assessment: case study of Hector's dolphin mortality due to gillnet entanglement. *Conservation Biology* **14**, 1264–1270.

Thackston, R. and M. Whitney (2001) The bobwhite quail in Georgia: history, biology, and management. Georgia Dept. of Natural Resources, Wildlife Resources Division, Game Management Section, Social Circle, GA.

U.S. Environmental Protection Agency (USEPA). (2010) Alabama 303(d) listed waters for Reporting Year 2010. Available at: <http://ofmpub.epa.gov/waters10/attains_impaired_waters.impaired_waters_list?p_state=AL&p_cycle=2010>.

US Fish and Wildlife Service (USFWS). (1992) Cahaba Shiner, Notropis cahabae, Recovery Plan. US Fish and Wildlife Service, Southeast Region, Atlanta GA.

US Fish and Wildlife Service (USFWS). (2007) Cahaba River National Wildlife Refuge, habitat management plan. US Fish and Wildlife Service, Jackson, MS.

PART IV. APPENDICES

Appendix A

Probability and Distributional Relationships

Probability axioms

1. The **probability** of an event E in the event space F is a nonnegative real number $P(E) \geq 0 \wedge P(E) \in \Re, \forall E \in F$.
2. The probability that some event Ω will occur in the entire sample space is 1: $P(\Omega) = 1$.
3. Any countable sequence of mutually exclusive events $E_1.E_2,\ldots$ satisfies

$$P(E_1.\cup E_2 \cup, \ldots) = \sum_{i-1}^{\infty} P(E_i) \qquad (A.1)$$

It follows that:

1. Monotonicity of probability: $P(A) \leq P(B)$ if $A \subseteq B$
2. Probability of the null (empty) set: $P(\{\varnothing\}) = 0$.
3. Numeric bound on probability: $0 \leq P(E) \leq 1$ for all $E \in F$.

Conditional probability

Conditional probability is defined as follows:

$$P(A \mid B) = \frac{P(A \cap B)}{P(B)} \qquad (A.2)$$

Decision Making in Natural Resource Management: A Structured, Adaptive Approach,
First Edition. Michael J. Conroy and James T. Peterson.
© 2013 John Wiley & Sons, Ltd. Published 2013 by John Wiley & Sons, Ltd.

where $P(A \cap B)$ (read as "probability of A and B") is the **joint probability** of events A and B occurring or the probability that both events will occur. Equivalently,

$$P(A \cap B) = P(A \mid B)P(B)$$

A and B are statistically independent iff

$$P(A \cap B) = P(A)P(B) \qquad (A.3)$$

Corresponding expressions can be formed for the distributions of discrete or continuous random variables x_1 and x_2. The **conditional distributions** are given as

$$f_1(x_1 \mid x_2) = \frac{f(x_1, x_2)}{f_2(x_2)}$$

and

$$f_2(x_2 \mid x_1) = \frac{f(x_1, x_2)}{f_1(x_1)} \qquad (A.4)$$

where $f(x_1, x_2)$ is the *joint distribution* of x_1 and x_2 and $f_1(x_1)$, $f_2(x_2)$ are the respective marginal distributions. The random variables are statistically independent iff

$$f(x_1, x_2) = f_1(x_1)f_2(x_2) \qquad (A.5)$$

Conditional independence

Two events are conditionally independent iff

$$P(A \cap B \mid C) = P(A \mid C)P(B \mid C) \qquad (A.6)$$

Conditional independence implies that, after conditioning on the event C is accounted for, knowledge of the event B adds no new information about the occurrence of A, and vice versa. Conditional independence applies to distributions as well; for instance

$$f(x_1, x_2 \mid x_3) = f_1(x_1 \mid x_3)f_2(x_2 \mid x_3) \qquad (A.7)$$

specifies that the random variables x_1 and x_2 are conditionally independent, with conditioning of each on x_3.

Expected value of random variables

The **expected value** of a function $g(x)$ of a discrete random variable X is

$$E[g(x)] = \sum_x g(x)f(x)$$

or when $g(x) = x$

$$E(x) = \sum_x xf(x) \tag{A.8}$$

where $f(x)$ is the **probability mass function** of x, defined as $f(x) = P(X = x)$.

The expected value of a function $g(x)$ of a continuous random variable x is

$$E[g(x)] = \int_x g(x)f(x)dx$$

or when $g(x) = x$

$$E(x) = \int_x xf(x)dx \tag{A.9}$$

where $f(x)$ is now the **probability density function** of x, such that $P(x_1 < X < x_2) = \int_{x_1}^{x_2} f(x)dx$.

Example: Suppose $x \in \{1,2,3,4,5\}$ with associated discrete probability mass function $f(\underline{x}) = (0.01, 0.05, 0.5, 0.4, 0.04)$. The expected value of x is given by

$$E(x) = \sum_x xf(x) = 1(0.01) + 2(0.05) + 3(0.5) + 4(0.4) + 5(0.04) = 3.41$$

Law of total probability

Consider 2 events, A and B, with A conditional on the value that B, an element of the event space $\{B_i : i = 1,2,3,......\}$. The probability $P(A)$ is then given by

$$P(A) = \sum_i P(A \mid B_i)P(B_i). \tag{A.10}$$

In the special case where the event space is simply B and its complement $\{B, \bar{B}\}$

$$P(A) = P(A \mid B)P(B) + P(A \mid \bar{B})P(\bar{B})$$

The law of total probability applies to distributions of random variables in an analogous way, in that if we define A as the event that $X_1 = x_1$ and B the event that $X_2=x_2$ where, for instance, X_i are discrete random variables, we have

$$f_1(x_1) = \sum_{x_2} f_1(x_1 \mid x_2) f_2(x_2)$$
(A.11)

For continuous variables the analogous expression is

$$f_1(x_1) = \int_{x_2} f_1(x_1 \mid x_2) f_2(x_2) dx_2$$
(A.12)

Bayes' theorem

Using the above properties we can calculate the following:

$$P(B \mid A) = \frac{P(A \mid B) P(B)}{P(A)}$$

and

$$P(B \mid A) = \frac{P(A \mid B) P(B)}{\sum_{i=1}^{k} P(A \mid B_i) P(B_i)}$$
(A.13)

Bayes' theorem forms a general relationship between **conditional** and **unconditional probabilities,** and has many useful applications. One very useful role for Bayes' theorem is in helping us to update knowledge (e.g., about a parameter value) from information (e.g., sample data). For example, if we let θ stand for the value of a parameter and x stand for sample data, we can rewrite Bayes' theorem as

$$P(\theta \mid x) = \frac{P(x \mid \theta) P(\theta)}{P(x)}$$

Using the above results for conditional distributions and total probability, Bayes' theorem extends to distributions in the same manner:

$$f_2(x_2 \mid x_1) = \frac{f_1(x_1 \mid x_2) f_2(x_2)}{f_1(x_1)}$$
(A.14)

where

$$f_1(x_1) = \sum_{x_2} f_1(x_1 \mid x_2) f_2(x_2)$$

or

$$f_1(x_1) = \int_{x_1} f_1(x_1 \mid x_2) f_2(x_2) dx_2$$

for continuous variables.

Again, if we let θ stand for the value of a parameter and x for sample data, then

$$f_\theta(\theta \mid x) = \frac{f_x(x \mid \theta) f_\theta(\theta)}{f_x(x)} \qquad (A.15)$$

where is the **posterior distribution** of θ and

$$f_x(x) = \int_\theta f_x(x \mid \theta) f_\theta(\theta) d\theta$$

is the weighted average of the **likelihood** $f_x(x \mid \theta)$ over the **prior distribution** of θ. Since this will usually be a constant for given data, this leads to the general statement

$$f_\theta(\theta \mid x) \propto f_x(x \mid \theta) f_\theta(\theta)$$

Distribution moments

Probability distributions can be summarized by their **moments.** The kth moment of a probability distribution is defined as

$$M_k = \sum_x x^k f(x)$$

for discrete distributions and

$$M_k = \int_x x^k f(x) dx$$

for continuous distributions. The two most commonly used distribution moments are the first central moment or **mean**

$$M_1 = E(x) = \mu = \int_x x f(x) dx \qquad (A.16)$$

and the second moment about the mean, known as the **variance**

$$M_2' = E[(x - \mu)^2] = \sigma^2 = \int_x (x - \mu)^2 f(x) dx \qquad (A.17)$$

The third central moment M_3 is often used as a measure of distribution skewness, with symmetric distributions such as the normal having $M_3 = 0$ (if the moment is defined). The fourth central moment M_4 provides a measure of kurtosis relative to a normal distribution (Appendix B).

The mean and variance play important roles in describing data and summarizing distributions. In addition to the mean and variance of the random variable x, we are often interested in means and variances of functions of x. Linear functions provide the most straightforward results. Thus, if

$$y = g(x) = b_0 + bx$$

then

$$E(y) = b_0 + bE(x)$$

and

$$\text{Var}(y) = b_0 + b^2 \text{Var}(x)$$

Note that $\text{Var}(b_0) = 0$, since b_0 is a constant.

The above extends to linear functions of several random variables. So, if

$$y = g(\underline{x}) = b_0 + b_1 x_1 + b_2 x_2 + \ldots\ldots + b_k x_k$$

then

$$E(y) = b_0 + \sum_{i=1}^{k} b_i E(x_i) \tag{A.18}$$

If the elements of $\underline{x}$ are mutually independent

$$\text{Var}(y) = \sum_{i=1}^{k} b_i^2 \text{Var}(x_i) \tag{A.19}$$

As a special case, the expectation and variance of a sum of independent random variables is given by the sum of the expectations and variances, respectively. That is, for

$$y = g(\underline{x}) = \sum_{i=1}^{k} x_i$$

$$E(y) = \sum_{i=1}^{k} E(x_i)$$

and

$$\text{Var}(y) = \sum_{i=1}^{k} \text{Var}(x_i)$$

Computation of the variance is more complicated when the elements of x are non-independent, and requires inclusion of a covariance term, defined below. If

$$y = g(\underline{x}) = b_0 + b_1 x_1 + b_2 x_2 + \ldots + b_k x_k$$

and $\underline{x}$ are non-independent, then

$$\text{Var}(y) = \sum_{i=1}^{k} b_i^2 \text{Var}(x_i) + 2 \sum_{i=1}^{k} \sum_{j>i}^{k} b_i b_j \text{Cov}(x_i, x_j) \qquad (A.20)$$

where $\text{Cov}(x_i, x_j)$ is the covariance between x_i and x_j.

The **covariance** of between two random variables, x_i and x_j, is defined as

$$\text{Cov}(x_i, x_j) = E\left[(x_i - E(x_i))(x_j - E(x_j))\right] = E(x_i x_j) - E(x_i) E(x_j). \qquad (A.21)$$

Note that $\text{Cov}(x_i, x_i)$ is simply $\text{Var}(x_i)$. The covariance can be scaled by the root product of the variances to produce the **product-moment correlation**

$$\rho(x_i, x_j) = \frac{\text{Cov}(x_i, x_j)}{\sqrt{\text{Var}(x_i) \text{Var}(x_j)}} \qquad (A.22)$$

or equivalently

$$\rho(x_i, x_j) = E\left[\left(\frac{x_i - E(x_i)}{\sqrt{\text{Var}(x_i)}}\right)\left(\frac{x_j - E(x_j)}{\sqrt{\text{Var}(x_j)}}\right)\right] = E\left[\left(\frac{x_i - \mu_{x_i}}{\sigma_{xi}}\right)\left(\frac{x_j - \mu_{x_j}}{\sigma_{xj}}\right)\right]$$

The correlation $\rho(x_i, x_j)$ ranges between -1 (perfect negative linear dependence) and $+1$ (perfect positive dependence). A correlation of 0 indicates absence of linear dependence; when x_i and x_j are jointly normally distributed (Appendix B) $\rho(x_i, x_j) = 0$ implies statistical independence.

Variances of nonlinear functions of random variables require more mathematical complexity. One approach, called the "delta method" invokes a Taylor-series approximation of the function to produce an approximate variance. Thus, if y is a continuous, differentiable function of $\underline{x}$,

$$y = g(\underline{x}) = g(x_1, x_2, \ldots, x_k) \qquad (A.23)$$

then

$$\text{Var}(y) \cong \sum_{i=1}^{k} \left(\frac{\partial g}{\partial x_i}\right)^2 \text{Var}(x_i) + 2 \sum_{i=1}^{k} \sum_{j>i}^{k} \left(\frac{\partial g}{\partial x_i}\right)\left(\frac{\partial g}{\partial x_j}\right) \text{Cov}(x_i x_j) \qquad (A.24)$$

Other methods for approximating variance from complex functions are based on simulation approaches (e.g., parametric bootstrapping) or resampling approaches such as the jackknife or nonparametric bootstrapping (Williams et al. 2002: Chapter 5, Appendix C).

Sample moments

Corresponding to distributions moments are **sample moments**. The kth sample is provided by

$$\hat{M}_k = \frac{1}{n} \sum_{i=1}^{n} X_i^k$$

where a random sample $X_1, X_2, \ldots, X_n$ is drawn from a distribution $f(x)$. The two most common sample moments are the sample mean and variance

$$\hat{M}_1 = \hat{\mu} = \bar{x} = \frac{1}{n} \sum_{i=1}^{n} X_i \tag{A.25}$$

and

$$\hat{M}_2 = \hat{\sigma}^2 = s^2 = \frac{1}{n} \sum_{i=1}^{n} (X_i - \mu)^2 \tag{A.26}$$

In a similar fashion, the sample moments corresponding to the covariance are computed for samples of the two random variables X_1 and X_2

$$\hat{M}_{12} = \hat{\sigma}_{12} = s_{12} = \frac{1}{n} \sum_{i=1}^{n} (X_{1i} - \mu_1)(X_{2i} - \mu_2) \tag{A.27}$$

The relationship between sample moments and distribution moments is exploited by the estimation method known as the **method of moments** (Appendix C).

Additional reading

Evans, M., N. Hastings, and B. Peacock. (2000) *Statistical Distributions*. 3rd edn. Wiley. New York.

Williams, B.K., J. D. Nichols, and M.J. Conroy. (2002) *Analysis and Management of Animal Populations*. Elsevier-Academic. New York.

Appendix B

Common Statistical Distributions

General distribution characteristics

We distinguish between two broad classes of distributions, based on whether the random variable X is **discrete** (taking on a finite number of values over its range) or continuous (taking on an infinite number of values over its range). For discrete random variables we can define the **probability mass function** (pmf) as

$$f(x) = P(X = x).$$

That is, we are assigning a finite probability to each specific outcome $X = x$. For continuous random variables we define a **probability density function** (pdf) that that the random variable is in a specific range

$$P(x_1 < X < x_2) = \int_{x_1}^{x_2} f(x)dx$$

In either case, the distribution operates over an admissible range for the random variable X known as the **support,** which is simply an interval or set whose complement has probability zero. By second and third probability axioms (Appendix A) the total probability over the support must sum (integrate) to 1; thus an essential property is

$$\sum_x f(x) = 1$$

Decision Making in Natural Resource Management: A Structured, Adaptive Approach,
First Edition. Michael J. Conroy and James T. Peterson.
© 2013 John Wiley & Sons, Ltd. Published 2013 by John Wiley & Sons, Ltd.

for pmfs and

$$\int_x f(x)dx = 1$$

for pdfs, where summation or integration is understood to be over the support.

The **cumulative distribution function** (cdf) or simply **distribution function** is closely related to the pmf or pdf. The cdf is defined as

$$F(x) = P[X \le x].$$

For discrete distributions the cdf is obtained by summation

$$F(x) = \sum_{x_L}^{x} f(x)$$

and for continuous by integration

$$F(x) = \int_{x_L}^{x} f(x)dx$$

where x_L is the lower bound of the support of x.

Example: Consider the example from Appendix A, where $x \in \{1,2,3,4,5\}$ with associated discrete pmf $f(\underline{x}) = (0.01, 0.05, 0.5, 0.4, 0.04)$. Suppose we are interested in

$$F(3) = P[X \le 3]$$

We can obtain this value by definition of the cdf

$$F(3) = \sum_{x_L}^{x} f(3) = 0.01 + 0.05 + 0.5 = 0.56$$

In like matter, we can obtain $F(1) = 0.01$, $F(2) = 0.06$, $F(4) = 0.96$, $F(5) = 1$.

Note that in this example we could obtain $P[X > 3]$ by observing that it is the complementary event to $P[X > 3]$, so

$$P[X \le 3] = 1 - F(3) = 0.44$$

More generally, we can obtain the probability that random variable is in a specific interval $[a, b]$ for discrete random variables as

$$P(x_1 < X \le x_2) = F(x_2) - F(x_1)$$

For the example, suppose we wish to find $P(1 < X \le 4)$ for the above pmf

$$P(1 < X \le 4) = F(4) - F(1) = 0.96 - 0.01 = 0.95$$

Of course, we could in this case easily have observed by direct inspection of the pmf that

$$P(1 < X \le 4) = f(2) + f(3) + f(4) = 0.95$$

By analogy, the probability over an interval for a pdf is obtained (as noted earlier) by

$$P(x_1 < X < x_2) = \int_{x_1}^{x_2} f(x)dx = F(x_2) - F(x_1)$$

Distribution parameters

Distributions are described in part by their moments, with the first moment (mean) describing central tendency, the second moment about the mean dispersion, etc. (Appendix A). However, a more concise and complete description is provided by the value of the distribution **parameters** $\underline{\theta}$, a vector of 1 or more constants in the pmf or pdf. Depending on the distribution, the parameter values control the location (central tendency), dispersion (scale), or shape of the distribution function. For this reason the distribution is more generally written as

$$f(x; \underline{\theta})$$

with the understanding that that x is the random variable and $\underline{\theta}$ the vector of constant parameters. As we will see, these roles are effectively reversed when we consider the problem of estimation (Appendix C).

Distribution quantiles

It is often useful to describe values of the random variable that are associated with specific cumulative probability levels. That is, for a specific value of the cdf, we seek the corresponding value of x that satisfies

$$F(x) = P[X \le x] = y$$

For example, we may desire the value of x that corresponds to $F(x_0) = 0.75$; that is, there is a 75% probability that a value of x selected at random will fall below x_0. This problem is solved by inversion of the cdf, that is by evaluating $x_0 = F^{-1}(y)$ or in this case $F^{-1}(0.75)$. The resulting value of x_0 is known as the $P = y$ **quantile** of the distribution. Standard quantiles ($y = 0.01, 0.05, 0.10, 0.5, 0.90, 0.95, 0.99$, etc.) of many distributions are tabulated in statistical appendices or can be obtained via inverse cdf functions built into most standard statistical software. We provide example code for deriving quantiles other statistics from specified distributions in Appendix G.

Generation of random numbers

A common problem is the generation of (pseudo)random number from given distributions. There are a number of approaches for random number generation, most of which depend at some point on the generation of pseudorandom numbers from the uniform $(0,1)$ distribution. A common practice is to draw a pseudorandom number U from this interval, and then to apply the inverse cdf for the relevant distribution. That is, given U we generate x as

$$x = F^{-1}(U)$$

This is based on the straightforward reversal of the fact that $F(x)$ is a mapping of the cumulative distribution of x on $(0,1)$. Thus, the problem of random variable generation devolves to the problem of the satisfactory generation of pseudorandom $U(0,1)$ numbers.

The inverse cdf approach can be applied directly for most continuous cdfs, but must be modified for discrete variables. We will describe more specific approaches for random variable generation for each of the distributions covered.

Continuous distributions

Uniform distribution

The **uniform (or rectangular) distribution** expresses equal probability of outcome over the real interval (a,b). Thus, support is defined as $a < x < b$; a and b are the distribution parameters (i.e., the minimum and maximum). When the uniform is referenced without parameters, implicitly one is referring to the uniform$(0,1)$. The probability density function of the uniform is given as

$$f(x; a, b) = \frac{1}{b-a}$$

and the cdf is

$$F(x; a, b) = \frac{x-a}{b-a}$$

The mean and variance are

$$E(x) = \frac{a+b}{2}$$

$$\mathrm{Var}(x) = \frac{(b-a)^2}{12}$$

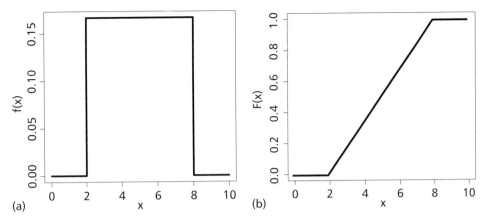

Figure B.1 Uniform distribution on interval $2 < x < 8$. (a) probability density function, (b) cumulative distribution function.

Because of the central nature of the uniform to random variable generation, uniform(0,1) pseudorandom number generators are routinely provided as part of statistical software (e.g., R).

Figure B1 illustrates the pdf and cdf for the uniform distribution.

Normal distribution

The **normal distribution** is perhaps the most familiar statistical distribution. It is symmetric about the mean, with the familiar bell-shaped curve, and is used to model continuous, real values with theoretical support from negative to positive infinity. It is the limiting distribution of many test statistics and functions and is commonly used as an approximation, even when the data are thought to follow some other distribution, often after transformation to reduce skewness or discontinuities in the data. The normal density is determined by the 2 parameters μ and σ^2 ($\sigma^2 > 0$) as

$$f(x; \mu, \sigma^2) = \frac{1}{\sqrt{2\pi\sigma^2}} \exp\left(\frac{-(x-\mu)^2}{2\sigma^2}\right), \quad -\infty \le x \le \infty$$

The mean and variance of the normal distribution are simply μ and σ^2. Often, the normal distribution is standardized by referring to standardized variates

$$Z_i = \frac{(X_i - \mu)}{\sigma}$$

If $X_i \sim N(\mu, \sigma^2)$ then $Z_i \sim N(0,1)$, which is referred to as a **standard normal distribution**. Figure B2 illustrates the standard normal distribution, and Figure B3 shows the influence of changes in the location and scale parameter.

Quantiles of the normal are easily obtained by inversion of the cdf; some typical quantiles of the standard normal are $F^{-1}(0.01) = -2.32, F^{-1}(0.05) = -1.64,$

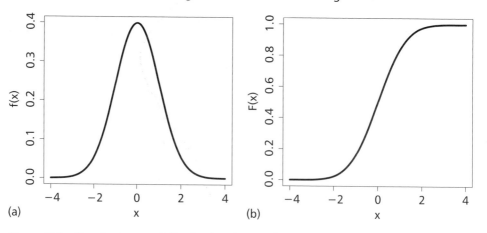

Figure B.2 Standard normal distribution ($\mu = 0, \sigma^2 = 1$). (a) probability density function, (b) cumulative distribution function.

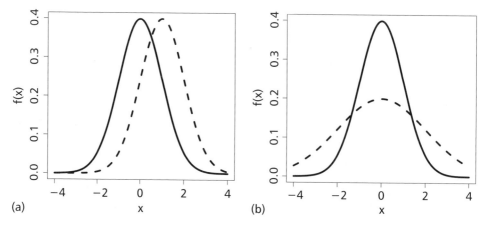

Figure B.3 Effect of change of location and scale parameters on normal distribution. Solid line is standard normal distribution ($\mu = 0, \sigma^2 = 1$). Dashed line indicates (a) $\mu = 1$, $\sigma^2 = 1$, (b) $\mu = 0, \sigma^2 = 2$.

$F^{-1}(0.25) = -0.67$, $F^{-1}(0.5) = 0$, with the remaining quantiles at probability levels of 0.75, 0.95, and 0.99 given by symmetry as 0.67, 1.64, and 2.32.

Random normal variates can be obtained by inversion of the cdf, but a more efficient method (Evans et al. 2000) proceeds as follows:

- Generate 2 independent uniform(0,1) variates U_1 and U_2.
- From these generate 2 independent standard normal variates
  - $Z_1 = \sqrt{-2\log(U_1)}\sin(2\pi U_2)$, $Z_2 = \sqrt{-2\log(U_1)}\cos(2\pi U_2)$.
- If needed, derive the normal (non-standard) variates as
  - $X_i = Z_i\sigma + \mu$.
- Repeat a sufficient number of times to obtain the desired sample.

The multivariate generalization of the normal distribution is the **multivariate normal distribution**. The multivariate normal response $\underline{x}$ is a vector of continuous attributes $\underline{x} = (x_1, \ldots, x_k)$ the individual elements of which may be correlated with one another. The multivariate normal probability density function is

$$f(\underline{x}; \underline{\mu}, \Sigma) = \frac{1}{\sqrt{2\pi}} |\Sigma|^{-1/2} \exp\left(-\frac{1}{2}(\underline{x} - \underline{\mu})' \Sigma^{-1}(\underline{x} - \underline{\mu}) \right)$$

where $\underline{\mu} = (\mu_1, \ldots, \mu_k)$ is a vector a means for each of the k attributes and Σ is a positive-definite $k \times k$ variance–covariance matrix elements $\Sigma_{ij} = \sigma_{ij}$, $i = 1, \ldots, k$, $j = 1, \ldots, k$, so

$$\Sigma = \begin{bmatrix} \sigma_1^2 & \sigma_{12} & \cdot & \cdot & \cdot & \cdot & \sigma_{1k} \\ \sigma_{21} & \sigma_2^2 & \sigma_{23} & \cdot & \cdot & \cdot & \sigma_{2k} \\ \cdot & \cdot & \cdot & \cdot & \cdot & \cdot & \cdot \\ \cdot & \cdot & \cdot & \sigma_i^2 & \cdot & \cdot & \cdot \\ \cdot & \cdot & \cdot & \cdot & \cdot & \cdot & \cdot \\ \cdot & \cdot & \cdot & \cdot & \cdot & \cdot & \cdot \\ \sigma_{k1} & \sigma_{k2} & \cdot & \cdot & \cdot & \cdot & \sigma_k^2 \end{bmatrix}$$

Thus, the variance terms appear down the principal diagonal and the covariance terms on the off-diagonal elements. In the case where the attributes are all uncorrelated, the off-diagonal elements are zero. In this case, the multivariate normal reduces to the product of k independent normal distributions, with means $\underline{\mu} = (\mu_1, \ldots, \mu_k)$ and variances $\underline{\sigma}^2 = (\sigma_1^2, \ldots, \sigma_k^2)$.

The bivariate normal distribution is a special case of the multivariate normal when there are two jointly distributed random variables, $\underline{x} = (x_1, x_2)$

$$f(x_1, x_2; \mu_1, \mu_2, \sigma_1^2, \sigma_2^2, \sigma_{12}) = \frac{1}{\sqrt{2\pi}} |\Sigma|^{-1/2} \exp\left(\frac{-(\underline{x} - \underline{\mu})' \Sigma^{-1}(\underline{x} - \underline{\mu})}{2} \right)$$

where

$$\underline{\mu} = (\mu_1, \mu_2)$$

$$\Sigma = \begin{bmatrix} \sigma_1^2 & \sigma_{12} \\ \sigma_{12} & \sigma_2^2 \end{bmatrix}$$

Alternatively, the distribution can be parameterized in terms of $\underline{\mu}$, σ_1^2, σ_2^2 and ρ_{12}, where

$$\sigma_{12} = \rho_{12} \sqrt{\sigma_1^2 \sigma_2^2}$$

In the absence of linear dependence $\rho_{12} = \sigma_{12} = 0$, so that Σ simplifies to

$$\Sigma = \begin{bmatrix} \sigma_1^2 & 0 \\ 0 & \sigma_2^2 \end{bmatrix}$$

It readily follows that

$$f(x_1, x_2; \mu_1, \mu_2, \sigma_1^2, \sigma_2^2) = f(x_1; \mu_1, \sigma_1^2)f(x_2; \mu_2, \sigma_2^2)$$

that is, x_1 and x_2 are statistically independent. Note that $\rho_{12} = 0$ is a **necessary** condition for independence, but is only **sufficient** for independence when x_1 and x_2 are jointly normally distributed.

Exponential distribution

The **exponential distribution** is typically used to model the distribution of time elapsed until an event (e.g., a machine component failure or a mortality) when the probability of such an event occurring over the next short time interval does not change through time. The exponential is thus the simplest model available for describing mortality or other events in terms of "failure times." The distribution has a single parameter, λ, and the random variable x is constrained to be nonnegative but otherwise may take any real value, $0 \le x < +\infty$. The probability density function of the exponential is given as

$$f(x; \lambda) = \lambda e^{-\lambda x}$$

and the cdf is

$$F(x; \lambda) = 1 - e^{-\lambda x}$$

The mean and variance are

$$E(x) = \frac{1}{\lambda}$$

$$Var(x) = \frac{1}{\lambda^2}$$

Random exponential numbers are easily generated from uniform(0,1) variates U by the transform

$$X = \frac{-\log(U)}{\lambda}$$

The **geometric distribution**, discussed below, is the discrete analogue to the exponential distribution. The exponential distribution is illustrated in Figure B4.

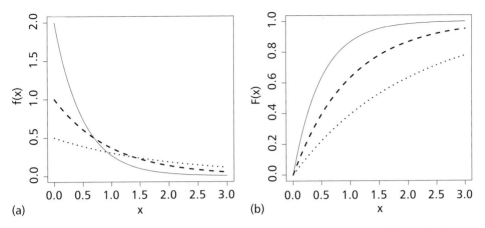

Figure B.4 Exponential distribution. (a) probability density function, (b) cumulative distribution function. $\lambda = 2$ solid line, $\lambda = 1$ dashed line, $\lambda = 0.5$ dotted line.

Gamma distribution

The **gamma distribution** generalizes the exponential distribution by the addition of a shape parameter; thus the exponential is a special case of the gamma (as is the chi-squared distribution). The distribution has two parameters: a scale parameter $b > 0$ and a shape parameter $c > 0$. As with the exponential, the gamma random variable x is constrained to be nonnegative but otherwise may take any real value, $0 \le x < +\infty$. The probability density function of the gamma is given as

$$f(x; b, c) = \frac{(x/b)^{c-1} e^{-(x/b)}}{b\Gamma(c)}$$

where $\Gamma(c)$ is the gamma function

$$\Gamma(c) = \int_0^\infty \exp(-u) u^{c-1} du$$

which in practice is obtained from a numerical function library. The cdf is easily tractable in the case where c is an integer.

$$F(x; b, c) = 1 - \left[e^{-(x/b)} \right] \sum_{i=0}^{c-1} \frac{(x/b)^i}{i!}$$

Example gamma distributions are plotted in Figure B5. The mean and variance of the gamma distribution are

$$E(x) = bc$$

$$\mathrm{Var}(x) = b^2 c.$$

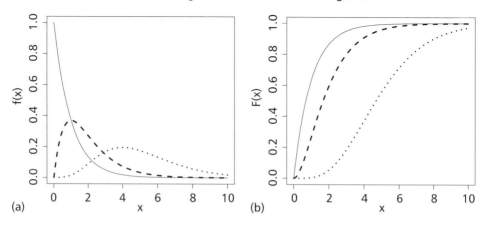

Figure B.5 Gamma distribution. (a) probability density function, (b) cumulative distribution function. Scale parameter $b = 1$, shape parameter $c = 1$ (solid line), $c = 2$ (dashed line), $c = 5$ (dotted line).

Gamma random variates can be generated from uniform (0,1) pseudorandom numbers. When c is an integer

$$X = \sum_{i=1}^{c} (-b \log U)$$

is distributed gamma(b,c), where $U \sim Uniform(0,1)$. Built-in random number functions (e.g., in R) provide random numbers and cdf values for the more general case of non-integer c; note that in the R function for the gamma the inverse scale parameter $1/b$ is used as an argument, rather than b (see Appendix G).

Besides being the distribution that leads to the exponential, chi-squared, and other important distributions as special cases, the gamma distribution is useful for modeling heterogeneity in density, resulting in the well-known Poisson–gamma mixture model. The gamma distribution is also the natural conjugate distribution for the Poisson likelihood in Bayesian analysis, and for the precision (inverse variance) parameter of a normal distribution with known mean (Appendix C).

Beta distribution

The **beta distribution** is important for modeling continuous random variables that are restricted to a finite range. In particular, the beta distribution is useful for modeling heterogeneity in probabilities. The beta also serves as the natural conjugate distribution for the Bernouilli and binomial likelihoods in Bayesian analysis.

We now take x to be confined to the unit interval, so as with the exponential and the gamma the random variable, x is constrained to be nonnegative but otherwise may take any real value, $0 < x < 1$.

The pdf of the beta is

$$f(x; \alpha, \beta) = \frac{x^{\alpha-1}(1-x)^{\beta-1}}{B(\alpha, \beta)}$$

where $B(\alpha, \beta)$ is the beta function

$$B(\alpha, \beta) = \int_0^1 u^{\alpha-1}(1-u)^{\beta-1} du$$

The two parameters $\alpha > 0$ and $\beta > 0$ control the location, shape and scale of the distribution. A beta with parameters $\alpha = 1$ and $\beta = 1$ is equivalent to a uniform(0,1). One interpretation of the beta parameters is as indicating α–1 successes in $\alpha + \beta$–2 trials; thus the beta can be useful for capturing "prior information" about a probability value. Example beta distributions are plotted in Figure B6. The mean and variance of the beta distribution are

$$E(x) = \frac{\alpha}{(\alpha + \beta)}$$

$$Var(x) = \frac{\alpha\beta}{\left[(\alpha + \beta)^2 (\alpha + \beta + 1)\right]}$$

These expressions can be solved for α and β as functions of $E(x)$ and $Var(x)$, using the method of moments (Appendix C).

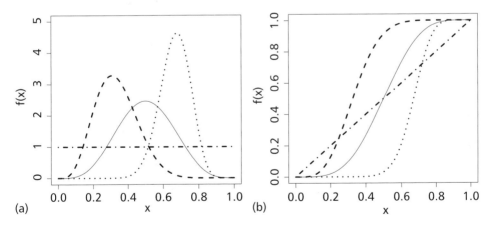

Figure B.6 Beta distribution. (a) probability density function, (b) cumulative distribution function. Parameters $\alpha = \beta = 5$ (solid line), $\alpha = 5$ $\beta = 10$ (dashed line), $\alpha = 20$ $\beta = 10$ (dotted line), $\alpha = 1$ $\beta = 1$ (dot-dashed line).

When α and β are integer, beta random variates can be generated from uniform (0,1) pseudorandom numbers as

$$X = \frac{a}{a+b}$$

where

$$a = -\log \sum_{i=1}^{\alpha} U_i, \; b = -\log \sum_{i=1}^{\beta} U_i$$

and

$$U_i \sim uniform(0,1)$$

Built-in random number functions (e.g., in R) provide random numbers and cdf values for the more general case of non-integer α and β.

The beta distribution generalizes to the multivariate case where x is now a vector response, with each of the $0 < x_i < 1$. The **Dirichlet** probability density function is

$$f(x_1, \ldots x_k; \alpha_1, \ldots \alpha_k) = \frac{\prod_{i=i}^{k} x_i^{\alpha_i - 1}}{B(\underline{\alpha})}$$

Similar to the beta, the Dirichlet serves as the natural conjugate distribution for the **multinomial** likelihood in Bayesian analysis (Appendix C).

Distributions of quadratic forms

Three important distributions arise from quadratic forms (sums of squares) of normal random variables; 2 of these are in turn derived from the first, the chi-squared distribution. The chi-squared distribution is also important as the asymptotic distribution of important test statistics such as the likelihood ratio test.

The **chi-squared distribution**, in turn is a special case of the gamma distribution previously considered. In fact, the chi-squared distribution with v degrees of freedom is simply a gamma distribution with scale parameter $b = 2$ and shape parameter $c = v/2$. Thus the pdf of the chi-squared is

$$f(x; v) = \frac{x^{(v-2)/2} e^{-(x/b)}}{2^{v/2} \Gamma(v/2)}$$

The mean and variance of the chi-squared distribution are

$$E(x) = v$$

$$\text{Var}(x) = 2v$$

The **Student's t distribution** and **Fisher's F distribution** are both derived as ratios of independent chi-squared variables. Student's t with v degrees of freedom is the square root of the ratio of a chi-squared random variable with $v = 1$ and a chi-squared $v = v$. Fisher's F variates with v, ω degrees of freedom are obtained as ratio of a chi-squared with v degrees of freedom to an independent chi-squared with ω degrees of freedom.

The **Wishart distribution** is a multivariate generalization of the chi-squared distribution that results from quadratic forms of multivariate normal variables. The main application of the Wishart distribution is as a conjugate prior distribution for the precision (inverse of the variance–covariance matrix) parameter of the multivariate normal (Appendix C).

Discrete distributions

Discrete uniform

The discrete uniform distribution describes the probability of an integer value over a specific range, conventionally 0 to n. The random variable $x \in \{0,1,2,....,n\}$. The probability mass function is given by

$$f(x; n) = \frac{1}{(n+1)}, \, x \in \{0, 1, 2, ..., n\}.$$

The cumulative distribution function is

$$F(x; n) = \frac{(x+1)}{(n+1)}$$

over the same support. The mean and variance are

$$E(x) = n/2$$

$$\text{Var}(x) = \frac{n(n+2)}{12}$$

The above pmf/cdf readily generalize to integers over any range $[a, n]$ by addition of a constant. For example, $y \sim \text{Disuniform}(a, n)$ if $y = a+x$ and $x \sim \text{Disuniform}(0, n-c)$.

Figure B7 illustrates the pdf and cdf for the discrete uniform.

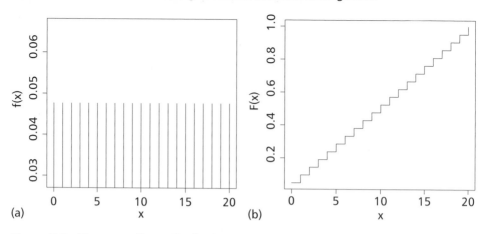

Figure B.7   Discrete uniform distribution $x \in \{0,1,2,....,20\}$ (a) probability mass function (b) cumulative distribution function.

Bernoulli and binomial distributions

The Bernoulli and binomial distributions are appropriate when responses are binary. Examples include "heads or tails" in a coin flip, survival or death of an individual over a time interval, captured or not captured for a marked animal. The **Bernoulli** random variable x takes on value of 1 or 0 representing "success" or "failure" respectively, with "success" depending on the experimental context. The probability mass function is written as

$$f(x; p) = p^x (1-p)^{1-x}, x = 0, 1$$

and simplifies to $f(0;p) = (1-p)$ and $f(1;p) = p$, where p is the probability of success. Note that we assume that there are only 2 possible outcomes, a success with probability p and a failure with probability $1-p$, and that by definition the probability that it is either a success or failure adds to 1. The mean and variance of the Bernoulli distribution are simply

$$E(x) = p$$

$$\mathrm{Var}(x) = p(1-p)$$

The **binomial distribution** is closely related, with the binomial variable x defined as number of success in n independent Bernoulli trials, each with probability p of success. The binomial thus has two parameters (n and p) though one of these (n) ordinarily is known and will not be estimated from data. The binomial mass function is

$$f(x; n, p) = \binom{n}{x} p^x (1-p)^{n-x}., x = 0, n$$

The binomial distribution function is

$$F(x; n, p) = \sum_{k=0}^{x} \binom{n}{k} p^k (1-p)^{n-k}, x = 0, n$$

The mean and variance are given by

$$E(x) = \mu = np$$

and

$$Var(x) = np(1-p)$$

See Figure B8 for examples of the binomial pdf and cdf.

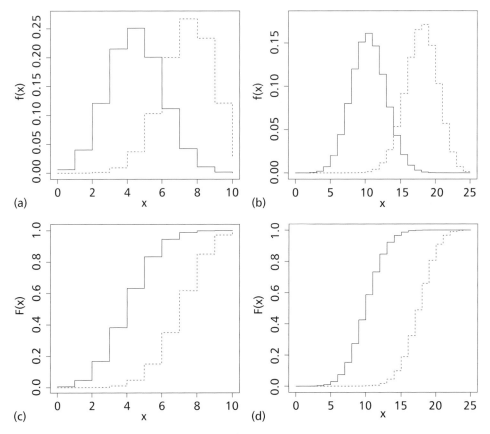

Figure B.8 Binomial distribution (a,b) probability mass function, (a) $n = 10$, (b) $n = 25$; (c,d) cumulative distribution function (c) $n = 10$, (d) $n = 25$. Parameter $p = 0.4$ (solid line), 0.7 (dashed line).

Bernouilli random variables are typically generated by the rejection technique, where a pseudorandom uniform(0,1) number U is obtained and compared to p. If $U < p$ then $x = 1$, otherwise $X = 0$. Binomial random variables can be generated in the same fashion, where n uniform(0,1) numbers are generated; the number of these that are $< p$ is x and the remainder is $n–x$. Alternatively, the **geometric distribution** (below) can be used to generate the number of failures before the first success, which by definition is $n–x$. The latter method is faster when p is small.

Multinomial distribution

The **multinomial distribution** extends the binomial to the situation where the events are k mutually exclusive outcomes; the binomial is a special case of the multinomial where $k = 2$. The multinomial variable $\underline{x}$ is defined as number of outcomes in each of k categories in n independent trials, with probability $\underline{p}$ of success. Thus, both $\underline{x}$ and $\underline{p}$ are defined as vectors

$$\underline{x} = [x_1, x_2, x_3, \ldots, x_k]$$

and

$$\underline{p} = [p_1, p_2, p_3, \ldots, p_k]$$

where p_i is the probability that in a single trial $x_i = 1$. Only $k–1$ of the outcomes are independent, with the kth by complementarity; thus these vectors reduce to

$$\underline{x} = \left[x_1, x_2, x_3, \ldots, x_{k-1}, n - \sum_{i=1}^{k-1} x_i \right]$$

and

$$\underline{p} = \left[p_1, p_2, p_3, \ldots, p_{k-1}, 1 - \sum_{i-1}^{k-1} p_i \right]$$

Thus, the multinomial has k parameters (n and $p_1, p_2, p_3, \ldots, p_{k-1}$) though again n ordinarily is known and will not be estimated from data. The multinomial mass function is

$$f(\underline{x}; n, \underline{p}) = \binom{n}{x_1, x_2, \ldots, x_k} \prod_{i=1}^{k} p_i^{x_i}$$

where again $\sum_{i=1}^{k} p_i = 1$ and $\sum_{i=1}^{k} x_i = n$. When $n = 1$ the multinomial reduces to a multidimensional version of the Bernoulli distribution; when $k = 2$ the multinomial reduced to the binomial. The cdf for the multinomial is the multivariate analog of the binomial cdf

$$F(\underline{x}; n, p) = \sum_{c_1=0}^{x_1} \sum_{c_2=0}^{x_2} \cdots \sum_{c_k}^{x_k} \binom{n}{c_1, c_2, \ldots c_k} \prod_{i=1}^{k} p_i^{c_i}$$

The means and variances are in terms of the individual outcomes and are given by

$$E(x_i) = \mu_i = np_i$$

and

$$Var(x_i) = np_i(1 - p_i)$$

Additionally, the covariance between outcomes i and j is given by

$$E[(x_i - \mu_{x_i})(x_j - \mu_{x_j})] = -np_i p_j, \, i \neq j$$

Poisson distribution

The **Poisson distribution** models outcomes that take on nonnegative integer values $(0, 1, 2, \ldots \ldots, n_\infty)$. Examples include counts of animal and plants, where the process generating the counts in space is "random" in the sense that counts are not clustered separated except by chance. The Poisson distribution is specified by the single parameter λ, which is equal to both the population mean and variance. Thus, sometimes the ratio of the sample mean to the variance is used as evidence (or lack thereof) of Poisson assumptions, with values of this ratio ~1 taken as support for a Poisson count model. The probability mass function of the Poisson is given by

$$f(x; \lambda) = \frac{\lambda^x e^{-\lambda}}{x!}, \, x = 0, 1, 2, 3, \ldots \ldots, n_\infty$$

where e is the base of the natural logarithm, $\lambda > 0$, and $x!$ denotes the factorial function $x(x-1)(x-2)\ldots..1$. The distribution function is simply given by summation over the discrete values of x of the density

$$F(x; \lambda) = \sum_{k=0}^{x} f(k; \lambda) = \sum_{k=0}^{x} \frac{\lambda^k e^{-\lambda}}{k!}, \, x = 0, 1, 2, 3, \ldots \ldots, n_\infty$$

As noted above, the mean and variance are given simply by

$$E(x) = \lambda$$

and

$$Var(x) = \lambda$$

The Poisson pmf and cdf are illustrated in Figure B9.

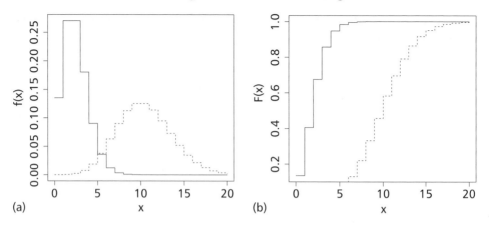

Figure B.9 Poisson distribution (a) probability mass function (b) cumulative distribution function. Parameter $\lambda = 2$ (solid line) and 10 (dashed line).

Poisson random variates are computed by calculating $F(x; \lambda)$ for specified λ up to an arbitrary cutoff $x_1, x_2, \ldots, x_N$, where N will be chosen to be sufficiently large so that higher values of x are improbable. Pseudorandom $U \sim \text{uniform}(0,1)$ numbers are drawn and compared to the (truncated) cdf. The value of x is then chosen such that

$$F(x) \le U < F(x+1)$$

Geometric distribution

The **geometric distribution** describes the number of independent Bernoulli trials that occur before the first success. It is thus the discrete analog to the exponential distribution, modeling trials rather than time elapsed until some event. The probability mass function of the geometric distribution is

$$f(x; p) = p(1-p)^x$$

where x is a nonnegative integer and p, the probability of success in each trial, is the single parameter. The cumulative distribution function is

$$F(x; p) = 1 - (1-p)^x$$

The mean and variance are given by

$$E(x) = \frac{(1-p)}{p}$$

$$\text{Var}(x) = \frac{(1-p)}{p^2}$$

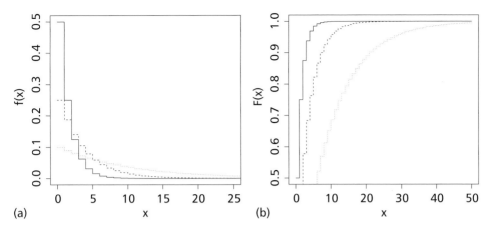

Figure B.10   Geometric distribution (a) probability mass function (b) cumulative distribution function. Parameter p = 0.5 (solid line), 0.25 (dashed line), 0.1 (dotted line).

The pmf and cdf for the geometric are illustrated in Figure B10.

Geometric random variates can be generated by first generating pseudorandom U~uniform(0,1) numbers. For each uniform variate U the value

$$\frac{\log(U)}{\log(1-p)} - 1$$

is rounded to the next largest integer to generate x. The geometric distribution can also provide a useful alternative for generating binomial random variates. The approach is to generate a series of geometric random variates $x_i, i = ,1,....,m$. until their sum exceeds $n–m$, where n is the (known) number of Bernoulli trials. The number m is now a binomial random variate. For example, if n =10 and $p = 0.01$, the sequence of random geometric numbers might be produced $x = \{5,5,5,12,39,34,5,33, 8,9\}$. The sum of these exceeds $100–m = 93$ when $m = 7$, returning a value of 7 for the binomial variate.

Negative binomial distribution

Similar to the geometric, the special case of the **negative binomial** arise as the number of failures x before the rth success. The distribution generalizes to non-integer values of r (so, "success" is not an integer); and has two parameters, r (referred to above) and p, the probability of success on each trial. The probability mass function for the general (non-integer) case is

$$f(x; r, p) = \frac{\Gamma(r+x)}{\Gamma(r)x!} p^r (1-p)^x, \, x = 0, 1, 2, ...$$

where Γ is the gamma function defined earlier.

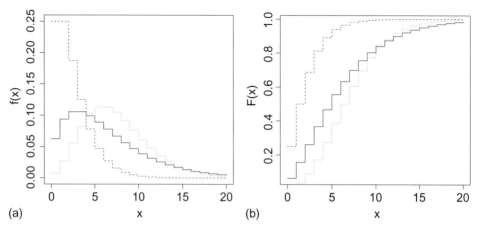

(a)   (b)

Figure B.11 Negative binomial distribution (a) probability mass function (b) cumulative distribution function. Parameters $r = 2$, $p = 0.25$ (solid line), $r = 2$, $p = 0.5$ (dashed line), $r = 7$, $p = 0.5$ (dotted line).

The mean and variance are given by

$$E(x) = \frac{r(1-p)}{p}$$

$$Var(x) = \frac{r(1-p)}{p^2}$$

The pmf and cdf for the negative binomial are illustrated in Figure B11.

The similarity of these to the mean and variance of the geometric distribution is not coincidental, since the geometric is a special case of the negative binomial with $r = 1$. Also, as r tends to infinity and p tends to 1 with fixed $r(1-p) = \lambda$, the negative binomial variate x tends to the Poisson(λ) variate. Finally, the negative binomial also arises as a mixture between the gamma and the Poisson distributions.

Negative binomial random variates can be generated several ways.

- Generate a sequence of uniform(0,1) numbers and keep track of the number of these $n < p$. When n reached r the number $> p$ is a negative binomial variate (this works for integer r).
- The sum of r geometric(p) random numbers produces a negative binomial variate (again for integer r); this is faster when p is small.
- Generate as a Poisson-gamma mixture
  - Draw λ as a random gamma(r, $p/(1-p)$).
  - Draw x as a random Poisson(λ).

Hypergeometric distribution

The **hypergeometric distribution** arises in sampling without replacement from a finite population with two or more categories. In the generalized hypergeometric (k categories), the random variable x_i is the frequency of individuals in category i from a random sample of n, $n = \sum_{i=1}^{k} x_i$; in this respect it resembles the multinomial outcome. The sample n is assumed to come from the finite population N without replacement. The population in turn is composed of M_i total individuals in each category i so that $N = \sum_{i=1}^{k} M_i$. The probability mass function for $\underline{x} = (x_1, x_2, \ldots, x_k)$ is then

$$f(\underline{x};, \underline{M}) = \frac{\prod_{i=1}^{k} \binom{M_i}{x_i}}{\binom{N}{n}}$$

where x_i $i=1, \ldots, k$ are nonnegative integers. When there are two categories the generalized hypergeometric reduces to the standard hypergeometric

$$f(x; n, N, M) = \frac{\binom{M}{x}\binom{N-M}{n-x}}{\binom{N}{n}}$$

with cumulative distribution function

$$F(x; n, N, M) = \sum_{k=0}^{x} \frac{\binom{M}{k}\binom{N-M}{n-k}}{\binom{N}{n}}$$

The mean and variance of the standard hypergeometric are

$$E(x) = n\frac{M}{N}$$

$$\mathrm{Var}(x) = n\left(\frac{M}{N}\right)\left(\frac{N-M}{N}\right)\left(\frac{N-n}{N-1}\right)$$

As the population size N becomes large relative to the finite sample n, the hypergeometric converges on the binomial distribution, with $p = M/N$ representing

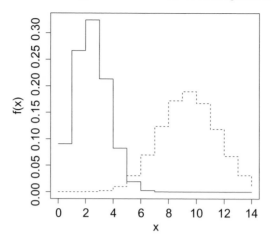

Figure B.12 Hypergeometric distribution (a) probability mass function (b) cumulative distribution function. Parameters $n = 8$, $M = 25$, $N = 100$ (solid line), $n = 25$, $M = 36$, $N = 100$ (dashed line).

the probability of "success" (occurrence in the first category). Examples of the hypergeometric pmf and cdf are given in Figure B12.

Hypergeometric random variates are easily generated by a rejection technique similar to that used for the binomial, except now accounting for depletion of the available sample units due to sampling without replacement. A simple approach (e.g., Evans et al. 2000) is to generate $i = 1, \ldots, n$ uniform(0,1) pseudorandom numbers, U_i. A success $z_i = 1$ is recorded if $U_i < p_i$, otherwise $z_i = 0$, where

- $p_1 = M / N$
- $p_{i+1} = \dfrac{[(N - i + 1)p_i - z_i]}{(N - i)}, i \geq 2$

The hypergeometric variate is then $x = \displaystyle\sum_{i=1}^{n} z_i$.

Reference

Evans, M., N. Hastings, and B. Peacock. (2000) *Statistical Distributions*. 3rd edn. Wiley. New York.

Additional reading

Williams, B.K., J.D. Nichols, and M.J. Conroy. (2002) *Analysis and Management of Animal Populations*. Elsevier-Academic. New York.

Appendix C

Methods for Statistical Estimation

General principles of estimation

Properties of estimators

An **estimator** $\hat{\theta}$ of a parameter θ is a function $\hat{\theta}(x)$ of the sample data x, and consequently inherits the distributional properties of the data. **Bias** refers to the average distance of the estimator $\hat{\theta}(x)$ from the parameter value θ

$$\text{Bias}\left(\hat{\theta}\right) = \text{E}\left[\hat{\theta}(x) - \theta\right]$$

which in turn is evaluated as

$$\text{Bias}\left(\hat{\theta}\right) = \int_{x} \hat{\theta}(x) f(x) dx - \theta$$

or for discrete distributions

$$\text{Bias}\left(\hat{\theta}\right) = \sum_{x} \hat{\theta}(x) f(x) - \theta. \tag{C.1}$$

Precision is measured by the **variance** of $\hat{\theta}$,

$$\text{Var}\left(\hat{\theta}\right) = \text{E}\left(\hat{\theta}(x) - \text{E}\left(\hat{\theta}\right)\right)^{2}. \tag{C.2}$$

Decision Making in Natural Resource Management: A Structured, Adaptive Approach,
First Edition. Michael J. Conroy and James T. Peterson.
© 2013 John Wiley & Sons, Ltd. Published 2013 by John Wiley & Sons, Ltd.

Notice that variance is a measure of the average quadratic distance of the estimate from its own mean. By contrast, **accuracy** represents the average quadratic distance of the estimate from the true value θ, conventionally referred to as **mean squared error** (MSE)

$$\text{MSE}(\hat{\theta}) = \text{E}\left[\left(\hat{\theta}(x) - \theta\right)^2\right].$$ (C.3)

Rearrangement of terms of this expression yields the equivalent

$$\text{MSE}(\hat{\theta}) = \text{E}\left[\left(\hat{\theta}(x) - \text{E}(\bar{\theta})\right)^2\right] + \left[\text{E}(\hat{\theta}(x)) - \theta\right]^2 = \text{Var}(\hat{\theta}) + \text{Bias}^2(\hat{\theta}).$$ (C.4)

Bias and precision can be evaluated a number of different ways. First, if an estimator is function of data that follows a known distribution, it may be possible to evaluate accuracy directly via evaluation of expressions (C.1)–(C.4). In practice, either the distribution of x is not a known form; the functional form of $\hat{\theta}(x)f(x)$ is too complex to permit analysis; or both. In many cases, we are specifically interested in evaluating estimator accuracy when assumptions of the underlying model are violated to varying degrees (e.g., we have heterogeneous rather than homogeneous p in a binomial model); in many of these cases it is not possible to analytically derive the distribution of $\hat{\theta}(x)$, eliminating the above analytical approach. If the distribution of $\hat{\theta}(x)$ can be simulated, then we can use simulation to generate large number of sample values of $\hat{\theta}(x)$ and evaluate their sample moments. For example, we could evaluate bias by

$$\text{Biâs}_{sim}(\hat{\theta}) = \sum_{i=1}^{n} \hat{\theta}(\underline{x}_i)/n - \theta$$ (C.5)

where n samples of data $\underline{x}$ are generated by any desired process (e.g., a mixture of several distributions) and each is used to calculate $\hat{\theta}(x)$; for n sufficiently large this can be a good approximation to bias. A similar approach could be used to evaluate variance or MSE.

A special case of simulation is termed **parametric bootstrapping**, where the parameters of the simulating distribution are estimates (e.g., by maximum likelihood) and then data are simulated, conditional on these as fixed values. Parametric bootstrapping can be very useful as an alternative to asymptotic or other variance approximations. Standard (or nonparametric) bootstrapping involves resampling with replacements from the sample data, on the assumption that the data itself represents a random sample from some underlying (but possibly unknown distribution).

Cross-validation is a data resampling method related to bootstrapping but conducted for the purpose of getting a measure of prediction error that is independent of the data used to fit the model. This is often defined as the expected error rate of the model. Cross-validation involves randomly partitioning the data into subsets, one of which is used to estimate model parameters, and the remaining used to assess the ability of the model to predict "outside the data", i.e., beyond the data used to fit the model.

Finally, **jackknifing** is another resampling method that involves passing through the data, sequentially leaving out observations and computing the estimate on the remaining observation. If we have an observed sample $\underline{x} = [x_1, x_2, .. x_n]$, the ith jackknife sample is given by $\underline{x}(i) = [x_1, x_2, .. x_{i-1}, x_{i+1}, ..., x_n]$. The ith jackknife replicate of $\hat{\theta}$ is then given by

$$\hat{\theta}(i) = \hat{\theta}(\underline{x}(i)).$$

A jackknife estimate of variance is provided by

$$\text{Var}\left(\hat{\theta}\right)_{jack} = \frac{n-1}{n} \sum_{i=1}^{n} \left(\hat{\theta}(i) - \hat{\theta}(.)\right)^2 \tag{C.6}$$

where $\hat{\theta}(.) = \frac{1}{n}\sum_{i=1}^{n} \hat{\theta}(i)$. A jackknife estimate of bias is provided by

$$\text{Bias}\left(\hat{\theta}\right)_{jack} = (n-1)\left(\hat{\theta}(.) - \hat{\theta}\right). \tag{C.7}$$

A simple example involves the reduction of bias in estimate of the ratio $\theta = y/x$. A standard estimate of θ is the ratio of the two sample means $\hat{\theta} = \bar{y}/\bar{x}$, but this is somewhat biased. Consider a sample of $n = 6$ observations of y and x, $x = (2,5,6,7,9,14)$, $y = (3,5,7,8,9,15)$. The 6 jackknife estimates are

$$\underline{\hat{\theta}} = (1.073171, 1.105263, 1.081081, 1.08333, 1.17647, 1.103448).$$

The jackknife estimate of variance and bias are

$$\text{Vâr}\left(\hat{\theta}\right)_{jack} = \frac{n-}{n} \sum_{i=}^{n} \left(\hat{\theta}(i) - \hat{\theta}(.)\right) = (0.0014898) = 0.0012415$$

$$\text{Biâs}(\hat{\theta})_{jack} = (n-1)(\hat{\theta}(.) - \hat{\theta}). = 5(1.093991 - 1.09302) = 0.0048.$$

The approach confirms that ratio of means is negatively biased but that the alternative mean of ratios estimator is unbiased.

We provide examples of bootstrapping, cross-validation, and jackknife estimation in Chapter 5.

Method of moments

The method of moments is very simple and can provide reasonable estimates of parameters in some situations. The basic steps are

- Determine the population moments (expected value, variance, etc.) as functions of the parameter(s).
- Set the population moments equal to the sample (data based) moments.
- Solve for the parameter(s) as functions of the data.

To take a very simple case, consider a binomial experiment where we have 10 independent Bernoulli trials, we observe 6 success, and we wish to estimate p the probability of success (assumed homogeneous among trials). The population moment is

$$E(x) = np.$$

Setting the population moment equal to the sample moment (in the case, simply x) provides

$$x = np$$

and solving for p provides

$$\hat{p} = x/n = 6/10 = 0.6. \tag{C.8}$$

A somewhat more complicated example involves the beta distribution and 2 moments: the mean and the variance. Recall that for the beta the mean and variance are

$$E(x) = \frac{\alpha}{\alpha + \beta}$$

and

$$Var(x) = \frac{\alpha\beta}{(\alpha + \beta)^2 (\alpha + \beta + 1)}.$$

Because there are two unknowns (α, β) and two equations, we can solve for the parameters as functions of the sample moments. First, we equate the expectation of the moments with the sample moments

$$\bar{x} = \frac{\alpha}{\alpha + \beta} \tag{C.9}$$

and

$$s^2 = \frac{\alpha\beta}{(\alpha+\beta)^2(\alpha+\beta+1)}. \tag{C.10}$$

Then, we solve for α and β

$$\hat{\alpha} = \bar{x}\{[\bar{x}(1-\bar{x})]/s^2 - 1\} \tag{C.11}$$

and

$$\hat{\beta} = (1-\bar{x})\{[\bar{x}(1-\bar{x})]/s^2 - 1\}. \tag{C.12}$$

For example, if we have a sample mean of 0.5 and SD of 0.1 in our data, C.11-C.12 provide estimates of $\hat{\alpha} = 12$, $\hat{\beta} = 12$. We can confirm that these correspond to the moments by substitution into Eqs. (C.9) and (C.10).

Least squares

Least squares estimation involves, in its simplest form (**ordinary least squares OLS**) the selection of an estimator that minimizes the quadratic distance of the estimate to the data. That is, if y_i, $i=1, \ldots, n$ are sample data and $\hat{y}_i$ are estimates or predictions for each observation, then OLS minimizes

$$\sum_{i=1}^{n}(y_i - \hat{y}_i)^2. \tag{C.13}$$

Thus, OLS can be viewed as a type of optimization (Chapter 8, Appendix E). Many familiar estimators can be derived via OLS. For example, an estimate of the expectation $E(y) = \mu$ can be derived as the value of μ that minimizes

$$g(\mu) = \sum_{i=1}^{n}(y_i - \mu)^2.$$

This is an example of unconstrained optimization (Appendix E) and can be solved by the calculus by solving the gradient

$$g'(\mu) = \frac{dg}{d\mu} = 0$$

$$\frac{d\sum_{i=1}^{n}(y_i - \mu)^2}{d\mu} = \sum_{i=1}^{n} d(y_i^2 - 2y_i\mu + \mu^2)/d\mu =$$

$$2\sum_{i=1}^{n}(-y_i + \mu) = 0$$

$$-\sum_{i=1}^{n} y_i + n\mu = 0$$

resulting in the familiar estimator of the sample mean

$$\hat{\mu} = \frac{\sum\limits_{i=1}^{n} y_i}{n}.$$ (C.14)

By definition, this estimator minimizes $\sum\limits_{i=1}^{n} (y_i - \hat{\mu})^2$, and the average of the quadratic distances is taken as an estimator of variance

$$\hat{\sigma}^2 = \frac{\sum\limits_{i=1}^{n} (y_i - \hat{\mu})^2}{n-1}.$$ (C.15)

For this reason, estimators that adhere to the assumptions of OLS are said to be **minimum variance estimators**.

Note that the denominator of Eq. (C.15) is n-1 rather than n. This reduction can be justified first on the grounds that 1 degree of freedom is lost in the estimation of μ, and secondly that Eq. (C.15) provides an estimator of variance with lower bias.

A second familiar example is provided by the case where the observations are predicted under a linear model

$$\hat{y}_i = \beta_0 + \beta_1 X_i.$$ (C.16)

In this case, OLS estimators of the parameters b_0 and b_1 are those that minimize

$$g(\beta_0, \beta_1) = \sum_{i=1}^{n} (y_i - \hat{y}_i)^2.$$

The solution to the gradient equations provides the OLS estimators

$$\hat{\beta}_1 = \frac{\sum\limits_{i=1}^{n} x_i y_i - \frac{1}{n} \sum\limits_{i=1}^{n} x_i \sum\limits_{i=1}^{n} y_i}{\sum\limits_{i=1}^{n} x_i^2 - \frac{1}{n} \left(\sum\limits_{i=1}^{n} x_i \right)^2} = \frac{\hat{\sigma}_{xy}}{\hat{\sigma}_x^2}$$ (C.17)

$$\hat{\beta}_0 = \bar{y} - \hat{\beta}_1 \bar{x}$$

with residual variance estimated as

$$\hat{\sigma}_e^2 = \frac{\sum\limits_{i=1}^{n} (y_i - \hat{y}_i)^2}{n-2}$$ (C.18)

where $\hat{y}_i$ is obtained by evaluating Eq. (C.16) at the OLS estimates from Eq. (C.17).

Least square estimation extends to the case involving multiple predictor variables, with matrix notation facilitating description. The response is now modeled as

$$\underline{y} = X\underline{\beta} + \underline{\varepsilon} \tag{C.19}$$

where $\underline{y}$ is the vector of responses $\underline{y}' = [y_1, y_2, \ldots, y_n]$

$$\begin{bmatrix} y_1 \\ \cdot \\ \cdot \\ \cdot \\ y_n \end{bmatrix} = \begin{bmatrix} 1 & x_{11} & x_{12} & \cdot & \cdot & \cdot & x_{1k} \\ 1 & \cdot & & \cdot & \cdot & \cdot & \cdot \\ 1 & \cdot & & \cdot & \cdot & \cdot & \cdot \cdot \\ 1 & \cdot & & \cdot & \cdot & \cdot & \\ 1 & x_{n1} & & \cdot & \cdot & \cdot & x_{nk} \end{bmatrix} \begin{bmatrix} \beta_0 \\ \beta_1 \\ \cdot \\ \cdot \\ \beta_{k-1+} \end{bmatrix} + \begin{bmatrix} \varepsilon_1 \\ \varepsilon_2 \\ \cdot \\ \cdot \\ \varepsilon_n \end{bmatrix}. \tag{C.20}$$

OLS now minimizes

$$(\underline{y} - X\underline{\beta})'(\underline{y} - X\underline{\beta}) \tag{C21}$$

and the OLS estimators are

$$\underline{\hat{\beta}} = (X'X)^{-1}(X'\underline{y}). \tag{C.22}$$

The OLS assumptions include

- proper specification of the regression model;

- the residuals $\varepsilon_i = (y_i - \hat{\mu})$ have a mean of zero, $\sum_{i=1}^{n} (y_i - \hat{\mu})/n = 0$;

- the predictors $\underline{X}$ are linearly independent;
- the variances of the observations are determined by $Var(\varepsilon_i) = \sigma_e^2 I_{nXn}$;
- the additional assumption of normality, $\varepsilon_i \sim N(0, \sigma_e^2 I_{nXn})$, is not required for OLS, but if satisfied justifies normal regression under MLE, as discussed in the next section).

OLS can be generalized to allow for non-homogeneous errors, via weighted least squares. Thus, if the errors are correlated but unequal and estimates are of the precision of each measurement are available, a **best linear unbiased estimator (BLUE)** can be obtained as

$$\underline{\hat{\beta}} = (X'WX)(X'W\underline{y}) \tag{C.23}$$

where W is a diagonal matrix of weights, each the reciprocal of variance.

$$W_{ii} = 1/\sigma_i^2.$$

Finally, least squares generalizes to **nonlinear least squares (NLS)**, where the response is modeled as

$$\hat{y}_i = f(\underline{X}_i, \underline{\beta}) \tag{C.24}$$

and $f(\underline{X}_i, \underline{\beta})$ is in general nonlinear.

The problem then becomes to minimize

$$\sum_{i=1}^{n} \varepsilon_i^2 = \sum_{i=1}^{n} (y_i - f(\underline{X}_i, \underline{\beta}))^2. \tag{C.25}$$

This is an example of nonlinear optimization and is potentially solvable via a gradient approach if $f(\underline{X}_i, \underline{\beta})$ is continuous and differentiable (Appendix E). However unlike OLS, the assumption of normal error in the OLS case does not lead to asymptotic normality of estimates, as it does under normal regression (see next section).

Maximum likelihood

Maximum likelihood methods have several advantages not necessarily shared by other approaches, and therefore are favored in much of statistics. Generally speaking maximum likelihood estimators (MLEs)

- are asymptotically (i.e., with large samples) unbiased;
- are asymptotically normally distributed;
- have minimum variance (i.e., have variance smaller than any other estimator);
- provide variance estimates directly as part of estimation.

The basic idea of MLE is simple: given the data, we consider the parameter(s) to be unknown variables; the density function now behaves instead as a likelihood function. We then solve for the parameter values that maximize the likelihood function, give the data values. There are several ways to do this:

- by graphing the likelihood function against candidate parameter values;
- by "brute force" searching over the parameter space;
- by exact solution using the calculus;
- by numerical optimization methods.

Example: binomial likelihood

We can illustrate all these approaches by taking a simple case involving the binomial distribution. Suppose we conduct 100 Bernoulli trials and observe 40 successes. For example, the 100 trials could be 100 nests that we have discovered and have followed from initiation to success (fledging) or failure. Because we know the number of trials ($n=100$) we will focus on estimating the probability of success. The statistical model is

$$f(x; n, p) = \binom{n}{x} p^x (1-p)^{n-x}.$$

However, we now know that $n = 100$ and $x = 40$ so we will recast this as a likelihood function

$$L(p; x = 40, n = 100) = \binom{100}{40} p^{40} (1-p)^{60}.$$

Now the task is to find a value for p that maximizes this function. Usually, it will be more convenient to work with the natural logarithm of the likelihood function. Because the logarithmic transformation is monotonic, if we find value of p that maximize $\log(L(p))$ we've also found the value that maximize $L(p)$. For this example the log of the likelihood is

$$\ln L(p; x = 40, n = 100) = \ln \binom{100}{40} + 40 \ln p + 60 \ln(1-p)$$

or in general (for any integers n and $x \leq n$)

$$\ln L(p; x, n) = \ln \binom{n}{x} + x \ln p + (n-x) \ln(1-p).$$

Graphical approach

When we plot $\ln L(p; x, n)$ vs. p we get a curve centered about a value of $p \sim 0.4$. (Figure C1.a). Similarly the log likelihood (Figure C1.b) seems to peak around 0.4. So, $p = 0.4$ appears to be a good candidate for the MLE.

Brute force

We can fairly easily find the maximum by "brute force" if 1) we have a single parameter (p) in this case and 2) the parameter is constrained over a reasonable range (0 to 1 here). In such a case we can simply compute the likelihood for a large number of values in the admissible range (e.g., values from 0 to 1 in increments of 0.001), rank the resulting likelihoods, and select the value of p corresponding to the maximal likelihood value. Unless there are discontinuities in the likelihood function, or we have overdiscretized the variable, this approach will work well for likelihood that have few parameters, with each confined to a manageable range. In the binomial example it is easy to enumerate $L(p)$ over the range (0,1) and confirm that for $x = 40$ and $n = 100$, the likelihood is maximized at approximately $p = 0.4$.

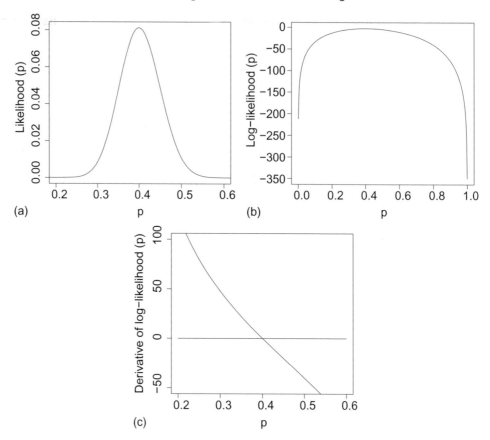

Figure C.1 Binomial likelihood for $x = 40$, $n = 100$. (a) likelihood function, (b) log-likelihood function, (c) derivative of log-likelihood with respect to p.

Exact approach using calculus

Calculus provides an exact solution to the likelihood maximization under certain conditions. In particular, if the likelihood is continuous and twice differentiable then a **necessary** condition that $L(p^*)$ is a maximum is that the first derivative with respect to p is zero (Appendix E). If the second derivative is negative, then this assures that $L(p^*)$ is a maximum and not a minimum. For the binomial likelihood this is best approached by operating with the log likelihood. The first derivative of the log likelihood is

$$\frac{d \ln L(p; x = 40, n = 100)}{dp} = \frac{40}{p} - \frac{60}{(1-p)}.$$

Setting this to zero yields

$$\frac{40}{p} = \frac{60}{(1-p)}.$$

and with a little algebra

$$\hat{p} = 40 / 100 = 0.4.$$

More generally

$$\frac{d \ln L(p; x, n)}{dp} = \frac{x}{p} - \frac{n-x}{(1-p)}$$

$$\frac{x}{p} = \frac{n-x}{(1-p)}$$

$$\hat{p} = x / n.$$

We can confirm graphically that the derivative becomes zero at $p = 0.4$ (Figure C1.c).

Direct solution of the log-likelihood equations by algebra is possible for many statistical models and their parameters. In addition to the binomial parameter p, the Poisson parameter λ can be estimated in this way, as can the normal parameters μ and σ, although analysis becomes more complicated when two or more parameters are involved. For example, estimation of the normal parameters μ and σ requires taking partial derivatives of the log-likelihood with respect to each parameter and setting each of these equations to zero. Solution of these two equations for μ and σ then provides the estimates

$$\hat{\mu} = \bar{x} = \sum_{i=1}^{n} x_i / n$$

$$\hat{\sigma}^2 = \sum_{i=1}^{n} (x_i - \bar{x})^2 / n.$$

Astute students will notice that the second formula differs slightly from the usual sample variance.

$$s^2 = \sum_{i=1}^{n} (x_i - \bar{x})^2 / (n-1).$$

The reason is that $\hat{\sigma}^2$ (the MLE) slightly biased for small samples, and use of s^2 reduces this bias; we can also make the argument, as with OLS, that a degrees of freedom was lost in the estimation of the mean.

Numerical methods

Explicit formulas for MLEs exist and are readily computed for many common statistical models. However, as models become more complex (more parameters and structure) it can be difficult or impossible to obtain algebraic solutions to the MLEs. Fortunately, high-speed computers are capable of solving the likelihood equations via numerical approaches. These approaches really are a special application of optimization approaches that we will consider in more detail later. They generally require the following:

- a mathematical expression (or computer code) for computing the log-likelihood for a given parameter value;
- an initial guess for the parameter value (sometimes based on simple statistics from the data);
- a means of "searching" to see if improvements (higher log-likelihood values) can be made by changing the parameter value;
- a stopping rule to determine that the parameter values has converged on the apparent MLE.

Gradient descent methods and **Newton's Method** are two of the more familiar (and simpler) optimization methods (Appendix E). Both require the ability to evaluate 1st and 2nd derivatives (partial derivatives if there is more than 1 parameter) with respect to each candidate parameter value (combination of values). The derivatives can be either explicitly written (i.e., algebraic) or computed via approximations.

The basic steps for Newton's Method are simple:

1. Start with an initial value for p, p_0.
2. Compute the gradient evaluated at the current value of $p(p_i)$
   $$g(p_i) = \frac{d \ln L(p_i)}{p_i}.$$
3. Compute $g'(p_i) = \dfrac{d^2 \ln L(p_i)}{dp_i^2}.$
4. Update p by $p_{i+1} = p_i - g(p_i) / g'(p_i)$
5. Return to Step 2 and repeat until convergence.

"Convergence" can be evaluated by evaluating how much (or little) p changes and/or by determining that $g(p_i)$ is sufficiently close to zero (i.e., differs from zero by some specified small amount). In the example code ($n = 100$, $x = 30$) p is initialized at 0.1 and converges rapidly to 0.3. R also has a built-in optimization function optimize() that performs maximization or minimization of specified function. We provide code for this example in Appendix G.

MLE for higher-dimensioned problems

In principle, exactly the same approaches used for single-parameter models extend to parameters with multiple parameters. However, both graphical and "brute force" approaches become cumbersome beyond about two parameters and are generally eschewed in favor of either direct or numerical solution of a system of likelihood equations.

Example – normal likelihood

We can take the example of the normal likelihood and a sample $\underline{x}$ of n observations. Assuming that the data are independent, the joint likelihood is formed by product of n likelihoods:

$$L(\mu, \sigma^2 \mid \underline{x}) = \prod_{i=1}^{n} \frac{1}{\sqrt{2\pi\sigma^2}} \exp\left(\frac{-(x_i - \mu)^2}{2\sigma^2}\right)$$

and the log-likelihood is

$$\log L(\mu, \sigma^2 \mid \underline{x}) = -n\log\left(\sqrt{2\pi\sigma^2}\right) + \sum_{i=1}^{n} \left(\frac{-(x_i - \mu)^2}{2\sigma^2}\right).$$

The likelihood is maximized by solving the system of equations

$$\underline{g}(\mu, \sigma^2; \underline{x}) = \underline{0} \tag{C.26}$$

where

$$\underline{g}(\mu, \sigma^2; \underline{x}) = \begin{bmatrix} \dfrac{\partial \log L(\mu, \sigma^2 \mid \underline{x})}{\partial \mu} \\ \dfrac{\partial \log L(\mu, \sigma^2 \mid \underline{x})}{\partial \sigma} \end{bmatrix}.$$

The partial derivatives of the log-likelihood with respect to the parameters simplify to

$$\frac{\partial \log L(\mu, \sigma^2 \mid \underline{x})}{\partial \mu} = 0 + \frac{-2n(\bar{x} - \mu)}{2\sigma^2}$$

$$\frac{\partial \log L(\mu, \sigma^2 \mid \underline{x})}{\partial \sigma} = -\frac{n}{\sigma} + \frac{\displaystyle\sum_{i=1}^{n}(x_i - \bar{x})^2 + n(\bar{x} - \mu)^2}{\sigma^3}.$$

It is easy to show that the analytical solution to $\underline{g}(\mu, \sigma^2; \underline{x}) = \underline{0}$ is provided by

$$\hat{\mu} = \bar{x} = \sum_{i=1}^{n} x_i / n$$

$$\hat{\sigma}^2 = \sum_{i=1}^{n} (x_i - \bar{x})^2 / n$$

the same result can be obtained by trial and error, gradient methods, Newton's Method, or other numerical methods. Application of Newton's Method and other derivative-based methods requires evaluation of the matrix of partial second derivatives

$$I = \begin{bmatrix} \dfrac{\partial^2 \ln L}{\partial \mu^2} & \dfrac{\partial \ln L}{\partial \mu \partial \sigma} \\[2ex] \dfrac{\partial \ln L}{\partial \sigma \partial \mu} & \dfrac{\partial^2 \ln L}{\partial \sigma^2} \end{bmatrix}. \qquad (C.27)$$

The matrix I is sometimes known as the **Hessian** or **information matrix**. The vector of the first partial derivates

$$G = \begin{bmatrix} \partial L / \partial \mu \\ \partial L / \partial \sigma \end{bmatrix}.$$

Solutions to the likelihood equations occur when $G = \underline{0}$; the inverse of F provides the estimated **variance–covariance matrix**, with the variances on the diagonal and the covariances on the off-diagonal. This same approach applies to any dimension MLE problem, with the sizes of G and F determined by the number of parameters (k, so G is length k and F is $k \times k$).

Maximization of multi-dimensional likelihoods is a particular application of the classical optimization solutions described in Appendix E. See also Appendix G for ML and general optimization software.

Normal regression

We can revisit the linear modeling (regression) problem in the context of MLE, by now requiring that the regression residuals assume a normal distribution. That is, we now model

$$y_i = \beta_0 + \beta_1 X_i + \varepsilon_i$$

where

$$\varepsilon_i \sim N(0, \sigma_e^2 I_{nXn}).$$

We thus replace μ in the normal likelihood by

$$\mu_i = \hat{y}_i = \beta_0 + \beta_1 X_i$$

leading to

$$L\left(\underline{\beta}, \sigma_e^2 \mid \underline{y}\right) = \prod_{i=1}^{n} \frac{1}{\sqrt{2\pi\sigma^2}} \exp\left(\frac{-(y_i - \hat{y}_i)^2}{2\sigma^2}\right). \qquad (C.28)$$

The log-likelihood function is

$$\log L\left(\underline{\beta}, \sigma_e^2 \mid \underline{y}\right) = n\log\left(\frac{1}{\sqrt{2\pi\sigma^2}}\right) + \sum_{i=1}^{n} \exp\left(\frac{-(y_i - \hat{y}_i)^2}{2\sigma^2}\right)$$

which is maximized when

$$\sum_{i=1}^{n}(y_i - \hat{y}_i)^2.$$

is minimized. However, this is exactly the condition for OLS. Thus, MLE and OLS are equivalent for linear models with the residuals are homogenously and normally distributed. In the general (multi-predictor) case, the inverse of the Hessian matrix in Eq. (C.27), when evaluated at the MLE, provides an estimate of the variance–covariance matrix. That is,

$$\Sigma\left(\underline{\hat{\beta}}\right) = I^{-1}\left(\underline{\hat{\beta}}\right). \tag{C.29}$$

Furthermore, under normal regression the ML estimates are unbiased and asymptotically normally distributed, with variance–covariance structure provided by Eq. (C.29).

Note that the equivalency between OLS and MLE does not generally hold for other error distributions, i.e., OLS and ML estimates could (and likely would) differ if ε_i follows another distribution such as Poisson or binomial.

Bayesian approaches

From a Bayesian viewpoint, all quantities in statistics are either known (e.g., our observed data), or unknown (parameters, predicted values, missing observations; Link and Barker 2010). Thus we can model the joint distribution of our unknown parameters θ and data as

$$f(\theta, x).$$

Using conditional distributions (Appendix A) we can rewrite this quantity as

$$f(\theta, x) = f_\theta(\theta \mid x)f_x(x).$$

Or equivalently as

$$f(\theta, x) = f_x(x \mid \theta)f_\theta(\theta).$$

Because these are equivalent, we can form the equality under Bayes' theorem

$$f_\theta(\theta \mid x) = \frac{f_x(x \mid \theta)f_\theta(\theta)}{f_x(x)}$$

where $f_\theta(\theta|x)$ is often referred to as the **posterior distribution** of θ and

$$f_x(x) = \int_\theta f_x(x \mid \theta) f_\theta(\theta) d\theta$$

is the weighted average of the **likelihood** $f_x(x|\theta)$ over the **prior distribution** of θ, $f_\theta(\theta)$. Since this integral will usually be a constant for given data, this leads to the general statement

$$f_\theta(\theta \mid x) \propto f_x(x \mid \theta) f_\theta(\theta).$$

Thus, Bayes' theorem provides a general, probability-based mechanism for expressing uncertainty in a parameter value before and after the collection of data. This relationship provides the basis for Bayesian "estimation", more properly referred to as Bayesian updating.

The general procedure is:

- Specify a model (parameter distribution and its parameter values) $f_\theta(\theta)$ that describes knowledge about the parameter before we collect data.
- Specify a model relating the data to the parameter values. This is the likelihood function, $f_x(x|\theta)$.
- Collect data and apply these to the likelihood function.
- Update the posterior distribution of the parameter with Bayes' theorem.

Binomial likelihood with a beta prior on p

We illustrate with an example where interest focuses on updating knowledge about the binomial parameter p. We will initially start by assuming that we have no prior information about p, apart from the obvious fact that $0 < p < 1$. In this case, a non-informative ("vague") prior would be

$$f(p) = U(0, 1)$$

which is the same as

$$f(p) = Beta(1, 1).$$

Both of these distributions can be readily evaluated for specific values of p, and we can readily obtain summary statistics such as means, variances and quantiles (Appendix B). Because it is more readily generalized to "informative" priors on p, we will use the beta.

$$f(p; \alpha, \beta) = \frac{p^{\alpha-1}(1-p)^{\beta-1}}{B(\alpha, \beta)}$$

and

$$f(p; 1, 1) = 1.$$

To update information on p we need the likelihood function

$$f(x \mid p) = Binomial(x \mid n, p)$$

or explicitly

$$f(x; n, p) = \binom{n}{x} p^x (1-p)^{n-x}$$

Bayes' theorem provides

$$f(p \mid x) = \frac{f(x \mid p)f(p)}{\int_p f(x \mid p)f(p)dp} = \frac{Binomial(x \mid n, p)Beta(1, 1)}{\int_p Binomial(x \mid n, p)Beta(1, 1)dp}$$

$$= \frac{\binom{n}{x} p^x (1-p)^{n-x}}{\int_p \binom{n}{x} p^x (1-p)^{n-x} dp}.$$

The denominator of this expression is a constant, as is the combinatorial terms, so we can rewrite

$$f(p \mid x) \propto p^x (1-p)^{n-x}$$

which is the kernel of a beta distribution with parameters $\alpha = x + 1$ and $\beta = n-x + 1$. Thus, it follows that

$$f(p \mid x) = Beta(x + 1, n - x + 1).$$

That is, the posterior distribution of p is simply another beta distribution, with parameters that contain the original prior parameters (1 and 1, known as **hyper-parameters**) and a summary of the data result (x, $n-x$). More generally, if the prior of p given by a beta(α, β) (so, with hyper-parameters α and β), and our sample experiment is has n trials with x success, we obtain the posterior of p by

$$f(p \mid x) = Beta(x + a, n - x + b). \tag{C.30}$$

Example – posterior inference with a vague (non-informative) prior

We can use the above result to obtain exact posterior inference under the binomial-beta model, once we have specified a prior and then obtained data.

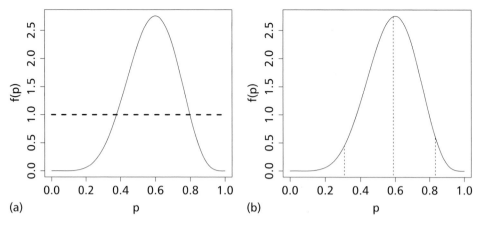

(a)

(b)

Figure C.2 Posterior distribution of p from beta(7,5) formed by conjugate prior beta(1,1) and binomial(10, 6) likelihood. (a) Posterior distribution superimposed on prior (dashed line), (b) 0.025, 0.5 and 0.975 quantiles of the posterior distribution.

Consider an example of survival of stream fishes. Suppose now that we mark a sample of 10 fish with radio tags, follow them for 1 year, and determine that 6 of the 10 survive. Given a vague (uniform) prior on p, the posterior distribution can be obtained exactly by

$$f(p \mid x = 6, n = 10) = Beta(6 + 1, 4 + 1) = Beta(7, 5).$$

The posterior is plotted over laid with the prior distribution (Figure C.2a). We can see that the observations have now given us some confidence that p is around 0.6 (and not just somewhere between 0 and 1).

We can readily obtain summary statistics such as quantiles from this distribution (e.g., using the R package; Chapter 5). For these example, we would obtain a median value of 0.588 and 0.0275 and 0.975 quantiles of 0.308 and 0.833 (Figure C.2b), which are analogous, respectively, to the MLE and 95% CI from a frequentist approach, with the latter more properly referred to as a **Bayesian credible interval** (BCI). A notable difference with CI is that the BCI is directly interpretable in terms of the probability that the parameter takes on a value in the interval. By contrast the CI only has meaning in the context of repeated samples, in $x\%$ of which the CI is predicted to include the parameter value.

We also could have simulated data from this posterior distribution, treated these as samples of the parameter value, and compute descriptive statistics. A sample of 10 000 from this distribution provides summary statistics very similar to those provided by direct examination of the distribution, with a histogram of simulated values shown in Figure C.3. Obviously, there is no purpose served in simulating distribution statistics when these can be obtained by direct analysis. However, more complex models often cannot be expressed as known distributions, but it may be possible still to sample from these distributions using

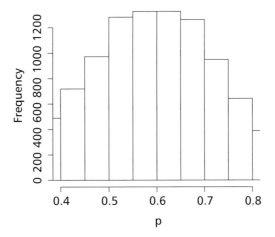

Figure C.3 Histogram of Monte Carlo simulations from the beta(7,5) posterior distribution formed by conjugate prior beta(1,1) and binomial(10, 6) likelihood.

simulation methods such as **Markov chain Monte Carlo (MCMC)**, considered below.

Informative prior (I) – prior based on x successes in n initial trials

Suppose that we in fact do have information about p, perhaps from a previous study. It may be appropriate to use this information to form an **informative prior** on p. The prior can then be updated with new data under a binomial likelihood, with the posterior representing a blending of the two studies.

The beta distribution can be used to form such a prior, since one interpretation of beta(α,β) is describing the posterior probability of success after having observed $\alpha-1$ success and $\beta-1$ failures. From that perspective, the previous study becomes the "updating" information used to revise prior of p from beta(1,1), and this informative prior is then used as the prior, to be combined with the new data likelihood. The mechanics are simple:

- Observe x_0 successes and n_0-x_0 failures in a preliminary study.
- Compute an informative prior on p as beta$(\alpha = x_0+1, \beta = n_0-x_0+1)$.
- Observe x successes and $n-x$ failures in the new study.
- Update p by beta$(x+\alpha, n-x+\beta)$.

Example – posterior inference with prior from a previous study

To return to the fish survival example, let us take the study just conducted with $x = 6$, $n = 10$ as a preliminary study. In a follow-up study, we observe $x = 55$, $n = 100$. The prior distribution is now beta$(\alpha = 6 + 1, \beta = 4 + 1) = $ beta(7,5), and the posterior distribution is given by beta$(55 + 7, 45 + 5) = $ beta(62,50). This distribution produces a median value of 0.553 and a 95% BCI of (0.463,0.644).

Informative prior (II) – prior estimate of mean and variance of *p*

Sometimes, information about the parameter will be based on multiple sources or studies. In such cases, there may be sufficient information to form an empirical prior distribution, but often there will be only summary information such as means, ranges, and perhaps variances on the prior estimates. If we have a prior estimate of the mean of *p* and its variance, we can use the method of moments (C.11) and (C.12) for the beta distribution to get a reasonable estimate of the parameters α and β. For example if we have prior information that leads us to believe that the mean of *p* is 0.6 and the sd is 0.1, we can use Eqs. (C.11) and (C.12) to produce MOM estimates of α and β of 13.8 and 9.2, respectively, and then use beta(13.8, 9.2) as an informative prior, together with whatever new data we collect.

At this point we offer some cautions. We must think carefully about potential consequence of heavily weighting prior information in comparison to our sample data. The second is a technical issue in that the MOM for the beta sometime fails (produces negative values for the parameters for example); this can happen if we use particularly large variances for *p*. In such cases, alternative approaches will be needed to obtain reasonable prior distributions. Finally, where concern arises about the sensitivity of results to the choice of priors, it is always good practice to perform analyses using both vague and informative priors.

Conjugate distributions

The beta–binomial pairing above is a pairing of statistical distributions known as conjugate distributions. **Conjugate distributions** are statistical distributions that form a natural or "conjugate" pair. They all have the basic property that the prior and posterior distributions are of the same distributional form. That is, we form a posterior distribution via Bayes' theorem (as above) by multiplication of the prior time the likelihood, divided by a normalizing constant. With conjugate distributions, the distribution we get back after this operating is the same distributional form as the prior.

So, in the case of the binomial-beta, the prior is

$$f_p(p; a, b) = \frac{\Gamma(\alpha+\beta)}{\Gamma(\alpha)\Gamma(\beta)} p^{\alpha-1}(1-p)^{\beta-1}$$

and the likelihood is

$$f_x(x; n, p) = \binom{n}{x} p^x (1-p)^{n-x}.$$

When we form the posterior distribution by invoking Bayes' theorem and simplify the result, we will get a posterior distribution where

$$f_p(p \mid x, n) \propto p^{\alpha+x-1}(1-p)^{\beta+n-x-1}.$$

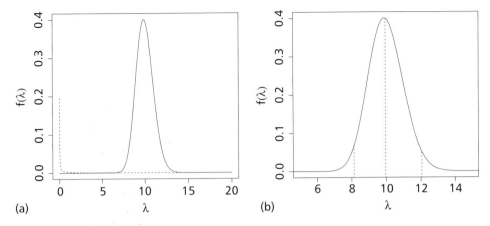

Figure C.4 Posterior distribution of λ from gamma(0.0999, 100.001), formed by conjugate prior gamma(1000,0.002) and Poisson likelihood with $\bar{x} = 10$, $n = 10$. (a) Posterior distribution superimposed on prior (dashed line), (b) 0.025, 0.5 and 0.975 quantiles of the posterior distribution.

Again, this is the kernel of a beta distribution; in order to integrate to 1 (required for a pdf) we must therefore have

$$f_p(p \mid x, n, \alpha, \beta) = \frac{\Gamma(\alpha + \beta + n)}{\Gamma(\alpha + x)\Gamma(\beta + n - x)} p^{\alpha + x - 1}(1 - p)^{\beta + n - x - 1}$$

thus,

$$f_p(p \mid x, n, \alpha, \beta) = Beta(\alpha + x, \beta + n - x).$$

There are several other important examples of conjugate prior-likelihood pairs in Bayesian statistics in addition to the binomial–beta shown above. We will discuss two additional pairings: the Poisson–gamma and normal–normal, which are summarized below along with the binomial-beta, and show how each can be used to provide posterior inference (Figure C.4). We summarize these three pairings and several others in Table C.1, with special attention to the prior and posterior hyper-parameters, which, respectively, model the prior information alone and prior updated by data using the likelihood model.

Poisson likelihood–gamma conjugate prior

Another important conjugate pair is the Poisson likelihood with a gamma prior. The Poisson likelihood can be appropriate when the outcome is a nonnegative integer value, such as the case with counts of animals, plants, or other objects that are randomly distributed in space.

$$f_x(x \mid \lambda) \sim Poisson(x; \lambda)$$

Table C.1 Some common conjugate distribution pairs.

Likelihood	Parameters	Conjugate Prior distribution	Prior hyper-parameters	Posterior	Posterior Hyper-parameters
Bernoulli Binomial	p	Beta	α, β	Beta	$\alpha + x,\ \beta + n - x$
Poisson	λ	Gamma	b, c	Gamma	$\dfrac{b}{bn+1},\ c + \bar{x}n$
Normal (known σ^2)	μ	Normal	μ_0, σ_0^2	Normal	$\left(\dfrac{\mu_0}{\sigma_0^2} + \dfrac{n\bar{x}}{s^2}\right)\Big/\left(\dfrac{1}{\sigma_0^2} + \dfrac{n}{s^2}\right)$, $\left(\dfrac{1}{\sigma_0^2} + \dfrac{n}{s^2}\right)^{-1}$
Normal (known μ)	$\tau = \sigma^{-2}$	Gamma	b, c	Gamma	$b + n/2$, $c + \displaystyle\sum_{i=1}^{n}(x_i - \mu)^2 / 2$
Multinomial	$\underline{p}$	Dirichlet	α	Dirichlet	$\alpha + \underline{x}$
Multivariate normal (known Σ)	μ	Multivariate normal	μ_0, Σ_0	Multivariate normal	$(\Sigma_0^{-1} + n\Sigma^{-1})^{-1}(\Sigma_0^{-1}\mu_0 + n\Sigma^{-1}\bar{x})$ $(\Sigma_0^{-1} + n\Sigma^{-1})^{-1}$

We will model uncertainty in the Poisson density parameter λ via a gamma distribution

$$f_\lambda(\lambda) = gamma(p \mid b, c)$$

that is,

$$f(\lambda; b, c) = \frac{(\lambda / b)^{c-1} e^{-(\lambda/b)}}{b\Gamma(c)}.$$

The likelihood is specified by

$$f_x(x \mid \lambda) = \frac{\lambda^x e^{-\lambda}}{x!}$$

Thus, the posterior distribution

$$f_\lambda(\lambda \mid x, b, c) \propto \frac{(\lambda / b)^{c-1} e^{-(\lambda/b)}}{b\Gamma(c)} \frac{\lambda^x e^{-\lambda}}{x!}.$$

With some algebraic simplification, this can be seen to be the kernel of a gamma distribution with parameters $\dfrac{b}{bn+1}$ and $c + \bar{x}n$. Thus, the posterior distribution

$$f_\lambda(\lambda \mid x, a, b) = gamma\left(\frac{b}{bn+1}, c + \bar{x}n\right) \tag{C.31}$$

where b and c are, respectively, the scale and shape parameters of the gamma and $\bar{x}$ and n are the mean and sample size from the data study. So once again we get the posterior distribution in simple form that allows us to include the summary results from our new study as updating information in the posterior.

Posterior inference with a vague (non-informative) prior

The question arises as to what should be used as a prior distribution for the gamma, and it turns out there is no easy answer, in part because λ has no upper bound (unlike p for the beta). An approach we will use for now is as follows, assuming we want a "vague" prior on λ

- Pick a "reasonable" value for the mean of λ, say $1 < \lambda < 10$.
- Specify a large variance, saw $var(\lambda) > 100mean\,(\lambda)$.
- Use MOM to obtain estimates of the hyper-parameters of for the gamma.

For example, we can specify a vague prior on λ, and then update λ from a study where we observe a mean of 10 counts over 10 samples (so, we count 100 total). The mean and variance of the gamma distribution (Appendix B) are

$$E(x) = bc$$

$$Var(x) = b^2 c$$

which leads to moment estimators of the parameters as

$$\hat{b} = \frac{s_0^2}{\bar{x}_0}$$

$$\hat{c} = \frac{(\bar{x}_0)^2}{s_0^2}$$

where $\bar{x}, s_0^2$ are the prior values of the mean and variance of x, chosen in this case to be deliberately "vague". For example $\bar{x}_0 = 2$, $s_0^2 = 2000$ produces $\hat{b} = 1000$, $\hat{c} = 0.002$, producing a prior distribution on λ of

$$f_\lambda(\lambda) = gamma(1000, 0.002).$$

We then conduct a study and observe $\bar{x} = 10$ in $n = 10$ sample counts. This leads to a posterior distribution

$$f_\lambda(\lambda \mid \bar{x}, n, a, b) = gamma\left(\frac{b}{bn+1}, c + \bar{x}n\right) = gamma(0.0999, 100.002)$$

This posterior pdf is superimposed on the "vague" prior, from which we can see that the prior has very little influence, relative to the likelihood.

As above, we can proceed with inference from this posterior distribution. For example, the median and 95% BCI from this distribution are, respectively, 9.965 and (8.136, 12.052).

Normal likelihood – normal conjugate prior

We will consider on more important conjugate pair, this time in which the data are continuous responses appropriately modeled by the normal likelihood, and both the prior and the posterior are also of normal form. Actually, the more general problem involving normal likelihoods is quite a bit more complicated, but if we simplify things a little we get a result that works nicely for illustration. The simplification assumes that we know the variance, and focuses on updating information on mean (things get quite a bit more complicated if we need to worry about both parameters).

Again, the statistical model for the data is normal

$$f_x(x \mid \mu, \sigma^2) = normal(x \mid \mu, \sigma^2).$$

We will model uncertainty in the mean μ via another normal distribution

$$f_\mu(\mu) = normal(\mu_0, \sigma_0^2).$$

The posterior distribution of λ is obtained by Bayes' theorem and conjugacy as another normal distribution

$$f_\mu(\mu \mid x) = normal(\mu_1, \sigma_1^2)$$

where

$$\mu_1 = \left(\frac{\mu_0}{\sigma_0^2} + \frac{n\bar{x}}{s^2}\right) \bigg/ \left(\frac{1}{\sigma_0^2} + \frac{n}{s^2}\right)$$

and

$$\sigma_1^2 = \left(\frac{1}{\sigma_0^2} + \frac{n}{s^2}\right)^{-1}.$$

Effectively this results in the posterior distribution being a mixing of the prior and the likelihood, with increasingly more weight given to the likelihood as sample sizes increase or sampling variance decreases.

Example – posterior inference with a vague (non-informative) prior

Again, because the normal has no upper (or lower) bounds it is not necessarily simple to form a vague prior. However, one reasonable approach is as follows.

- Set $\mu_0 = 0$, effectively centering the prior on the real line (it may be necessary to transform the response first for this to make sense, for example using logarithms to eliminate negative responses.
- Set σ_0^2 as very large (say 10^6); this will provide a very diffuse normal distribution.
- Proceed as above with incorporating data to the posterior.

As an example, we will use $\mu_0 = 0$ $\sigma_0^2 = 1000^2$ as normal hyper-parameters. Then, we take a sample of 100 and observe a mean of 10 with sd $= 10$ (so, not a particularly precise result, but an adequate sample). We could obtain posterior inference on μ as follows

$$f_\mu(\mu \mid x) = normal(\mu_1, \sigma_1^2)$$

where

$$\mu_1 = \left(0 + \frac{1000(10)}{10^2}\right) \bigg/ \left(\frac{1}{1000^2} + \frac{1000}{10^2}\right) = 9.999$$

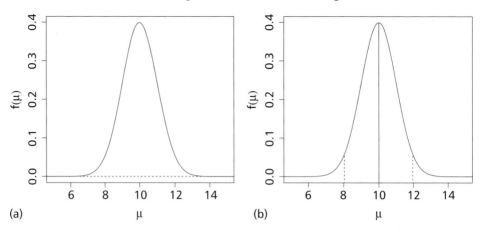

Figure C.5 Posterior distribution of μ from normal(9.999, 0.999), formed by conjugate prior normal(0, 1000) and normal likelihood with $\bar{x} = 10$, $s = 10$, $n = 100$. (a) Posterior distribution superimposed on prior (dashed line), (b) 0.025, 0.5 and 0.975 quantiles of the posterior distribution.

and

$$\sigma_1^2 = \left(\frac{1}{1000^2} + \frac{100}{10^2} \right)^{-1} = 0.9999.$$

The plot of the prior superimposed on the posterior illustrates its "vagueness" (Figure C.5)

We obtain posterior inference on μ from the posterior distribution

$$f_\mu(\mu \mid x) = normal(9.999, 0.9999)$$

for instance, the median and 95% BCI from this distribution are, respectively, 9.999 and (8.04, 11.95).

Monte Carlo methods for Bayesian inference

The above examples exploit the conjugacy relationship between certain distributions (some of which are summarized in Table C.1), which leads to direct inference by examination of that posterior. However, there are many instances in which the posterior distribution cannot be directly analyzed, possible because it does not conform to a known distribution. For example, the posterior distribution may arise via a complex combination of conditional or hierarchical relationships (Chapters 5, 6). The problem stems from the fact that, while we can always combine the prior and posterior distributions via Bayes' theorem

$$f_\theta(\theta \mid x) = \frac{f_x(x \mid \theta) f_\theta(\theta)}{\displaystyle\int_\theta f_x(x \mid \theta) f_\theta(\theta) d\theta}.$$

the resulting posterior frequently cannot be derived analytically, because of difficulty in solving the integral in the denominator. However, it is still possible to obtain samples from this distribution (i.e., without necessarily knowing its mathematical properties), then posterior inference is still possible. Monte Carlo methods provide this means via simulating values, which are then summarized. We briefly describe three important approaches, the first of which we have already seen (Appendix B); Link and Barker (2010) provide a more comprehensive (but very readable) coverage of these topics. The first two Monte Carlo methods provide simulated samples that are independent, whereas the last (MCMC) provides a dependent sequence of samples that nevertheless behaves, under the right conditions, like an independent sample from the posterior distribution.

Standard Monte Carlo simulation

We saw in Appendix B that the problem of generating random variables from a known distribution $F(x)$ generally reduces to the problem of the satisfactory generation of pseudorandom $U(0,1)$ numbers. For example, a standard approach is the generate a random number

$$U \sim uniform(0, 1)$$

and then generate the random variable x by applying the inverse cdf for the specified distribution.

$$x = F^{-1}(U).$$

As noted in Appendix B, this approach just reverses the fact that $F(x)$ is a mapping of the cumulative distribution of x on $(0,1)$. For some distributions (e.g., the Bernoulli), the inversion is even simpler: one draws $U \sim uniform(0,1)$ and specifies $x = 1$ if $U < p$ and $x = 0$ otherwise. In general, Monte Carlo sampling requires the ability to specify $F(x)$, produce its inverse (or the equivalent), and the ability to generate independent uniform values. Standard Monte Carlo algorithms produce the sample of n independent variates with a single step. By comparison, the next methods considered discard (reject) some of the sample variates generated in the process.

Rejection methods

Sometimes, it is inconvenient to sample from the distribution of interest, but fairly easy to sample from another distribution. Suppose that the pdf of interest, called the "target distribution", is $t(x)$, but there exists another distribution $c(x)$ such that

$$t(x) \leq Mc(x)$$

for the entire range of x, where M is a finite constant. The approach is to draw a sample $\underline{x}$ of size n from $c(x)$ and calculate

$$w(x) = t(x) / Mc(x)$$

for the sample. For each element we draw a Bernoulli trial with probability $w(x)$ and select the corresponding x value for the sample if the outcome is 1 and reject the value otherwise. This will result in an acceptance sample with $n' < n$ elements, since some $(n-n')$ will have been rejected. Link and Barker (2010) demonstrate this approach where the target distribution $t(x)$ is a standard normal, so that

$$t(x) = \frac{1}{\sqrt{2\pi}} \exp(-x^2 / 2).$$

The enveloping distribution is a triangular distribution, which arises as the sum of two independent uniform distributions; specifically

$$c(x) = (2/7) - 4|x| / 49.$$

The condition $t(x) \le Mc(x)$ is met over the range of x if $M = \dfrac{15}{4\sqrt{2\pi}}$, which is confirmed by graphing an overlay of $t(x)$ and $Mc(x)$ (Figure C.6a). We can easily generate random samples U from $c(x)$ by

$$U_1 = uniform(-3.49, 3.49)$$
$$U_2 = uniform(-3.49, 3.49)$$
$$U = U_1 + U_2$$

These are then used to evaluate

$$w(U) = t(U) / Mc(U)$$

with U restricted to the range $|U| \le 3.49$ (to avoid issues with the otherwise infinite tails of the normal). Finally, Bernoulli trials with these probabilities are conducted, with $x = U$ for the successes and U discarded otherwise. We wrote a small algorithm in R to perform these computations from 100 000 independent samples of U generated as above. Approximately 2/3 of these were accepted; a histogram of the accepted values suggests that the retained sample closely matches a standard normal distribution, with sample mean -0.0004 and variance 0.9992 (Figure C.6b).

Markov chain Monte Carlo

Markov chain Monte Carlo (MCMC) is a form of **rejection sampling** that involves the generation of a sequence of random variables $X_1, X_2, \dots X_n$ that has the property that given all values to that point, the probability of the next value is conditional on only the last k values. That is

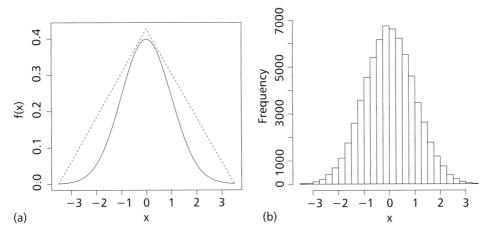

Figure C.6 Illustration of envelope sampling distribution (a) triangular sampling distribution (dashed line) and standard normal target distributions (solid line). (b) histogram of 10 000 rejection samples.

$$P[X_t \mid X_{t-1}, X_{t-2}, \ldots\ldots, X_1] = P[X_t \mid X_{t-1}, X_{t-2}, \ldots\ldots, X_{t-k}]$$

MCMC methods depend on convergence to a **stationary distribution**, that is, one in which the above conditional probability relationship does not depend on time, that is

$$P[X_t \mid X_{t-1}, X_{t-2}, \ldots\ldots, X_1] = P[X_t \mid X_{t-1}, X_{t-2}, \ldots\ldots, X_{t-k}] = f$$

MCMC methods provide a fairly general alternative for the (common) situation in which inverse cdf or rejection sampling methods will not work or are very inconvenient (e.g., an appropriate enveloping distribution cannot be found or results in very inefficient sampling). If it is still possible to derive a Markov chain with a stationary distribution, then although the resulting sample values will not (by definition) be independent, they should still constitute a valid sample of the target distribution.

Metropolis–Hastings algorithm

The **Metropolis–Hastings** (MH) algorithm (Metropolis et al. 1953, Link and Barker 2010) is an MCMC method that will work with a wide range of distributions. The MH algorithm requires only the capacity to generate a Markov chain and to evaluate the pdf at each point in the chain. The basic steps of MH are:

1. Obtain an initial value X_0; this can be a specified constant or draw from the prior distribution.
2. Generate a X_t, $X_{cand} \sim j\,(x \mid X_{t-1})$, where $j\,(x \mid X_{t-1})$ is a conditional distribution of the state X_t given state X_{t-1}.

3. Calculate $r = \dfrac{t(X_{cand})j(X_{t-1} \mid X_{cand})}{t(X_{t-1})j(X_{cand} \mid X_{t-1})}$.

4. Generate a uniform(0,1) variable u.
5. If $u < r$ set $X_t = X_{cand})$, otherwise set $X_t = X_{t-1}$.
6. Return to step 2 and repeat.

Example

Link and Barker (2010) illustrate the Metropolis–Hastings algorithm by using it to create samples from a normal (0,1) distribution. They used a symmetric distribution for j $(x \mid X_{t-1})$, so that

$$j(X_{cand} \mid X_{t-1}) = j(X_{t-1} \mid X_{cand}).$$

Thus, in step 3 of the algorithm r simplifies to

$$r = \frac{t(X_{cand})}{t(X_{t-1})}.$$

The specific algorithm used is:

1. Generate two independent uniform(0,1) variables u_1 and u_2.
2. Calculate a candidate value for X_t, $X_{cand} = X_{t-1} + 2A(u_1 - 1/2)$.
3. Calculate $r = \dfrac{t(X_{cand})}{t(X_{t-1})}$ where $t(x) = \exp(-0.5x^2)$.
4. If $u_2 < r$ set $X_t = X_{cand}$, otherwise set $X_t = X_{t-1}$.
5. Return to step 2 and repeat.

We have written simple code in R to implement the above example; like Link and Barker (2010) we tried different values of the "tuning" constant $0.5 < A < 20$ (Appendix G). We found, as did these authors, that $A \approx 3.5$ seems to work well in producing a random-looking MCMC trace and resulting posterior distribution from 10 000 samples (Figure C.7). Note that the target distribution $t(x)$ above is technically not even a pdf, since it lacks the normalizing constant $1/\sqrt{2\pi}$. However, this constant is irrelevant to the MH algorithm, which should produce values that converge on the normalized target.

Finally, in the above description of rejection methods for convenience we have used $t(x)$, etc. to stand for a generic distribution that from which we wish to generate samples, x. In our particular application to Bayesian analysis $t(x)$ will be the posterior distribution $f(\theta \mid x)$, from which we wish to generate sample values of θ.

Gibbs sampling

Gibbs sampling is an MCMC method specifically designed for sampling from multivariate posterior distributions, where we have a vector of unknown

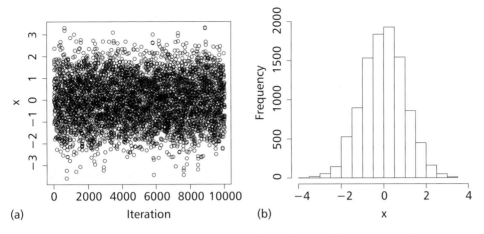

Figure C.7 10 000 Metropolis–Hastings samples using a uniform proposal distribution, targeting a standard normal posterior distribution. (a) posterior trace and (b) histogram of posterior distribution.

quantities ("parameters") conditioned on a data outcome. That is, interest centers on sampling from

$$f(\theta_1, \theta_2, \ldots, \theta_k \mid \underline{x}) \propto f(\underline{x} \mid \theta_1, \theta_2, \ldots, \theta_k) f(\underline{\theta})$$

Gibbs sampling works by sequential sampling from the **full conditional distributions** of the unknowns

$$f(\theta_j \mid \underline{\theta}_{(-j)}, \underline{x})$$

where θ_j is the jth element of $\underline{\theta}$ and $\underline{\theta}_{(-j)}$ is the vector of the remaining unknowns. The Gibbs sampling steps are (Link and Barker 2010):

1. Initialize $\underline{\theta}^{(0)} = \left(\theta_1^{(0)}, \theta_2^{(0)}, \theta_3^{(0)}, \ldots, \theta_k^{(0)}\right)$.
2. Sample $\theta_1^{(t)}$ from $f(\theta_1 \mid \underline{\theta}_{(-1)}^{(t-1)}, \underline{x})$.
3. Sample $\theta_2^{(t)}$ from $f(\theta_2 \mid \underline{\theta}_{(-2)}^{(t-1)}, \underline{x})$.
...

k. Sample $\theta_k^{(t)}$ from $f(\theta_k \mid \underline{\theta}_{(-k)}^{(t-1)}, \underline{x})$
k+1. Set $\underline{\theta}^{(t)} = \left(\theta_1^{(t)}, \theta_2^{(t)}, \theta_3^{(t)}, \ldots, \theta_k^{(t)}\right)$ and return to step 2, $t+1$.

Alternatively, the individual elements of $\underline{\theta}^{(t)}$ can be updated at each of the steps 2....k and the updated values used in the subsequent full conditional.

Example – closed population capture–recapture

We illustrate Gibbs sampling with an example from closed population capture–recapture, using the basic model M_0 that stipulates a homogeneous capture probability (Williams et al. 2002). The unknown parameters of this model are abundance (N) and capture probability (p). The data are $n\cdot$, the total number of captures summed over $k \geq 2$ capture occasions, and $u\cdot$, the total number of unique (initially unmarked) animals captured over the course of the study.

The posterior distribution we are interested is

$$f(p, N \mid n\cdot, u\cdot) \propto f(n\cdot, u\cdot \mid p, N)f(p, N)$$

which assuming independence in the priors is

$$f(p, N \mid n\cdot, u\cdot) \propto f(n\cdot, u\cdot \mid p, N)f(p)f(N).$$

The obvious prior distribution for p is a beta distribution

$$f(p) \propto p^{\alpha-1}(1-p)^{\beta-1}$$

A prior for N is less obvious; it will turn out to be convenient to use a *uniform*(a,b) prior, where a and b are upper and lower bounds to abundance. Because the pdf is a constant, the actual values for these limits are irrelevant. That is, irrespective of a and b

$$f(N) = c.$$

The data distribution (likelihood) is

$$f(n\cdot, u\cdot \mid N, p) = \frac{N!}{(N-u\cdot)!}p^{n\cdot}(1-p)^{kN-n\cdot}$$

or

$$f(n\cdot, u\cdot \mid U, p) = \frac{(U+u\cdot)!}{U!}p^{n\cdot}(1-p)^{k(U+u\cdot)-n\cdot}$$

where $U = N-u\cdot$ is the unknown number of unmarked animals in the population after sampling. The joint distribution of the unknowns and data is

$$f(U, p, n\cdot, u\cdot) \propto f(n\cdot, u\cdot \mid U, p)f(p)f(N) = \frac{(U+u\cdot)!}{U!}p^{n\cdot}(1-p)^{k(U+u\cdot)-n\cdot}p^{\alpha-1}(1-p)^{\beta-1}c$$

$$\propto \frac{(U+u\cdot)!}{U!}p^{n\cdot+\alpha-1}(1-p)^{k(U+u\cdot)-n\cdot+\beta-1}.$$

The full conditional distribution for p is

$$f(p \mid U, n\cdot, u\cdot) \propto p^{n\cdot+\alpha-1}(1-p)^{k(U+u\cdot)-n\cdot+\beta-1}$$

which is the kernel for a beta distribution $Beta(n\cdot + \alpha, k(U + u\cdot) - n\cdot + \beta)$.

The full conditional distribution for U

$$f(U \mid p, n\cdot, u\cdot) \propto \frac{(U+u.)!}{U!}(1-p)^{k(U+u\cdot)-n\cdot+\beta-1} \propto \frac{(U+u.)!}{U!}(1-p)^{kU} = \frac{(U+u.)!}{U!}\pi_0^U$$

where $\pi_0 = (1-p)^k$. This is the kernel of a negative binomial distribution $NB(u. + 1, 1 - \pi_0)$

We implemented the algorithm as follows, for $u\cdot = 54$, $n\cdot = 121$, $k = 5$, and specifying a vague prior on p ($\alpha = \beta = 1$):

1. Specify initial $N^{(0)} = u\cdot = 54$.
2. Sample $p^{(1)}$ from $Beta(121 + 1, 5(54) - 121 + 1) = Beta(122, 150)$.
3. Using $p^{(1)}$ calculate $\pi_0 = 1 - (1 - p^{(1)})^5$.
4. Sample U from $NB(u. + 1, 1 - \pi_0)$.
5. Calculate $N^{(1)} = u\cdot + U$.
6. Return to step 2 replacing 54 with $N^{(1)}$ to calculate $p^{(2)}$ and continue.

The algorithm produces reasonably "smooth" looking posterior distributions for p and N (Figure C.8) based on 10 000 Gibbs samples. The median value for p is 0.418 with 95% BCI of (0.350,0.485); the median value for N is 58 with a 95% BCI of (54,64).

When feasible, Gibbs sampling can be far more efficient than MH or other rejection sampling, in that no "burn in" (discarding of initial samples) should be required. However, frequently some of the full conditional distributions will not themselves be kernels of known distributions, in which case MH or other rejection sampling will be needed. The freely available software BUGS (Bayesian inference under Gibbs sampling) provides a Metropolis-within-Gibbs sampler that allows for tremendous flexibility in model specification (Appendix F).

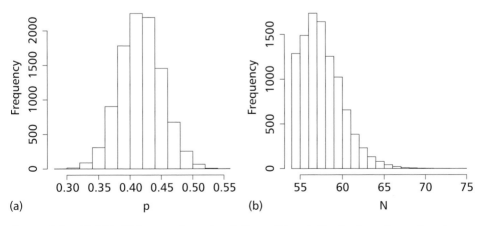

(a)   (b)

Figure C.8   10 000 Gibbs samples using full conditional distributions for a closed population capture-mark-recapture problem. Posterior histograms of (a) capture probability and (b) abundance.

There are numerous technical issues that users need to be aware of before using MCMC. In particular, caution must be exercised before assuming that MCMC has provided a valid sample of the posterior distribution. Several standard diagnostic procedures are available and will reveal obvious problems in convergence failure. Usually it is a good idea to run multiple MCMC chains with different initial values, to verify that ergodicity (lack of dependence on initial conditions) is fulfilled. Finally, it is standard procedure to "burn in", that is, discard an initial fraction of the MCMC sample, and some recommend "thinning" MCMC samples to reduce autocorrelation (but see Link and Barker 2010 on this topic). Again, freely available software such as WinBUGS, CODA, and PyMC exists to perform MCMC sampling and conduct diagnostic tests. Alternatively, users can build their own customized MCMC samplers using programming language such as MATLAB or R. See Appendix F for a discussion of software available to perform MCMC sampling.

Empirical Bayes estimation

The term **empirical Bayes (EB)** is sometimes applied to an approach in which parameter estimates are obtained from a procedure such as MOM or MLE, and then combined in a joint distribution. For example, we might take 10 binomial samples each with $n_i = 10$ trials and observe $x_i = (4,5,2,3,4,7,9,1,5,6)$. We could then obtain 10 ML estimates of p_i as $\hat{p}_i = (0.4, 0.5, 0.2, 0.3, 0.4, 0.7, 0.9, 0.1, 0.5, 0.6)$. Finally, we could describe the distribution of p across the samples via fitting these as data to a beta distribution and estimating the beta parameters, and take the resulting beta distribution as the posterior distribution for p. A **full Bayesian (FB)** approach would, by contrast, specify a prior distribution for $\underline{p}$, a joint likelihood for the data $\underline{x}$, and form the posterior distribution via Bayes' theorem, with inference as described above obtained either directly from the posterior distribution under conjugacy, or via MCMC. The EB approach is a convenient "plug in" approximation to FB that works well in certain cases where full Bayes is difficult. The term "empirical Bayes" is frowned up by some authors (e.g., Gelman et al. 1995) in that it incorrectly implies that EB is "empirical" (i.e., based on data) and FB is not.

References

Gelman, A.B., J. S. Carlin, H.S. Stern, and D. B. Rubin. (1995) *Bayesian Data Analysis*. Chapman and Hall, Boca Raton, FL.

Link, W.A. and R.J. Barker. (2010) *Bayesian Inference with Ecological Applications*. Elsevier-Academic. New York.

Metropolis, N., A.W. Rosenbluth, M.N. Rosenbluth and A.H. Teller. (1953) Equations of state calculations by fast computing machines. J. Chem. Phys. **21**, 1087–1092.

Williams, B.K., J. D. Nichols, and M.J. Conroy. (2002) *Analysis and Management of Animal Populations*. Elsevier-Academic. New York.

Appendix D

Parsimony, Prediction, and Multi-Model Inference

General approaches to multi-model inference

We are frequently faced with the situation where more than one model is capable of explaining the data, and the question arises: which model, or combination of models, should we use to explain and predict the data? Below we consider several criteria that may be considered in building and comparing models. We use as our motivating example the generalized linear model example consider in Box 5.10, but the principles extend to more general (e.g., nonlinear) models.

A general expression for the model in Box 5.6 is

$$Y_i = \log\left(\frac{p_i}{1 - p_i}\right) = \beta_0 + \sum_{i=1}^{k} \beta_i X_i$$

where p_i is the probability of nest success, X_i $i = 1, \ldots, k$ are predictors (X_1 = Julian date of nest discovery, X_2 = elevation in m) including quadratic or other polynomial terms ($X_i^2, X_i^3, , , , X_i^n$), interactions ($X_i X_j$) or other transformations, and β_i $i = 1, \ldots, k$ are model coefficients associate with each of the predictors and the intercept β_0. In the example, six alternative models were expressed and described by corresponding code in the glm procedure in R; using the above notation the models are:

$$H_0 : Y_i = \beta_0$$

$$H_1 : Y_i = \beta_0 + \beta_1 X_1$$

Decision Making in Natural Resource Management: A Structured, Adaptive Approach, First Edition. Michael J. Conroy and James T. Peterson.

$$H_2 : Y_i = \beta_0 + \beta_2 X_2$$

$$H_3 : Y_i = \beta_0 + \beta_1 X_1 + \beta_2 X_2$$

$$H_4 : Y_i = \beta_0 + \beta_2 X_2 + \beta_3 X_2^2$$

$$H_5 : Y_i = \beta_0 + \beta_1 X_1 + \beta_2 X_2 + \beta_3 X_2^2$$

Maximum likelihood is used to estimate the parameters β_i $i = 0, \ldots, k$ together with the variance–covariance structure (Appendix C).

Model plausibility

We suggest that the first step in the appropriate use of alternative models for inference, is in the proper selection of a set of models that is, a priori, plausible. That is, we are strong advocates of building models that make biological sense and have behind them a plausible biological hypothesis. It is very easy to construct complex models involving many factors, but many of these models are likely to be nonsensical (or at least there is no plausible mechanism behind the model) and difficult to interpret. In the nest success example, we have plausible hypotheses behind why nest success is predicted to increase with elevation (having to do with timing of food availability relative to nest phenology) and decrease with Julian date of initiation (general decrease in food availability and increase in predators as the season progresses), which motivate models H_1–H_3. We might also hypothesize that there is an optimal elevation for nest survival, motivating a quadratic relationship (models H_4 and H_5). We could also make a good case for an interaction between elevation and Julian date, for example

$$H_5 : Y_i = \beta_0 + \beta_1 X_1 + \beta_2 X_2 + \beta_4 X_1 X_2$$

On the other hand, we cannot envision a plausible hypothesis involving the interaction of elevation and Julian date in the absence of the main effect of either factor. Thus, the model

$$H_6 : Y_i = \beta_0 + \beta_4 X_1 X_2$$

makes (to us) no sense. We would not build this model, because for one thing we have no idea how to interpret its estimates or predictions.

Assuming that we have confined ourselves to a set of plausible models, we have a number of criteria that we can use to evaluate individual models and compare among models. One criterion we could use is **model fit** as measured by R^2, the proportion of total variance accounted of total variance in the response accounted for by the model. However, it is easy to show (Appendix C) that model fit can be increased to 100% simply by adding more parameters For instance, increasing the number of polynomial terms or interactions terms to the point where $k \to n$. Since this conflicts with both the goal of parsimony and that of prediction accuracy (see Appendix C), we caution against this approach.

Another approach is based on the comparison of alternative models a **likelihood ratio hypothesis testing** (LRT). Although LRT can be appropriate for model selection, particularly when under experimentation to test a specific hypothesis, there are a number of issues, not the least of which is what order is the testing done (simplest to most complex, the reverse, or something else). Null hypothesis testing can also lead to overfitting of models. Finally, LRT obviously is limited to cases where the simpler model is nested as a null hypothesis within the more complex model. However, there are many instances where the simpler model is not nested in the more complex one. For example, above we considered

$$H_3 : Y_i = \beta_0 + \beta_1 X_1 + \beta_2 X_2$$

$$H_4 : Y_i = \beta_0 + \beta_2 X_2 + \beta_3 X_2^2$$

Neither of these can be formed as a null of the other (both have the same number of parameters) and so LRT will not work, although both are null hypotheses when compared to

$$H_2 : Y_i = \beta_0 + \beta_2 X_2$$

We instead advocate the approach developed by Akaike (1973) and described in detail in the excellent book by Burnham and Anderson (2002). Instead of viewing model selection as a hypotheses testing exercise, it is instead viewed from an **optimization** perspective, in which the goal is to maximize predictive accuracy, given the tradeoff between variance and bias (Appendix C). The approach uses information theory to develop a statistic known as **Akaike's information criterion** (AIC). AIC is computed from the log-likelihood and the number of model parameters by the formula

$$AIC = -2 \ln \log L + 2k$$

where k is the number of parameters estimated in the model (e.g., $k = 3$ for a simple regression: b_0, b_1 and σ_e^2. We recommend adjusting AIC for sample size (especially for small samples) by the correction

$$AICc = AIC + \frac{2k(k+1)}{n-k-1}$$

the adjustment will be especially important as k becomes large relative to n (as in the polynomial example). Notice too that as n becomes large AICc $\rightarrow$ AIC. Returning to that example we can compute AIC_c for each model. The actual value of AIC is not important, rather is the value of a model's AIC value relative to that of the model with the smallest AIC. We thus form the difference between the models AIC (or AICc)

$$\Delta AICc_i = AICc_i - \min(AICc)$$

where min(AIC) is the minimum AIC value over the set of models under consideration. Finally, $\Delta AICc_i$ can be used to calculate an "information weight" for each model, essentially representing how much that model should contribute to inference, given the data. Finally, ΔAIC_i can be used to calculate an "information weight" for model, essentially representing how much that model should contribute to inference, given the data.

$$w_i = \frac{\exp(-0.5\Delta AICc_i)}{\sum_{j=1}^{m} \exp(-0.5\Delta AICc_j)}$$

where the denominator sums the term $\exp(-0.5\Delta AIC_j)$ over the $j = 1, \ldots, m$ models under consideration. Below, we calculate adjusted AIC weights for the nest success example and add this information to the table.

Example

Generalized linear models (Box 5.7) were used to estimate the parameters under each of the alternative models

Model	2log(L)	K	AIC	AICc	Δ AICc	wt
H_0	−24.88	1	51.75	51.81	3.65	0.07
H_1	−21.99	2	47.98	48.16	0.00	0.41
H_2	−24.57	2	53.13	53.31	5.15	0.03
H_3	−21.03	3	48.06	48.43	0.26	0.36
H_4	−24.53	3	55.06	55.42	7.26	0.01
H_5	−21.01	4	50.01	50.63	2.47	0.12

Models H_1 and H_3 have the most support, with nearly 80% of the overall model weight accounted for by these two models. We return to this example below, when we calculate model-averaged estimates.

Multi-model inference and model averaging

In the above, the emphasis was on comparing among models with the idea of selecting a single model (or perhaps one or two models) on which to base inference. However, it is often the case that a parameter or prediction occurs in several models. If our primary focus is estimation or prediction, then we are going to care more about appropriate estimation of that parameter (or prediction) than about what model or models it is based on. Suppose that θ is a parameter or a prediction that we are interested in estimating, and that estimates are available under several candidate models. Burnham and Anderson (2002) describe how we can combine the estimates over the alternative models using Akaike weights:

$$\bar{\theta}_{Modlavg} = \sum_{i=1}^{m} w_i \hat{\theta}_i$$

where $\hat{\theta}_i$ is the estimate under each model and w_i are the Akaike weights (adjusted as appropriate for sample sizes).

Inference on θ also requires an estimate of estimator precision. Under each candidate model we have an estimated standard error (based on either the least-squares deviance or maximum likelihood), $\hat{SE}(\hat{\theta}_i)$. However, this estimate by definition assumes that the estimating model is the true model. To account for uncertainty in the parameter estimate due to model uncertainty, Burnham and Anderson (2002) advocate computation of an "unconditional standard error".

$$SE_{uncond}(\bar{\theta}) = \sum_{i=1}^{m} w_i \sqrt{SE(\hat{\theta}_i)^2 + (\hat{\theta}_i - \bar{\theta}_{Modlavg})^2}$$

In general, use of the unconditional standard error will produce larger estimated SEs and correspondingly wider confidence intervals, which is appropriate given that inference no longer rests on belief in a single model.

Example

Returning to the nest success example, interest centered on predictions of nest success at 1200 m elevation and a Julian date of 176 days, both values near the averages observed for nests in the study. That is, $\hat{\theta}_i = \tilde{Y}_i \mid X_1 = 176, X_2 = 1200$. We computed $\hat{\theta}_i$ for each model, along with its estimated conditional standard error $SE(\hat{\theta}_i)$.

Model	$\hat{\theta}_i$	$SE(\hat{\theta}_i)$	wt
H_0	−2.04769	0.375665	0.06633675
H_1	−2.39096	0.507361	0.41115311
H_2	−2.09178	0.390573	0.03134593
H_3	−2.5431	0.561755	0.36023813
H_4	−2.23314	0.655753	0.01092694
H_5	−2.65445	0.749055	0.11999913

These values were used along with the AICc weights calculated earlier to provide a model averaged estimate and unconditional standard error for the prediction. Thus,

$$\bar{\theta}_{Modlavg} = -2.04769(0.06633675) + (-2.39096)(0.41115311)$$
$$+ \ldots + (-2.65445)(0.11999913) = -2.44351$$

$$SE_{uncond}(\bar{\theta}) = 0.06633675\sqrt{0.375665 + (-2.04769 + 2.45184)^2}$$
$$+ \ldots + 0.11999913\sqrt{0.749055^2 + (-2.65445 + 2.45184)^2}$$
$$= 0.5688231$$

We computed a 95% approximate normal confidence interval on this prediction from

$$\bar{\theta}_{Modlavg} \pm SE_{uncond}(\bar{\theta})Z_{1-\alpha/2} = -2.45184 \pm 0.577(1.96) = (-3.56, -1.33)$$

where $Z_{0.975}$ is the upper standard normal deviate associated with $\alpha = 0.05$. After back transformation to the probability scale, this produces a 95% confidence interval for θ of (0.028, 0.2094).

AIC for least squares

As seen above, AIC is computed from the log-likelihood under each model, and thus requires distributional assumptions (e.g., normality). However, Burnham and Anderson (2002) describe an approximation for computing AIC based least-squares estimation:

$$AIC_{ls} = n\log(\sigma_e^2) + 2k$$

The AIC value based on least squares is then used for computing model weights and model-averaged estimates and predictions, as above.

In general, we recommend using an appropriate likelihood to compute AIC, rather than the least square approach. However, in cases where no ML model is appropriate and LS estimation must be used, we advocate the above approach for approximating AIC for model comparison and model-averaged estimation.

Adjustment for overdispersion (quasi-likelihood) for count data

In some cases, the variance of count data exceeds the variance expected under the proposed sampling distribution. For example, the variance of x for Poisson count data is predicted to be λ (the same as the mean). Likewise, the variance of x for binomial count data is predicted to be

$$np(1-p)$$

where the expectation would be evaluated at the MLEs of the parameters (λ for the Poisson or p for the binomial). A realized variance, x in excess of this value could occur for a number of reasons, such as lack of independence (e.g., animals travel in social groups and thus their fates are at least partially co-dependent). The ratio of

$$c = \frac{\text{Var}(x)}{\sigma_0^2}$$

represents **overdispersion,** with $c = 1$ indicating perfect agreement between the realized and model variance, and $c > 1$ indicating overdispersion. Burnham and Anderson (2002) recommend adjusting AIC for overdispersion when $1 < c < 4$, with larger values suggesting structural problems with the model (e.g., the Poisson is not the appropriate statistical model). The adjustment creates a "quasi-likelihood" value and a corresponding value for AIC:

$$QAIC = \frac{-2\ln\log L}{\hat{c}} + 2k$$

which can be adjusted for small sample sizes as

$$QAICc = QAIC + \frac{2k(k+1)}{n-k-1}$$

Various approaches exist for obtaining estimates of c. If the study has replication, it may be possible to provide direct, empirical estimates of variance for comparison to the model-based estimates. Another approach is to use the ratio of the chi-square goodness of fit statistic to its degrees of freedom

$$\hat{c} = \chi^2 / df$$

A third approach uses the deviance ($-2\ln L$) divided by its degrees of freedom

$$\hat{c} = -2\ln L / df$$

Finally, parametric bootstrapping (Chapter 5, Appendix C) can be used to compute the deviance or other measure of fit from data simulated under the model assuming the ML parameter estimates. Then, the ratio of the corresponding data-based statistic to the simulated values is used to evaluate whether the sample results are within that expected under the model, or require quasilikelihood adjustment (with c estimated as, e.g., the sample value divided by the mean value from the simulations).

Multi-model Bayesian inference

Multi-model inference under the Bayesian paradigm is a natural extension of Bayesian inference (Appendix C) where now the "unknowns" include the model state, M on which the parameter value θ is conditioned. The joint distribution of unknowns and data is then

$$f(M, \theta, x) = f_x(x \mid \theta, M) f_\theta(\theta \mid M) f_M(M)$$

where sequential conditioning is of the data on the parameters and model in the likelihood, parameters conditional on the model in model-specific priors, and prior probabilities of the model states. As before, posterior inference is achieved by conditional arguments under Bayes' theorem.

For example, the model-specific posterior distribution of a parameter is

$$f_\theta(\theta \mid M, x) = \frac{f(M, \theta, x)}{f_{x,M}(x, M)} = \frac{f(M, \theta, x)}{f_x(x \mid M) f_M(M)} = \frac{f(x \mid \theta, M) f_\theta(\theta \mid M) f_M(M)}{f_x(x \mid M) f_M(M)}$$

where the data distribution is integrated over the prior distribution of the parameter given the model:

$$f_x(x \mid M) = \int_\theta f_x(x \mid \theta) f_\theta(\theta \mid M) d\theta.$$

Similarly, the "model averaged" posterior distribution is obtained by further integration (or for finite model states by summation)

$$f(\theta \mid x) = \int_M \frac{f(M, \theta, x)}{f_x(x)} dM = \frac{\int_M f_x(x \mid \theta, M) f_\theta(\theta \mid M) f_M(M) dM}{f_x(x)}$$

where

$$f_x(x) = \int_M \int_\theta f(x \mid \theta) f(\theta \mid M) f(M) d\theta dM$$

that is, the data distribution now must be integrated (or for finite model states summed) over the prior distribution of model states and the parameter values for each model.

Posterior probabilities of the model states, analogous to the AIC model weights considered earlier, are likewise computed as

$$f_M(M \mid x) = \int_\theta \frac{f_M(M, \theta, x)}{f_x(x)} d\theta = \frac{\int_\theta f_x(x \mid \theta, M) f_\theta(\theta \mid M) f_M(M) d\theta}{f_x(x)}$$

$$f_x(x) = \int_M \int_\theta f(x \mid \theta) f(\theta \mid M) f(M) d\theta dM$$

Computational approaches for Bayesian MMI

Although it is relatively straightforward to express model uncertainty in the Bayesian paradigm, multi-model Bayesian inference is complicated by the involvement of the uncertain model states, especially in the integrals of the posterior pdf. A number of approaches have been suggested, and we consider some of these, starting with the simplest.

Link and Barker (2010) describe use of the **Bayes' factor (BF)** for comparing alternative models. The BF for two alternative models *i* and *j* is the ratio of the posterior marginal likelihoods averaged over the parameters

$$f_x(x \mid M) = \int_\theta f(x \mid \theta, M) f(\theta \mid M) d\theta$$

$$BF_{ij} = \frac{f_x(x \mid M_i)}{f_x(x \mid M_j)}$$

$$\frac{f_M(M_i \mid x)}{f_M(M_j \mid x)} = BF_{ij} \frac{f_M(M_i)}{f_M(M_j)}$$

where $\dfrac{f_M(M_i)}{f_M(M_j)}$ is the prior odds and $\dfrac{f_M(M_i \mid x)}{f_M(M_j \mid x)}$ the posterior odds of model *i* relative to *j*. Computation of BF does not depend on prior model probabilities, but is influenced by the choice of prior parameters distributions under each model, thus complicating their use for MMI. Various authors have suggested using what is termed a "posterior Bayes factor" (PBF) that uses posterior parameter distributions for the parameter. Link and Barker provide an example using WinBUGS of PBF for comparison of Geometrics and Poisson models for a simple data set, and we have reproduced their example in Appendix F.

Computational issues notwithstanding, BF is attractive because it provides a means of allowing for sequential updating of model probabilities given acquisition of data. That is, suppose we acquire x_1 in a study. We compute the BF

$$BF_{ij}(x_1) = \frac{f_x(x_1 \mid M_i)}{f_x(x_1 \mid M_j)}.$$

Upon acquisition of a second data set x_2 we repeat the updating, using $BF_{ij}(x_1)$

$$BF_{ij}(x_2 \mid x_1) = \frac{f_x(x_2 \mid x_1, M_i)}{f_x(x_2 \mid x_1, M_j)}.$$

In this manner,

Reversible jump MCMC

Reversible jump MCMC (RJMCMC) is a generalized Metropolis–Hastings procedure that is based on the idea of sampling both over the model space and the parameter space conditioned on the model. That is, we envision a state-space (combination of model states and parameter values)

$$S = \{(M_i, \theta_{M_i}), M_i \in \{M\}, \theta_{M_i} \in \Theta_{M_i}\}$$

Then the posterior distribution factors as

$$f(M_i, \theta_{M_i} \mid x) = f_\theta(\theta_{M_i} \mid M_i) f_M(M_i)$$

The RJMCMC procedure is implemented by specifying a "bijection function" $g_M(\psi) = (\theta_M, u_M)$ that maps the parameters of (M, θ_M) into a "palette of parameters" that governs all the models in the model set (Link and Barker 2010). Sampling proceed via Gibbs sampling, alternating between the full conditional distribution of $f_\psi(\psi \mid M, x)$ and $f_\psi(M \mid \psi, x)$ with sampling from the full conditional of M via Metropolis-Hastings, with the sample from the full conditional of M either a proposed value M^* or the current value by a Bernoulli trial with probability $\min(r,1)$ where

$$r = \frac{f_x(x \mid \psi, M^*) f_\psi(\psi \mid M^*)}{f_x(x \mid \psi, M) f_\psi(\psi \mid M)} \times \frac{f_M(M^*) J(M \mid M^*)}{f_M(M) J(M^* \mid M)}$$

where model M^* is selected at random with probability $J(M^* \mid M)$.

Bayesian information criteria

The **Bayesian information criterion (BIC)**

$$\text{BIC}_M = -2\log\left(f_x\left(x \mid \hat{\theta}_M, M\right) + k\log(n)\right)$$

where k is the number of parameters and n is sample size, has been proposed as an approximation to $-2 \log(f_x(x \mid M)$ (Schwarz 1978) and as a large-sample approximation of the Bayes factor (Link and Barker 2010). Thus, BIC has been proposed as a model selection criterion, analogous to AIC considered earlier in a frequentist context. Alternatively, AIC based on the posterior likelihood has also been considered as a model selection criterion for Bayesian models. A problem arises with either statistic, however, in that the factor k (the number of parameter) is ambiguous when there is hierarchical structuring among parameters (i.e., hyper-parameters), as is the case with random effects and full hierarchical models. An alternative, termed the **deviance information criterion (DIC)**, has been proposed by Spiegelhalter et al. (2002) to deal with this more general situation:

$$\text{DIC}_M = \bar{D}(\theta_M) + p_D$$

where p_D is an "effective number of parameters" calculated as

$$p_D = \bar{D}(\theta_M) - D(\bar{\theta}_M)$$

and $D(\theta_M) = -2 \log f_x(x \mid M, \theta_m)$; the quantities $\bar{D}(\theta_M)$ and $D(\bar{\theta}_M)$ are, respectively, the mean value of $D(\theta_M)$ and $D(\bar{\theta}_M)$ evaluated at the mean of the parameters for model M. The DIC statistic is implemented in the WinBUGS software (Appendix F).

References

Akaike, H. (1973) Information theory and an extension of the maximum likelihood principle . pp. 267-281 in. B.N. Petran and F. Csàaki (Eds). *Second International Symposium on Information Theory.* Akadèemiai Kiadi. Budapest, Hungary.

Burnham, K.P. and D.R. Anderson. (2002) *Model Selection and Inference: A Practical Information-Theoretic Approach.* 2nd edn. Springer. New York.

Link, W.A. and R.J. Barker. (2010) *Bayesian Inference with Ecological Applications.* Elsevier- Academic. New York.

Schwarz , G. (1978) Estimating the dimension of a model. *Ann. Stat.* **6**, 461–464.

Spiegelhalter, D.J., N.G. Best, B.P. Carlin, and A. van der Linde. (2002) Bayesian measures of model complexity and fit. *J.R. Stat. Soc.*, Ser. B. **64**, 583–639.

Appendix E

Mathematical Approaches to Optimization

In Chapter 8, we provided an overview of optimization approaches, as illustrated by problems in natural resource management. Here, we provide further mathematical details of optimization methods. These details should allow readers a basic understanding of the mechanics of how optimization methods work, which is key to understanding why they sometimes do *not* work. Using these principles, readers should be able to construct their own optimization algorithms or modify existing ones for specific needs. Readers are referred to Williams et al. (2002) for a more detailed coverage of these topics.

As discussed in Chapter 8, the hierarchy of traditional optimization approaches is determined by the interplay between the mathematical complexity of the objective function and the involvement of constraints, if any (see Figure 8.1). The simplest optimization problems involve a straightforward relationship between the decision variables and the resulting objective function. More complex problems involve the incorporation of dynamic relationships, stochasticity, or both. Likewise, the simplest problems involve no or a few simple (e.g., boundary) constraints, while more complex problems involve inequality, linear, or nonlinear constraints on the decision space. More complex constraints can include the incorporation of dynamics and even stochastic relationships of certain types, although these are typically better handled by approaches, such as the calculus of variations and dynamic programming (for dynamic problems) and stochastic dynamic programming (for dynamic, stochastic problems).

Decision Making in Natural Resource Management: A Structured, Adaptive Approach,
First Edition. Michael J. Conroy and James T. Peterson.
© 2013 John Wiley & Sons, Ltd. Published 2013 by John Wiley & Sons, Ltd.

Review of general optimization principles

General statement of the optimization problem

Verbally, the optimization problem (Chapter 8) sounds straightforward:

- "Given a range of possible decisions that can be chosen, chose the combination of decisions that provides the best result in terms of the objective".

Here, we invoke a more formal statement of the optimization problem that allows us to address the issue rigorously and quantitatively. First we will define a set of 1 or more **controls** or **decision variables**.

$$\underline{d}' = (d_1, d_2, \ldots, d_n)$$

For example, $\underline{d}' = (d_1, d_2)$ where d_1 is the daily bag limit (e.g., 0–6) and d_2 is the maximum season length (e.g., 0–75).

A second component is a set of constraints (if any) on the decision space due to costs, logistics, or other constraints. For example, decisions outside $0 \leq d_1 \leq 6$, $0 \leq d_2 \leq 75$ are not considered, nor are certain combinations (e.g., $d_1 = 0$ with $d_2 > 0$ or the reverse); we can collectively refer to the permissible decisions as the "opportunity set" $\underline{D}$ and the specified conditions as **constraints**.

Finally, we define an **objective**, $F(\underline{d})$, to achieve as a function of the decision variables. For example, our objective could be to maximize yield from the population over the next 10 years.

Our formal statement of the optimization problem then is:

- Select $\underline{d}$ to maximize $F(\underline{d})$ subject to $\underline{d} \in \underline{D}$.

Note that although the problem is stated as "maximize" we can easily convert it to a problem where we are minimizing $F(\underline{d})$ (for instance, if $F(\underline{d})$ is cost or probability of extinction):

- Select $\underline{d}$ to maximize $-F(\underline{d})$ subject to $\underline{d} \in \underline{D}$.

This still leaves us to deal with the 2 main problems of classical optimization:

- How to find the decisions that maximize the objective.
- How to deal with constraints.

We will put aside until later 2 even thornier issues:

- How to deal with dynamics.
- How to deal with stochastic uncertainty.

Necessity and sufficiency conditions

We will at various points in the discussion that follows refer to conditions for optimization as fulfilling either necessity or sufficiency conditions. By **necessary** we mean that a condition must be fulfilled for a solution to be optimal; however, its fulfillment does not guarantee that a solution is optimal. By **sufficient** we ordinarily mean that fulfillment of a condition assures optimality of the solution.

Unconstrained Optimization

We start with the case where there are no constraints imposed on the decision space, so that the task is to select a value of $\underline{d}$ from $\underline{D}$ that optimizes (maximizes or minimizes) $F(\underline{d})$. Optimal solutions exist and are sufficient if they fill convexity or concavity requirements, the details of which are provided in Williams et al. (2002:745). Briefly, an **objective function** $F(\underline{d})$ is concave over a convex set $\underline{D}$ if

$$F[a\underline{d_1} + (1-a)\underline{d_2}] \leq aF(\underline{d_1}) + (1-a)F(\underline{d_2})$$

for all $\underline{d_1} \in \underline{D}$, $\underline{d_2} \in \underline{D}$ and $a \in [0, 1]$. These conditions are illustrated graphically in Figures E.1 and E.2. Demonstration that the function is concave over a convex set is sufficient to guarantee that local maximum is in fact global, which is important because ordinarily we are not interested in maxima (or minima) that are local (apply only to a specified subset of the decision set).

Single decision controls

Concavity (and thus the existence of a global optimum) can be evaluated by applying the rules of the calculus, in cases where $F(\underline{d})$ is a twice-differential, continuous function. This can be shown easily for the case of a single decision variable d. A local optimum d^* by definition exists if $F(d^* + \Delta d) \leq F(d^*)$ for small values of Δd. It can be shown by Taylor series expansion that this condition is met by $F'(d^*)$. That is, a **necessary** (but not sufficient) condition for d^* to be a maximum (or minimum) is that the first derivative of the function with respect to d equals zero when evaluated at d^*. **Sufficiency** that d^* provides a maximum (minimum) is provided by evaluation of the second derivate, with $F''(d^*) < 0$ (>0) ensuring a maximum (minimum; see Figure E.2). Finally, sufficiency that the optimum is **global** rather than local is provided by demonstrating (as above) that $F(\underline{d})$ is concave for every value d in the convex set D.

Example- MSY

We will start with a "classical" problem in conservation, determining maximum sustainable yield (MSY). This is solvable as a relatively simply optimization problem given a few assumptions:

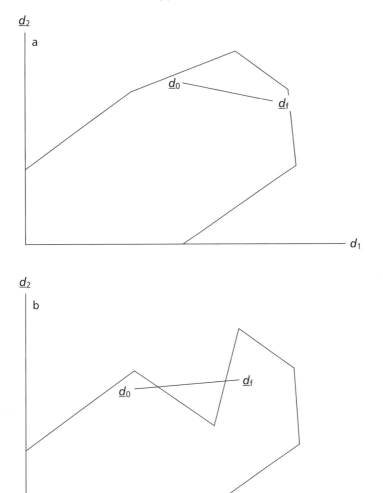

Figure E.1 Convex and concave decision sets for a 2-dimensional decision problem (a) the line segment connecting the decision points d_0 and d_f is completely contained in the opportunity set; (b) the line segment connecting the decision points d_0 and d_f is partially outside the opportunity set (Reproduced with permission from Williams, B.K., J.D. Nichols, and M.J. Conroy; *Analysis and management of animal populations*; Elsevier-Academic, 2002).

- The population is growing deterministically according to the logistic growth model.
- We can remove a specified harvest from the population (perfect controllability) that perfectly balances population growth (by definition this is **sustainable yield**).

Our management control consists of specifying the desired equilibrium population level (N) at which to achieve MSY.

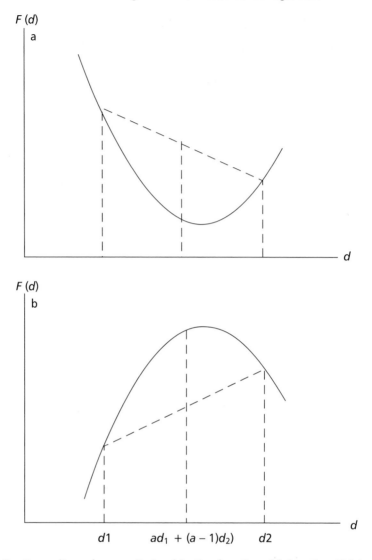

Figure E.2 Concavity and convexity in objective functions.(a) function $F(d)$ is convex if for all d if $F(ad_1 + (a - 1)d_2) \leq aF(d_1) + (1 - a)F(d_2)$, $0 < a < 1$. (b) the function $F(d)$ is concave if $-F(d)$ is convex (Reproduced with permission from Williams, B.K., J.D. Nichols, and M.J. Conroy; *Analysis and management of animal populations*; Elsevier-Academic, 2002).

The logistic growth assumption results (in discrete form) in population growth as

$$N(t+1) = N(t) + r_{max}N(t)[1 - N(t)/K]$$

where r_{max} is the maximum per capita growth rate and K is the carrying capacity (an upper limit to N). If we take a harvest from this population each year we have

$$N(t+1) = N(t) + r_{max} N(t)[1 - N(t)/K] - H(t)$$

and we can exactly balance population growth by taking

$$H(t) = r_{max} N(t)[1 - N(t)/K]$$

at which point we have $N(t+1) = N(t) = N$, so the population is at equilibrium, so we can rewrite H as a function of N as

$$H(N) = r_{max} N[1 - N/K]$$

Our optimization problem is to find the value of N than maximizes this function. Because this is a continuous, (twice) differentiable function, we'll use the exact approach based on calculus. As above, the necessary condition for a maximum will be fulfilled in general by

$$\frac{dF(x)}{dx} = 0$$

or in this case

$$\frac{dH(N)}{dN} = 0$$

$$r_{max} - \frac{2N r_{max}}{K} = 0$$

$$N^* = \frac{K}{2}$$

To show that this is a maximum (and not a minimum) we need to show that $\frac{d^2 H(N)}{dN^2} < 0$; i.e., that the function is concave down rather than up. The second derivative is

$$\frac{d^2 H(N)}{dN^2} = -2r_{max}/K,$$

which is negative for all $r_{max} > 0$, $K > 0$. Therefore, MSY is obtained at $N^* = \frac{K}{2}$.

Finally, we get the yield at this optimum by substituting $N^* = \frac{K}{2}$ back into the yield equation.

$$H(N^*) = r_{max} N^*[1 - N^*/K] = r_{max} \frac{K}{2}[1 - 1/2] = r_{max} \frac{K}{4}.$$

This provides simple, analytical expressions for MSY that require us only to specify r_{max} and K (Figure E.3).

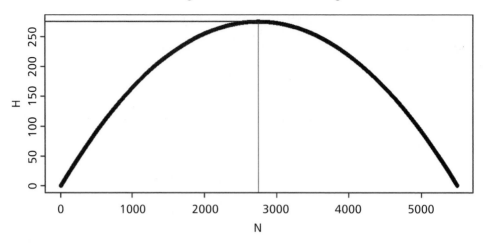

Figure E.3 Relationship between yield and abundance under logistic model with $r_{max} = 0.2$, $K = 5500$. Maximum sustainable yield (MSY) of $H(N^*) = 275$ occurs at $N^* = K/2 = 2750$.

Multiple decision controls

Unconstrained optimization extends in a straightforward way to multi-variate decision problems, by invoking multivariate calculus. For a bivariate decision problem, $\underline{d}' = [d_1, d_2]$. Sufficiency for an optimum is provided by

$$\frac{\partial F(d_1^*, d_2^*)}{\partial d_1} \text{ and } \frac{\partial F(d_1^*, d_2^*)}{\partial d_2} = 0.$$

The partial derivatives are usually stacked in a vector referred to as the **gradient**, to make it clear that the conditions must jointly hold:

$$G(\underline{d}) = \begin{bmatrix} \dfrac{\partial F(d_1^*, d_2^*)}{\partial d_1} \\ \dfrac{\partial F(d_1^*, d_2^*)}{\partial d_2} \end{bmatrix} = \begin{bmatrix} 0 \\ 0 \end{bmatrix}. \tag{E.1}$$

Sufficiency conditions for a local maximum (minimum) are provided by additionally showing that the **Hessian** matrix

$$H(\underline{d}) = \begin{bmatrix} \dfrac{\partial^2 F(d_1^*, d_2^*)}{\partial d_1^2} & \dfrac{\partial F(d_1^*, d_2^*)}{\partial d_1 \partial d_2} \\ \dfrac{\partial F(d_1^*, d_2^*)}{\partial d_2 \partial d_1} & \dfrac{\partial^2 F(d_1^*, d_2^*)}{\partial d_2^2} \end{bmatrix} \tag{E.2}$$

is negative-definite (positive-definite).

The approach generalizes in the obvious way to multiple (>2) decision problems. Again, the optimum will be fulfilled by

$$\frac{\partial F(\underline{d}^*)}{\partial \underline{d}} = \underline{0}$$

where now we are taking partial derivatives with respect to each of the individual controls and finding where they are simultaneously equal to zero, resulting in the gradient

$$G(\underline{d}) = \frac{\partial F(\underline{d}^*)}{\partial \underline{d}} = \begin{pmatrix} \partial F / \partial d_1 \\ \partial F / \partial d_2 \\ . \\ . \\ . \\ \partial F / \partial d_n \end{pmatrix} = \begin{pmatrix} 0 \\ 0 \\ . \\ . \\ . \\ 0 \end{pmatrix} \tag{E.3}$$

and Hessian

$$H(\underline{d}) = \frac{\partial^2 F(\underline{d}^*)}{\partial \underline{d}^2} = \begin{pmatrix} \partial^2 F / \partial d_1^2 & \partial^2 F / \partial d_1 d_2 & . & . \; . \; . & \partial^2 F / \partial d_1 d_n \\ \partial^2 F / \partial d_2 d_1 & \partial^2 F / \partial d_2^2 & \partial^2 F / \partial d_2 d_3 & . & . \\ . & . & . & . \; . \; . & . \\ . & . & . & . \; . \; . & . \\ \partial^2 F / \partial d_n d_1 & . & . & . \; . & \partial^2 F / \partial d_n^2 \end{pmatrix}$$

$$\tag{E.4}$$

Approaches for identifying $\underline{d}^*$ fall into 2 basic camps: non-differential and differential methods. Non-differential (derivative-free) methods work by taking initial (starting) values for $\underline{d}$ successively repeated searching in different directions for values that result in maximum change in $F(\underline{d})$. Differential methods require specification (or approximation) of the first (and sometimes second) derivatives of $F(\underline{d})$, that is, the gradient and the Hessian. In a few cases, the gradient results in a system of linear equations that can be solved by linear algebra. In most practical cases, iterative methods such as the method of steepest ascent, Newton's Method, or conjugate gradient methods are used that hopefully converge on optima. To illustrate one of these, Newton's Method employs an iterative formula

$$\underline{d}_{k+1} = \underline{d}_k - H(\underline{d})^{-1} G(\underline{d})'$$

where $\underline{d}_k$ is the decision value at iteration k; obviously this approach, in addition to requiring starting values for $\underline{d} = \underline{d}_0$, necessitates the inverse of the Hessian exists (and thus that the Hessian is non-singular).

Algorithms to solve for unconstrained optima via Newton's Method and similar approaches are relatively easily constructed, particularly for small (low dimensioned) programs. To illustrate, we provide R code to implement Newton's Method in Appendix G. Built-in procedures are also readily available for many software packages. For example, the optim() procedure in R uses a numerical method known as BFGS (Broyden 1970) to approximate the point at which the gradient (first derivative vector) is zero. We also provide examples in R code implementing the BFGS procedure in Appendix G.

Example – conservation of 2 species

This example was used by Williams et al. (2002: 594) to illustrate constrained optimization with 2 controls; we will start out with an unconstrained problem. The objective is to sustain 2 species that are in partial competition. The objective function reflects potential for population sustainability and relative value of the 2 species. The values d_1 and d_2 are population levels for each species, over which we assume that we have perfect control (i.e., the decision is the level (d_1, d_2) at which to maintain the 2 species.

The objective is to maximize $F(\underline{d})$ where

$$F(\underline{d}) = 224d_1 + 84d_2 + d_1d_2 - 2d_1^2 - d_2^2.$$

The gradient is

$$\frac{\partial F(\underline{d})}{\partial \underline{d}} = \begin{pmatrix} \partial F / \partial d_1 \\ \partial F / \partial d_2 \end{pmatrix} = \begin{pmatrix} 224 + d_2 - 4d_1 \\ 84 + d_1 - 2d_2 \end{pmatrix} = \underline{0},$$

which is fairly easily solved for $(d_1^*, d_2^*) = (76, 80)$.

We can also produce the Hessian matrix for this problem

$$\frac{\partial^2 F(\underline{d})}{\partial \underline{x}^2} = \begin{pmatrix} \partial^2 F / \partial d_1^2 & \partial^2 F / \partial d_1 d_2 \\ \partial^2 F / \partial d_2 d_1 & \partial^2 F / \partial d_2^2 \end{pmatrix} = \begin{pmatrix} -4 & 1 \\ 1 & -2 \end{pmatrix}.$$

This can be shown to be negative-definite (all the eigenvalues are negative) demonstrating that we have found a maximum.

The calculations to produce the objective function and gradient for this problem are easy in R (Appendix G).

Classical programming

Classical programming extends the above approaches used for unconstrained optimization to allow for equality constraints on the decision variables. Now the problem is

Maximize $F(\underline{d})$

subject to constraints

$$
\begin{bmatrix}
g_1(\underline{d}) = a_1 \\
g_2(\underline{d}) = a_2 \\
\\
\\
\\
g_n(\underline{d}) = a_n
\end{bmatrix}
\tag{E.5}
$$

where each $g_i(\underline{x})$ can involve some or all of the decision variables, and a_i is a specified vector of constants. The constraint function allow for tremendous generality, the only proviso being that they are (along with the objective) differentiable. Examples of constraints include cost and resource availability, system dynamics, and linkages among the decision variables.

One approach is to use constraints to express some decision variables in terms of others, effectively eliminating some decision variables. As a simple case, suppose that harvest in 2 areas in described by the decision variables d_1, d_2 but that harvest in area 2 is always proportional to that in area 1 by a predetermined factor c. We can make the substitution $d_2 = cd_1$, eliminate d_2, and optimize solely as a function of d_1.

A more general approach is the use of a device known as the **Lagrangian multiplier** to incorporate the constraints into a generalized objective function. The Lagrangian approach modifies the objective function by including the constraint as part of an augmented "Lagrangian function". We illustrate this approach first for a bivariate (2-decision) problem and then extend it to the more general multivariate case (first introduced above).

Bivariate classical programming

The bivariate programming problem involves 2 decision variables d_1, d_2 and a single constraint $g(d_1, d_2) = c$.

The Lagrangian function incorporates the constraint as

$$
L(d_1, d_2, \lambda) = F(d_1, d_2) + \lambda[c - g(d_1, d_2)]
\tag{E.6}
$$

Differentiating $L(d_1, d_2, \lambda)$ with respect to d_1, d_2, and λ provides the system of 3 equations, which are equated to zero

$$
\frac{\partial L}{\partial d_1} = \frac{\partial F}{\partial d_1} - \lambda \frac{\partial g}{\partial d_1} = 0
$$

$$
\frac{\partial L}{\partial d_2} = \frac{\partial F}{\partial d_2} - \lambda \frac{\partial g}{\partial d_2} = 0
$$

$$
\frac{\partial L}{\partial \lambda} = c - g(d_1, d_2) = 0
$$

Solutions to these equations provide the necessity conditions for the optimum as

$$\frac{\partial F}{\partial d_1}(d_1^*, d_2^*) = \lambda^* \frac{\partial g}{\partial d_1}(d_1^*, d_2^*)$$

$$\frac{\partial F}{\partial d_2}(d_1^*, d_2^{**}) = \lambda^* \frac{\partial g}{\partial d_2}(d_1^*, d_2^*)$$

$$g(d_1^*, d_2^*) = c$$

with a maximum assured if the Hessian of $F(d_1^*, d_2^*)$ is negative-definite. Thus, in a manner analogous to unconstrained optimization, the objective function is optimized subject to the equality constraint on $g(\underline{d})$.

Example – conservation of 2 species

We can return to the 2 species conservation example to illustrate optimization with constraints.

The original objective reflects potential for population sustainability and relative value of the 2 species. The objective is to maximize $F(\underline{d})$ where

$$F(\underline{d}) = 224d_1 + 84d_2 + d_1d_2 - 2d_1^2 - d_2^2.$$

We now add a resource constraint reflecting the higher demand for resources of species 1:

$$3d_1 + d_2 = 280$$

$$L(\underline{d}, \lambda) = 24d_1 + 84d_2 + d_1d_2 - 2d_1^2 - d_2^2 + \lambda(280 - 3d_1 - d_2)].$$

Optimization then proceeds by finding the value of d_1, d_2, and λ that satisfies

$$\partial L(\underline{d}, \lambda) / \partial d_1 = 224 - 4d_1 + d_2 - 3\lambda = 0$$
$$\partial L(\underline{d}, \lambda) / \partial d_2 = 84 + d_1 - 2d_2 - \lambda = 0$$
$$\partial L(\underline{d}, \lambda) / \partial \lambda = 280 - 3d_1 - d_2 = 0.$$

These equations can be fairly easily solved for $d_1^*, d_2^*, \lambda^* = (69, 73, 7)$. Substitution back into the augmented gradient produces $\underline{0}$. If we substitute the optimal decision and Lagrangian values into the objective we would obtain a value of 11 774, which is lower than the value of 11 872 that was obtained under unconstrained optimization. In general, constraints will result in an lower objective return than would be obtained by releasing the problem from such constraints.

Multivariate classical programming

Classical programming extends readily to the general case a k-dimension decision with $m < k$ equality constraints. That is, the problem is now to maximize $F(\underline{d})$ subject to

$$
\begin{bmatrix}
g_1(\underline{d}) = c_1 \\
g_2(\underline{d}) = c_2 \\
\\
\\
\\
g_m(\underline{d}) = c_m
\end{bmatrix}
$$

or

$$
\underline{g}(\underline{d}) = \underline{c}
$$

where $\underline{d}' = (d_1, d_2, \ldots, d_k)$. Applying the method of Lagrangian multipliers now yields the augmented function

$$
L(\underline{d}, \underline{\lambda}) = F(\underline{d}) + \underline{\lambda}[\underline{c} - \underline{g}(\underline{d})] \tag{E.7}
$$

where $\underline{\lambda} = (\lambda_1, \lambda_2, \ldots, \lambda_m)$. Differentiation with respect to $\underline{d}$ and λ yields

$$
\frac{\partial L}{\partial \underline{d}} = \frac{\partial F}{\partial \underline{d}} - \lambda \frac{\partial \underline{g}}{\partial \underline{d}}
$$

$$
\frac{\partial F}{\partial \underline{d}} = \underline{c} - g(\underline{d})
$$

providing the necessary conditions for local optimum

$$
\frac{\partial F}{\partial \underline{d}}(\underline{d}^*) = \underline{\lambda}^* \frac{\partial \underline{g}}{\partial \underline{d}}(\underline{d}^*)
$$

$$
\underline{g}(\underline{d}^*) = \underline{c}
$$

Sufficiency for the multivariate case is evaluated by computation of the Hessian matrix of the Lagrangian function

$$
H_L(\underline{d}, \underline{\lambda}) = \begin{bmatrix} 0 & J_g(\underline{d}) \\ J_g(\underline{d})' & H_F(\underline{d}) \end{bmatrix} \tag{E.8}
$$

where

$$
H_F(\underline{d}) = \begin{bmatrix}
\dfrac{\partial^2 F(\underline{d})}{\partial d_1^2} & \dfrac{\partial F(\underline{d})}{\partial d_1 \partial d_2} & \cdot & \dfrac{\partial F(\underline{d})}{\partial d_1 \partial d_2} \\
\dfrac{\partial F(\underline{d})}{\partial d_2 \partial d_1} & \dfrac{\partial^2 F(\underline{d})}{\partial d_2^2} & & \cdot \\
\cdot & & & \cdot \\
\dfrac{\partial F(\underline{d})}{\partial d_n \partial d_1} & & & \dfrac{\partial^2 F(\underline{d})}{\partial d_n^2}
\end{bmatrix}
\tag{E.9}
$$

is the Hessian of the objective function and

$$
J_g(\underline{d}) = \begin{bmatrix}
\dfrac{\partial g_1(\underline{d})}{\partial d_1} & \cdot & \dfrac{\partial g_1(\underline{d})}{\partial d_n} \\
& & \\
\dfrac{\partial g_m(\underline{d})}{\partial d_1} & & \dfrac{\partial g_m(\underline{d})}{\partial d_n}
\end{bmatrix}
\tag{E.10}
$$

is the **Jacobian** matrix of partial derivatives of the constraint function (n decisions and $m < n$ constraints). See Williams et al. (2002: 750–751 for a more complete discussion of these issues).

Sensitivity analysis

A useful result of the Lagrangian development above is that the optimal values of the Lagrangian multiplier are expressions of the **sensitivity** of decision problems to constraints via the relationship

$$
L(\underline{d}, \underline{\lambda}) = F(\underline{d}) + \underline{\lambda}[\underline{c} - \underline{g}(\underline{d})]
$$

Under the necessity conditions described above

$$
\partial F = \underline{\lambda}^* \partial \underline{g}
$$

and thus

$$
\underline{\lambda}^* = \Gamma(\underline{g}) = \frac{\partial F}{\partial \underline{g}}
$$

so that the Lagrangian coefficients can be interpreted as sensitivities of the objective to the constraint function. It is usually preferable to express these in terms of proportional sensitivity or **elasticity** (Chapter 8), so that

$$
e(\underline{g}) = \Gamma(\underline{g}) \frac{\underline{g}}{F} = \underline{\lambda}^* \frac{\underline{g}}{F}
$$

For example, in the 2-species problem above, the value $\lambda^* = 7$ providing $F^* = 11\ 774$ on the constraint of $g(d_1, d_2) = 280$ results in an elasticity of

$$e(\underline{g}) = 7\left(\frac{280}{11\,774}\right) = 0.166$$

Sensitivity and elasticity analysis can be useful, among other purposes, for exploring the implications of constraints on achieving resource objectives. In the above example, the elasticity value gives us a measure of the proportional marginal gain we could expect in our objective, for a proportional relaxation in the resource constraint (e.g., through habitat improvement).

Nonlinear programming

Nonlinear programming generalizes the classical programming approach to allow for inequality constraints. The general problem is to maximize (or minimize) $F(\underline{d})$ subject to

$$\underline{g}(\underline{x}) \leq \underline{a}$$
$$\underline{x} \geq 0. \tag{E.11}$$

Notice that the direction of the constraints is easily reversed by changing signs (e.g., $\underline{a}' = -\underline{a}$) and that equality constraints can be effectively imposed by 2 inequality constraints (e.g., $x \leq 100 \cap x \geq 100 \Rightarrow x = 100$. Thus, both classical programming (equality constraints) and unconstrained optimization (no constraints) can be readily cast as special cases of nonlinear programming.

Inequality constraints are introduced by modifying the classical optimization framework by means of "slack" variables, one variable for each inequality constraint. This essentially converts the problem into a classical programming problem, with the modified Lagrangian function

$$L(\underline{d}, \underline{\lambda}, \underline{s}) = F(\underline{d}) + \underline{\lambda}[\underline{c} - \underline{g}(\underline{d}) - \underline{s}]. \tag{E.12}$$

Nonnegativity constraints on the decisions and the slack variables produce the optimality conditions, collectively known as the **Kuhn–Tucker conditions**:

$$\frac{\partial L(\underline{d}^*, \underline{\lambda}^*)}{\partial \underline{d}} = \frac{\partial F(\underline{d}^*)}{\partial \underline{d}} - \underline{\lambda}^* \frac{\partial \underline{g}(\underline{d}^*)}{\partial \underline{d}} \leq \underline{0}'$$

$$\frac{\partial L'(\underline{d}^*, \underline{\lambda}^*)}{\partial \underline{\lambda}} = \underline{c} - \underline{g}(\underline{d}^*) \geq \underline{0}$$

$$\frac{\partial L(\underline{d}^*, \underline{\lambda}^*)}{\partial \underline{d}} \underline{d}^* = \left[\frac{\partial F(\underline{d}^*)}{\partial \underline{d}} - \underline{\lambda}^* \frac{\partial \underline{g}(\underline{d}^*)}{\partial \underline{d}}\right] \underline{d}^* = 0$$

$$\underline{\lambda}^* \frac{\partial L(\underline{d}^*, \underline{\lambda}^*)}{\partial \underline{\lambda}} = \underline{\lambda}^* \left[\underline{c} - \underline{g}(\underline{d}^*) \right] = \underline{0}$$

$$\underline{d}^* \geq \underline{0}$$

$$\underline{\lambda}^* \geq \underline{0}. \qquad \text{(E.13)}$$

If the constraints define a convex decision (opportunity) set and the objective is concave (convex), it can be shown that the Kuhn–Tucker conditions are sufficient to guarantee a global maximum (minimum). In a few cases, it is possible to identify an optimal solution by finding a solution that fulfills the Kuhn–Tucker conditions (e.g., Williams et al. 2002:599). In most cases, the solution will be found by sequential searching methods and the Kuhn–Tucker conditions evaluated for candidate solutions to verify sufficiency.

Example: conservation of 2 species with inequality constraints

Again, return to the 2-species conservation example to illustrate optimization with constraints, but now with inequality constraints. The objective is to maximize $F(\underline{d})$ where

$$F(\underline{d}) = 224d_1 + 84d_2 + d_1 d_2 - 2d_1^2 - d_2^2$$

We now impose the constraint

$$3d_1 + d_2 \leq 290,$$

so instead of a fixed total resource limitation, we require an upper limit on resource use. The Lagrangian (excluding the slack variables) is

$$L(\underline{d}, \lambda) = 224d_1 + 84d_2 + d_1 d_2 - 2d_1^2 - d_2^2 + \lambda(290 - 3d_1 - d_2)$$

Solutions for this problem can be found by iterative methods (e.g., the BFGS procedure in R) to provide optimal values of $\underline{d}^* = (71.5, 75.5)$, $\lambda^* = 4.5$. The Kuhn–Tucker conditions are readily evaluated as

$$\frac{\partial F(\underline{d}^*)}{\partial \underline{d}} - \underline{\lambda}^* \frac{\partial g(\underline{d}^*)}{\partial \underline{d}} = \begin{bmatrix} 13.5 \\ 4.5 \end{bmatrix} - 4.5 \begin{bmatrix} 3 \\ 1 \end{bmatrix} = \begin{bmatrix} 0 \\ 0 \end{bmatrix}$$

$$c - g(\underline{d}^*) = 290 - 290 = 0$$

$$\left[\frac{\partial F(\underline{d}^*)}{\partial \underline{d}} - \underline{\lambda}^* \frac{\partial g(\underline{d}^*)}{\partial \underline{d}} \right]' \underline{d}^* = \left[\begin{bmatrix} 13.5 \\ 4.5 \end{bmatrix} - 4.5 \begin{bmatrix} 3 \\ 1 \end{bmatrix} \right]' \begin{bmatrix} 71.5 \\ 4.5 \end{bmatrix} = [0 \quad 0] \begin{bmatrix} 71.5 \\ 4.5 \end{bmatrix} = 0$$

$$\lambda^* [b - g(\underline{d}^*)] = 4.5(290 - 290) = 0$$

$$\underline{d}^* = \begin{bmatrix} 71.5 \\ 75.5 \end{bmatrix} > \begin{bmatrix} 0 \\ 0 \end{bmatrix}$$

$$\lambda^* = 4.5 > 0.$$

This solution provides an objective value of $F(\underline{d}^*) = 11\,831.5$, which is higher than the value of $11\,774$ obtained under the equality constraint of $3d_1 + d_2 = 280$. Removal of the constraint returns us to the original, unconstrained solution with objective return of $11\,872$; these results are consistent with the general feature that relaxing (or removing) constraints increases objective return. They also make clear that both classical programming and unconstrained optimization can be viewed as special cases of **nonlinear programming.**

Linear programming

Linear programming is yet another special case of nonlinear programming with the following features

- The objective can be expressed as a linear function of the decision variables; that is

$$F(\underline{d}^*) = \underline{b}\underline{d}$$

where

$$\underline{b} = [b_1, b_2, b_3 \ldots . b_n]$$

Explicitly, the objective function is formed by the row-column product

$$F(\underline{d}) = [b_1 \quad b_2 \quad . \quad . \quad . \quad b_n] \begin{bmatrix} d_1 \\ d_2 \\ . \\ . \\ . \\ d_n \end{bmatrix}$$

The constraints are likewise a linear function of the decision space

$$\underline{A}\underline{d} \le \underline{c}$$

where $\underline{A}$ is n X m matrix of constants

$$\underline{A} = \begin{bmatrix} a_{11} & a_{12} & & a_{1n} \\ a_{21} & a_{22} & & a_{2n} \\ & & & \\ & & & \\ a_{m1} & a_{m2} & & a_{mn} \end{bmatrix}$$

The optimization is now expressed as

maximize

$$\underline{b}\underline{d}$$

subject to

$$\underline{A}\underline{d} \le \underline{c}$$
$$\underline{d} \ge \underline{0}. \tag{E.14}$$

Geometrically, the nonnegativity constraints $\underline{d}$ restrict solutions to a nonnegative orthant of n-dimensioned Euclidian space. The constraints impose upper-bounding hyperplanes, and the intersection of these spaces determines the opportunity set for decisions (Williams et al. 2002: 757).

Because linear programming is a special case of nonlinear programming, Lagrangian multipliers can be used to derive the Kuhn–Tucker conditions for optimality. Obviously the Lagrangian function is simpler, due to linearity

$$L(\underline{d}, \lambda) = \underline{b}\underline{d} + \underline{\lambda}(\underline{c} - \underline{A}\underline{d})$$

as are the resulting Kuhn–Tucker conditions:

$$\frac{\partial L(\underline{d}^*, \underline{\lambda}^*)}{\partial \underline{d}} = \underline{b} - \underline{\lambda}\underline{A} \le \underline{0}'$$

$$\frac{\partial L'(\underline{d}^*, \underline{\lambda}^*)}{\partial \underline{\lambda}} = \underline{c} - \underline{A}\underline{d}^* \ge \underline{0}$$

$$\frac{\partial L(\underline{d}^*, \underline{\lambda}^*)}{\partial \underline{d}} \underline{d}^* = [\underline{b} - \underline{\lambda}\underline{A}]\underline{d} = 0$$

$$\underline{\lambda}^* \frac{\partial L(\underline{d}^*, \underline{\lambda}^*)}{\partial \underline{\lambda}} = \underline{\lambda}^*[\underline{c} - \underline{A}\underline{d}] = 0$$

$$\underline{d}^* \ge \underline{0}$$
$$\underline{\lambda}^* \ge \underline{0} \tag{E.15}$$

An interesting consequence of linearity in the objective function is that a second linear programming problem, called the **dual,** can be formed by switching the roles of the Lagrangian multipliers and decision variables. That is, the problem now is

minimize

$$\underline{\lambda}\underline{c}$$

subject to

$$\underline{\lambda}\underline{A} \geq \underline{b}$$
$$\underline{\lambda} \geq \underline{0}'.$$

The Lagrangian function of the dual is

$$L(\underline{\lambda}, \underline{d}) = \underline{\lambda}\underline{c} + (\underline{b} - \underline{\lambda}A)\underline{d}. \tag{E.16}$$

It is easy to show that the Kuhn–Tucker conditions for the dual problem are identical to the original (or **primal**) problem, indicating that solution of either problem will result in the same (unique) values for $\underline{d}^*$ and $\underline{\lambda}^*$. Sometimes it is more efficient to solve $\underline{d}^*$ via the primal problem and $\underline{\lambda}^*$ via the dual problem; the above result guarantees that these values jointly satisfy optimality, including values for sensitivity $\underline{\lambda}^*$.

The linear nature of the objective function and constraints permits efficient searching for optimal solutions to the linear programming problem. The **simplex** algorithm takes advantage of the fact that solutions have to lie along intersection of constraint hyperplanes or in the boundary region of a hyperplane. Because there are a finite number of these vertices, finding the optimal decision reduces to searching among candidate values at these intersections for the maximum or minimum. Thus, linear programming tends to rapidly converge to solutions even for a large number of decision variables and constraints, making it attractive for highly dimensioned problems.

Linear programming represents a compromise that favors number and complexity of constraints over objective realism. On the one hand, we can rapidly find the solution to an optimization problem that involves a large number of constraints, as long as we are willing or able to cast the objective as a linear function of the controls, and if the constraints can be expressed as a linear equalities or inequalities. For this reason (and in spite of the necessary sacrifices to realism), linear programming has proved valuable for a wide range of planning problems, e.g., those involving allocation of resources among units in space.

Example – Reserve design

Suppose we have 3 biological reserves, and we need to decide how much habitat to conserve on each. Let d_1, d_2, d_3 represent the areas of habitat conserved on each reserve. Suppose further that it has been established that the reserved differ in their efficacy as far as the resource outcome, predicted probability of persistence according to the model:

$$\text{logit(persistence)} = Y = -5 + 0.001d_1 + 0.0012d_2 + 0.0009d_3$$

and that implementation costs differ among reserves at \$5, \$7, and \$6.50 per ha, respectively. We also assume that we have a total cost constraint of \$5000 and an area constraint that the maximum conservation area is dictated by the existing reserve size of 1000, 2500, and 2000 ha, respectively. Finally, for "political" reasons we have to conserve at least 250h in each reserve. We can express the objective function as the linear objective

$$F(\underline{d}) = Y + 5 = 0.001d_1 + 0.0012d_2 + 0.0009d_3$$

(note that the intercept 5 has no bearing since it is not a function of any of the decisions), which we wish to maximize subject to the constraints

$$5d_1 + 7d_2 + 6.5d_3 \leq 5000$$

$$d_1 \leq 1000$$
$$d_2 \leq 2500$$
$$d_3 \leq 2000$$

$$d_1 \geq 250$$
$$d_2 \geq 250$$
$$d_3 \geq 250.$$

Application of the simplex algorithm (e.g., using lp() in R) results in a solution of $d^* = [325,250,250]$ with an objective value of $F(d^*) = 0.85$, which can be converted to the probability of persistence 0.9971 by the inverse logit transform. Thus, to fulfill the objective of maximizing species persistence subject to the stated constraints, we should allocate conservation to each reserve as 325 ha, 250 ha, and 250 ha, respectively. Note that we could have undoubtedly achieved a higher probability of persistence by discarding the cost constraints, allowing expansion of the reserves, or ignoring politics. For example, if we eliminate the 250 ha minimum constraint, we achieve a different optimal decision of $d^* = [1000,0,0]$. If we also remove the cost constraint, we get $d^* = [1000,2500,2000]$, which lies on the boundary of the area constraint: we can conserve no more habitat than exists. In each case, relaxing a constraint results in somewhat higher objective value. Of course if we remove this constraint, the problem has no bounded solution: given a linear objective function maximizing the function means maximizing (or minimizing) the controls without bound.

Dynamic decision problems

Up to this point we have dealt with problems that involve implementation of a decision at a single point in time followed by realization of a single objective value for the candidate decision. More general problems involved 2 features that we have not explicitly dealt with

- decisions that are made sequentially through time; and
- decisions that are linked because earlier decisions may affect the transition of the resource state, which in turn affects the opportunity for future decisions.

A general formulation of the optimization problem in discrete time is

$$\text{Maximize} \atop [\underline{d}(t) \in \underline{D}] \sum_{t=t_0}^{t_f} F(\underline{x}, \underline{d}, \underline{z}, t) + F_T[\underline{x}(t_f)]$$

subject to

$$\underline{x}(t+1) = \underline{x}(t) + f(\underline{x}, \underline{d}, z, t)$$
$$\underline{x}(t_0) = \underline{x}_0 \tag{E.17}$$
$$\underline{x}(t) \in \underline{X}$$

where $\underline{x}(t)$, is a vector of system states evolving through time, $\underline{d}(t)$ is a vector of time-specific decisions (controls), $\underline{z}(t)$ is a vector of random variables influencing dynamics but not under decision control, $F()$ is a time-specific utility function, $F_T()$ assigns a terminal to the system state, and $f()$ describes system dynamics. The last 2 conditions specify an initial ($t = t_0$) state for the system and a feasible set of state trajectories (e.g., forbidding negative population values).

The parallel formulation for continuation time is

$$\text{Maximize} \atop [\underline{d}(t) \in \underline{D}] \int_{t_0}^{t_f} F(\underline{x}, \underline{d}, \underline{z}, t) + F_T[\underline{x}(t_f)]dt$$

subject to

$$\dot{\underline{x}} = d\underline{x}(t)/dt = f(\underline{x}, \underline{d}, z, t)$$
$$\underline{x}(t_0) = \underline{x}_0 \tag{E.18}$$
$$\underline{x}(t) \in \underline{X}.$$

Under either formulation, the objective is to maximize the accumulation of utilities over the decision timeframe (t_0, t_f), subject to (for the moment) a very general statement about dynamics. Note also that in this formulation the constraining dynamics include, in general, a stochastic element.

Several approaches are available for solving the dynamic decision problem, and their appropriateness will depend on several factors:

- the mathematical form of the objective, $F()$ and $F_T()$;
- the complexity (dimensionality) of the system state $\underline{x}(t)$;
- the mathematical form of the dynamics $f()$ and any additional constraints;
- the treatment of time and/or system states as continuous vs. discrete;
- whether stochastic terms $\underline{z}(t)$ are present or absent.

In the discussion to follow we will typically start with problems that have simple objectives, low-dimensioned system states with simple (e.g., linear), deterministic dynamics, and evolve toward problems with more complex objectives and higher-dimensioned states that include nonlinear and stochastic dynamics.

Nonlinear programming

Although not specifically designed for dynamic problems, nonlinear programming (and sometimes linear programming) can be formulated to handle dynamics. This is accomplished essentially be treating the dynamic model of decision impacts as another constraint and then optimizing the objective function that now includes

- consideration of time (present and future value) in the objective;
- constraints on initial conditions;
- constraints on state transition;
- other constraints as needed.

Example

Suppose we have a population that is currently at $N = 100$, and is assumed to be growing deterministically according to the discrete logistic model with $r_{max} = 0.3$ and $K = 125$. We wish to optimize the annual, per-capita harvest rate $d(t)$ to maximize total harvest over the next 10 years (including this year), plus the size of the stock remaining after 10 years. Finally, annual harvest rate is to be restricted to levels less than $r_{max} = 0.3$ to avoid depleting growth potential in any year.

Writing the objective function and Lagrangian for this problem explicitly as function of the decisions is a little involved, because of the recursive nature of the dynamic equation. However we show (Appendix G) that this problem is fairly straightforward to solve numerically via the use of user-defined functions for the constraining dynamics combined with a numerical procedure, such as BFGS as implemented in R.

The problem can then be stated as

$$\text{Maximize}_{[d(t)]} \sum_{i=1}^{10} d(t)N(t) + N(11)$$

subject to

$$N(t+1) = N(t) + r_{max}N(t)[1 - N(t)/K] - d(t)N(t)$$
$$0 \le d(t) \le r_{max}$$
$$N(0) = 100.$$

Solution of this problem via nonlinear programming is fairly straightforward, leading to an optimal harvest decision over the 10 years of $d^*(t) = [0.30000, 0.29524, 0.15000, 0.15001, 0.14999, 0.15000, 0.14999, 0.15000, 0.14999, 0.0999]$.

Although we can use nonlinear programming (and in some cases linear programming) to solve simple dynamic problems, this approach has a number of drawbacks. An obvious one is that as the timeframe lengthens (try doing this for a 100- or 500-year timeframe) computational demands become prohibitive.

Another is that the optimization is confined to a single observe current state, and we would have to repeat the exercise for other state possibilities. Finally, and most seriously, this approach does not really feedback anticipation of where the future state of the system could go, particularly under stochastic assumptions. For these reasons we emphasize modern approaches come back to the dynamic optimization problem using modern approaches, in particular **the calculus of variations** and **dynamic programming.**

The calculus of variations

The calculus of variations is analogous to classical programming considered earlier, in which we had an objective function that we sought to maximize or minimize, subject to equality constraints. Because we are seeking an analytical, calculus-based approach, both time and the system state will be considered as continuous but for now we will deal with univariate system states $x(t)$ and controls $d(t)$. An initial formulation of the calculus of variations approach is

$$\underset{[x(t)]}{\text{Maximize}} \int_{t_0}^{t_f} F(x, \dot{x}, t)dt$$

subject to

$$x(t_0) = x_0$$
$$x(t_f) \in x_f. \tag{E.19}$$

In this formulation, the influence of system controls $d(t)$ (if present) are contained in $\dot{x}$; we will make the connection more explicit below. The problem is to find a piecewise-differentiable function $x(t)$ that maximizes this integral subject to the initial value constraints, where $x(t)$ is deterministic and time is continuous. In the cases where the rate of change is under direct control, so that $\dot{x} = d(t)$, then the above can be reformulated as

$$\text{maximize} \int_{t_0}^{t_f} F(x, d, t)dt$$

subject to

$$\dot{x} = d(t)$$
$$x(t_0) = x_0$$
$$x(t_f) = x_f$$

so that now the controls are explicitly part of the objective and constraint.

In a manner analogous to classical optimization, a necessary condition for optimality is stated by Euler's equation

$$\frac{\partial F}{\partial x} - \frac{d}{dt}\left(\frac{\partial F}{\partial \dot{x}}\right) = 0.$$

Therefore, a search for the optimal trajectory (control) reduces to a search for solutions to the above equation, with the boundary conditions $x(t_0) = x_0, x(t_f) = x_f$. These constraints also force transversality conditions in addition to convexity and continuity requirements; see Williams et al. (2002: 758) for a more complete discussion.

Example – Pest control

A simple example of univariate calculus of variations is provided by Williams et al. (2002). The problem is to reduce a pest population from an initial state of $x(t_0) = 3$ to a terminal state of $x(t_f) = 0$ after 2 periods of control. The rate of change $\dot{x}$ is assumed to be under direct control, so $\dot{x} = d(t)$, with control costs a quadratic function of effort $d(t)$. The objective reducing the pest population is assured by the terminal condition $x(2) = 0$ and the objective of minimizing costs is provided by

$$\underset{[d(t)]}{\text{Maximize}} \quad -\int_0^2 \left[x(t) + d(t)^2 \right] dt$$

subject to

$$\dot{x} = d(t)$$
$$x(0) = 3$$
$$x(2) = 0$$

(we could equivalently have minimized the unsigned integral). Substituting dynamics into the equation we get

$$\underset{[x(t)]}{\text{Maximize}} \quad -\int_0^2 \left[x(t) + \dot{x}^2 \right] dt$$

subject to

$$x(0) = 3$$
$$x(2) = 0.$$

Euler's equation provides

$$\frac{\partial F}{\partial x} - \frac{d}{dt}\left(\frac{\partial F}{\partial \dot{x}} \right) = -1 + 2\ddot{x} = 0$$
$$\ddot{x} = 1/2$$

where

$$\ddot{x} = \frac{d^2 x(t)}{dt^2}.$$

The solution obtained from the definite integral over the timeframe for the optimal trajectory $x(t)$ and control $d(t)$ is then

$$x(t)^* = t^2 / 4 - 2t + 3$$
$$d(t)^* = \dot{x} = t / 2 - 2.$$

The calculus of variations extends to multivariate cases in a manner analogous to classical programming. The general multivariate problem is

$$\text{maximize} \int_{t_0}^{t_f} F(\underline{x}, \dot{\underline{x}}, t) dt$$

subject to

$$\underline{x}(t_0) = \underline{x}_0$$
$$\underline{x}(t_f) = \underline{x}_f.$$

Implicit in the above is the fact that controls can also be multivariate, for example mapping directly into state transitions, i.e., $\dot{\underline{x}} = \underline{d}(t)$. Optimality conditions are directly analogous to those for the univariate case, with application of the multivariate Euler equation as

$$\frac{\partial F}{\partial \underline{x}} - \frac{d}{dt}\left(\frac{\partial F}{\partial \dot{\underline{x}}}\right) = \underline{0}'.$$

Transversality, continuity, and convexity requirement hold as in the univariate case; see Williams et al. (2002) 614, 759).

Up to this point, we have only considered constraints (in addition to the dynamics implicitly incorporated) on the initial an terminal conditions. Imposition of equality and inequality constraints can be allowed by a Lagrangian approach very similar to the one employed in classical programming. Thus, an augmented objective is provided by

$$L(\underline{x}, \dot{\underline{x}}, \underline{\lambda}, t) = F(\underline{x}, \dot{\underline{x}}, t) + \underline{\lambda}\left[\underline{c} - \underline{g}(\underline{x}, \dot{\underline{x}}, t)\right]$$

and the problem become maximizing

$$\int_{t_0}^{t_f} L(\underline{x}, \dot{\underline{x}}, \underline{\lambda}, t) dt.$$

Euler's equation provides the optimality condition

$$\frac{\partial L}{\partial \underline{x}} - \frac{d}{dt}\left(\frac{\partial L}{\partial \dot{\underline{x}}}\right) = \underline{0}'.$$

Inequality constraints $g(\underline{x}, \underline{\dot{x}}, t) \leq \underline{c}$ can be allowed by analogy to nonlinear programming, with resultant optimality conditions

$$\frac{\partial L}{\partial \underline{x}} - \frac{d}{dt}\left(\frac{\partial L}{\partial \underline{\dot{x}}}\right) = \underline{0}'$$

$$g(\underline{x}, \underline{\dot{x}}, t) \leq \underline{c}$$

$$\underline{\lambda} \geq \underline{0}$$

$$\underline{\lambda}[\underline{c} - g(\underline{x}, \underline{\dot{x}}, t)] = 0.$$

The first of these is simply Euler's equation, the remainder correspond to the Kuhn–Tucker conditions described for nonlinear programming.

Pontryagin's maximum principle

Pontryagin's maximum principle extends the dynamic, multivariate case beyond casting it as a calculus of variations problem, by means of the incorporation "co-state" variables that allow for complex constraints on the decision space. It also generalized the problem to the case where change in the system state is influenced but not directly controlled by the decision variables. The problem is now

$$\underset{[\underline{d}(t) \in \underline{D}]}{\text{Maximize}} \int_{t_0}^{t_f} F(\underline{x}, \underline{d}, t)dt + F_T[\underline{x}(t_f)]$$

subject to

$$\underline{\dot{x}} = f(\underline{x}, \underline{d}, t)$$
$$\underline{x}(t_0) = \underline{x}_0 \qquad\qquad\qquad\qquad (E.20)$$
$$\underline{x}(t_f) = \underline{x}_f.$$

Constraints are introduced by means of the co-state variables, essentially a vector of time-varying Lagrangian multipliers

$$\underline{\lambda}(t) = [\lambda_1(t), \lambda_2(t), \ldots\ldots, \lambda_m(t)] \qquad\qquad (E.21)$$

one for each constraint

$$\underline{\dot{x}} - f(\underline{x}, \underline{d}, t) = \underline{0}$$

The **Hamiltonian** is then formed as

$$H(\underline{x}, \underline{d}, \underline{\lambda}, t) = F(\underline{x}, \underline{d}, t) + \lambda(t)f(\underline{x}, \underline{d}, t) \qquad\qquad (E.22)$$

and an augmented objective function

$$\int_{t_0}^{t_f} [H(\underline{x}, \underline{d}, \underline{\lambda}, t) - \underline{\lambda}\underline{x}]dt + F_T[\underline{x}(t_f)]. \tag{E.23}$$

Optimality conditions for $[\underline{x}(t)], [\underline{d}(t)], [\underline{\lambda}(t)]$ must satisfy

$$\frac{\partial \underline{H}}{\partial \underline{d}} = \underline{0}, \quad t_0 \le t \le t_f$$

$$-\frac{\partial \underline{H}}{\partial \underline{x}} = \underline{\lambda}, \quad t_0 \le t \le t_f \tag{E.24}$$

$$\left[\underline{\lambda} - \frac{\partial F_T}{\partial \underline{x}}\right]\delta\underline{x} = 0, \quad t = t_f$$

where at $t = t_f$ either the small variation $\delta\underline{x} = 0$, or $\underline{\lambda} - \dfrac{\partial F_T}{\partial \underline{x}} = 0$ (Williams et al. 2002: 618).

Williams et al. (2002) provide several examples to illustrate application of the maximum principle. Notably, all of these involved relatively short timeframes and fairly simple forms for the dynamics and constraints. These authors also describe a discrete-time version of the above equations, and note that in practice these equations must be solved by iterative methods (not in closed form as in the earlier, simple example or others presented by Williams et al. 2002).

It should be obvious from the above development that, while it is possible to obtain a mathematically "elegant" solution to the problem of dynamic optimization, the mathematics can be daunting for anything other than trivially simple problems. Furthermore, none of the above deals with this issue of stochastic influences on dynamics, to which we will return later in the Appendix.

Dynamic programming

Dynamic programming starts with the identical framing of the decision problem stated by the maximum principle. That is, the problem remains

$$\underset{[\underline{d}(t) \in \underline{D}]}{\text{Maximize}} \int_{t_0}^{t_f} F(\underline{x}, \underline{d}, t)dt + F_T[\underline{x}(t_f)]$$

subject to

$$\dot{\underline{x}} = f(\underline{x}, \underline{d}, t)$$
$$\underline{x}(t_0) = \underline{x}_0$$
$$\underline{x}(t_f) = \underline{x}_f$$

In discrete form the problem is

$$\underset{[\underline{d}(t)\,\in\,\underline{D}]}{\text{Maximize}} \sum_{t=t_0}^{t_f} F(\underline{x}, \underline{d}, t) + F_T[\underline{x}(t_f)]$$

subject to

$$\underline{x}(t+1) = \underline{x}(t) + f(\underline{x}, \underline{d}, z, t)$$
$$\underline{x}(t_0) = \underline{x}_0 \qquad\qquad\qquad (E.25)$$
$$\underline{x}(t) = \underline{x}_f.$$

For now, we formulate the problem with deterministic dynamics; we will later make a straightforward extension to stochastic state dynamics by replacing the objective function with its expected value, averaged over the uncertain factors that affect the system (in addition to the controls).

Despite the common roots with the maximum principle, dynamic programming takes a very different approach to finding solutions. Whereas the maximum principle obtains solutions via incorporation of system dynamics as co-state variables, dynamic programming exploits the **principle of optimality** (Bellman 1957) to reformulate the optimization problem by means of backwards induction. The principle of optimality states that an optimal policy has the property that, regardless of initial conditions and decisions from time t_0 to time t, all remaining decisions from t to t_f must be also be optimal. That is, the optimal behavior of the system from (t, t_f) is independent of how the system arrives at t, which can be considered as a new starting point for optimization. Dynamic programming is thus appropriate to systems where the future behavior of the system depends only on the current system state and controls, and independent of its past trajectories – in other words, the system is **Markovian**. In fact, most systems we are familiar with in ecology, are reasonably accurately representing by Markovian dynamics, including essentially all standard population dynamic models (Puterman 1994).

We present dynamic programming by starting with deterministic dynamics and assuming that we have optimized the objective function. That is we have an optimal value for Eq. (E.25) above, starting at time t (any arbitrary time in the timeframe (t_0, t_f). By the principle of optimality

$$J^*[\underline{x}(t), t] = \underset{[\underline{d}(t)\,\in\,\underline{D}]}{\text{Max}} [F(\underline{x}, \underline{d}, t)\Delta t + J^*[\underline{x}(t) + \Delta x, t + \Delta t]] \qquad (E.26)$$

which simply states that to be optimal, at time t, all remaining decisions must be optimal. Given the optimal decision at $t+\Delta t$, the optimal decision at time t is that which maximizes the sum of this value $J^*[]$ with the objective index value $F[]$ at time t, and is now optimal from that time forward. Given continuous differentiability of $J[\underline{x}(t), t]$ leads to the Hamilton–Jacobi–Bellman (HJB) equation as

$$\frac{-\partial J^*[\underline{x}(t), t]}{\partial t} = \frac{\text{Max}}{[\underline{d}(t) \in \underline{D}]} \left[F(\underline{x}, \underline{d}, t) + \frac{\partial J^*[\underline{x}(t), t]}{\partial \underline{x}} F(\underline{x}, \underline{d}, t) \right]. \quad \text{(E.27)}$$

This equation and the boundary condition

$$J^*[\underline{x}(t_f), t_f] = F_T[\underline{x}(t_f)]$$

form the analytical basis for the solving the control problem in continuous systems. The framework requires examination of all possible states of the system at each point in time. In practice, both the system state and time are typically discretized to facilitate analysis, although William et al. (2002: 629–630) illustrate simple applications in continuous time.

Discrete time dynamic programming

Discrete time dynamic programming starts with the HJB optimality relationship

$$J^*[\underline{x}(t), t] = \frac{\text{Max}}{[\underline{d}(t) \in \underline{D}]}[F(\underline{x}, \underline{d}, t)\Delta + J^*(\underline{x}(t + \Delta t), t + \Delta t)].$$

Conventionally, the time interval is usually taken as $\Delta t = 1$, recognizing that the units can vary among applications (i.e., 1 may signify 1 year, several years, or a fraction of a year).

$$J^*[\underline{x}(t), t] = \frac{\text{Max}}{[\underline{d}(t) \in \underline{D}]}[F(\underline{x}, \underline{d}, t) + J^*(\underline{x}(t + 1), t + 1)]. \quad \text{(E.28)}$$

The effect of this representation is to cast the problem of optimizing decisions over the entire timeframe $[t_0, t_f]$ as a series of optimizations over single time steps (stages). The optimal state-specific policy trajectory $[\underline{d}^*(\underline{x}(t))]$ is then found by a backwards inducting progress as follows:

- Starting at the terminal time t_f, determine an action that maximizes $F_T[\underline{x}(t_f), \underline{d}(\underline{x}(t_f))]$. Often, this value will not be affected by the actions, in which case the value function is simple the value of the terminal state itself, that is $J^*[\underline{x}(t_f)] = F_T[\underline{x}(t_f)]$. In some cases (e.g., many harvest problems), the terminal state has no value, so that $J^*[\underline{x}(t_f)] = 0$.
- Using the values obtained at t_f, determine the state-specific decision to take at $t_f - 1$ in order to maximize $J[\underline{x}(t_{f-1})] = F[\underline{x}(t_{f-1}), \underline{d}(t_{f-1})] + J^*[\underline{x}(t_f)]$.
- Using the values obtained at $t_f - 1$, determine the state-specific decision to take at $t_f - 2$ to maximize $J[\underline{x}(t_{f-2})] = F[\underline{x}(t_{f-2})] + J^*[\underline{x}(t_{f-1})]$.
- Continue the backwards iteration process until t_0 has been reached.

On completion, the dynamic programming algorithm provides a complete trajectory of time- and state-specific optimal decisions $[\underline{d}^{*}(\underline{x}(t))]$. The dynamic programming algorithm requires

- a current value function $F[\underline{x}(t), \underline{d}(\underline{x}(t))]$;
- a value $J^{*}[\underline{x}(t_{t+1})]$ for the optimal decisions at $t+1$;
- a model for transition from t to $t+1$;
- specification of the terminal value.

Because of the decomposition of the problem into single-stage problems, and the sequential elimination by backwards induction of suboptimal decisions under the principle of optimality, dynamic programming can result in extraordinary savings in computational time compared to forward-time enumeration of all possible decision pathways. Additional efficiency is introduced by recognizing that many dynamic decision problems converge on **stationary policies**, in which the state-specific optimal decision remains unchanged over time. When this occurs, the optimal policy is now state dependent but independent of time, so that in principle the optimal policy $[\underline{d}^{*}(\underline{x})]$ can be applied to any future observed state without recomputation of the optimal policy. Such a strategy is said to be **closed loop** in that it implicitly involves feedback from future (as of yet unobserved) realized states in the calculation of the current optimal decision. Significantly, the problem has now been converted from a finite-time optimization over $[t_0,t_f]$ to an infinite-time problem over $[t_0,t_\infty]$. Thus, stationary dynamic programming solutions are particularly applicable to the problem of sustainable harvest, where interest is in the ability to extract resources of a population while maintaining stock, both over an arbitrarily long time horizon.

Example

We illustrate dynamic programming by returning to the deterministic harvest problem considered in Box 8.6

    Maximize

$$\sum_{t=t_0}^{t_f} F[N(t) \times h(t)] + F_T[x(t_f)]$$

where x is discrete abundance ranging from 1 to 100 and the decision is the harvest rate $d \in D = \{0, 0.1, 0.2, 0.3, 0.4, 0.5\}$.

    Dynamics are provided by

$$x(t+1) = x(t) + 0.3x(t)[1 - x(t)/125] - d(t)x(t)$$

with outcomes rounded to the nearest integer. Because we are assigning no terminal value to the population state, we have

$$F_T[x(t_f)] = 0$$

so that under the optimality principle we have

$$J^*[x(t), t] = \frac{\text{Max}}{[d(t) \in D]}[F(x, d, t) + J^*(x(t+1), t+1)]$$

with

$$J^*[\underline{x}(t_f)] = 0.$$

Effectively, this means that we can start backward induction at t_{f-1}. To illustrate, we start at t_{f-1} and assume that the state value is $x(t_{f-1}) = 50$. The optimal state-specific decision at that time is that which maximizes

$$J^*[50, t_{f-1}] = \frac{\text{Max}}{[d(t) \in D]}[F(50, d, t_{f-1}) + J^*(x(t_f), t_f)]$$

$$J^*[50, t_{f-1}] = \frac{\text{Max}}{[d(t) \in D]}[50 \times d(t_{f-1}) + 0].$$

Fairly obviously, the decision that maximizes this value is the maximal harvest rate over the admissible range, so $d(t_{f-1}) = 0.5$, providing a value for J^* of 25; for other state values the optimal decision likewise will be 0.5, with value equal to $0.5x(t_{f-1})$. We now step back one time interval to t_{f-2}, now examining the system at $x(t_{f-2}) = 50$. Here, the optimal state-specific decision maximizes

$$J^*[50, t_{f-2}] = \frac{\text{Max}}{[d(t) \in D]}[F(50, d, t_{f-2}) + J^*(x(t_{f-1}), t_{f-1})]$$

where $J^*(x(t_{f-1}), t_{f-1})$ will (under the principle of optimality) be the optimal value of the state arrive the following time step, which in turn is determined by the decisions made at time t_{f-2} and the dynamic model. For the initial state $x(t_{f-2}) = 50$ the 6 possible harvest decisions result in states, respectively, at t_{f-1} of $x(t_{f-1}) = \{59, 54, 49, 44, 39, 34\}$ with respective values $J^*(x(t_{f-1}, t_{f-1}) = \{29.5, 27.0, 24.5, 22, 19.5, 17\}$. The immediate objective return at t_{f-2} is $50 \times d(t_{f-2}) = \{0, 5, 10, 15, 20, 25\}$. The optimal decision is that that maximizes the sum of these 2 indices, $50 \times d(t_{f-2}) + J^*(x(t_{f-1}), t_{f-1}) = \{29.5, 32.34.5, 37.39.5, 42\}$ which again is $d(t_{f-2}) = 0.5$ with value $J^* = 42$. Similar calculations are performed for each of the other states, resulting in the state-specific optimal decision d and values J^* at t_{f-2}. The algorithm continues stepping back through time, with state-specific decisions eventually changing to reflect the forward-looking nature of the algorithm. For example, if we consider $x = 50$ at t_{f-4}, the 6 possible harvest decisions result in states respectively at t_{f-3} of $x(t_{f-3}) = \{59, 54, 49, 44, 39, 34\}$, which will already have been shown to have optimality values of $J^*(x(t_{f-3}), t_{f-3}) = \{63, 58, 53, 48.5, 44.38.1\}$, resulting in $50 \times d(t_{f-4}) + J^*(x(t_{f-3}), t_{f-3}) = \{63, 63, 63, 63.5, 64, 63.1\}$, so that the optimal harvest decision is now $d(t_{f-4}) = 0.4$.

Eventually (after approximately 20 iterations) the algorithm reaches stationarity, with state-specific harvest decisions ranging from 0 (for low abundance states) to >0.3 (for states approaching 100).

In the implementation of this example in R, we assumed discrete states of nature (x capable of only integer values 0 to 100). More generally, continuous states of nature will be discretized, necessitating the interpolation of value functions between states, something that is accomplished in more sophisticated software such as program SDP (Lubow 1995). Although introducing computational complexity, such generalizations follow directly from the backwards iterative procedure illustrated above.

Stochastic dynamic programming

Stochastic dynamic programming (SDP) extends dynamic programming to allow stochastic uncertainty in the objective and dynamics. That is, the system now evolves over time according to

$$\underline{x}(t+1) = \underline{x}(t) + \underline{f}(\underline{x}(t), \underline{d}(t), \underline{z}(t), t). \tag{E.29}$$

Additionally, the value index $F()$ may itself involve stochastic influences (e.g., partial control; Chapter 7)

$$F(\underline{x}(t), \underline{d}, \underline{z}(t), t).$$

The objective is now evaluated by maximizing the expected value of the HJB relationship

$$J^*[\underline{x}(t), t] = \underset{[\underline{d}(t) \in \underline{D}]}{\text{Max}} \{E[F(\underline{x}(t), \underline{d}, t) + J^*(\underline{x}(t+1), t+1)]\}. \tag{E.30}$$

The expectation for this is calculated over the distribution(s) of the random variables $\underline{z}(t)$, typically discretized, with probability p_i for each discrete value $\underline{z}_i$:

$$J^*(\underline{x}(t), t) = \max\left(\sum_i^k F[\underline{x}(t), \underline{d}(t), t, \underline{z}_i(t)]p_i + J^*(\underline{x}(t+1), t+1, \underline{z}_i(t+1)p_i)\right). \tag{E.31}$$

Likewise, the expected single time step dynamics are also typically part of the state transition function, so that

$$E[\underline{x}(t+1)] = \underline{x}(t) + \sum_{i=1}^k p_i f(\underline{x}, \underline{d}, \underline{z}_i, t). \tag{E.32}$$

Example – Harvest of a stochastically growing population

Here we return to the example of Box 8.8 but generalize it slightly to allow for stochastic dynamics. As before, the problem is stated as

Maximize

$$E\left(\sum_{t=t_0}^{t_f}[N(t)\times h(t)]\right)$$

subject to

$$N(t+1) = E\{N(t) + r(t)N(t)[1 - N(t)/K] - h(t)N(t)\}$$

where partial controllability is represented by a discrete random variable distribution $\underline{h}(t) = [0.75d(t), d(t), 1.25d(t)]$ indicating that realized harvest rate is 75%, 100%, or 125% of intended harvest rate $d(t)$, with probabilities associated with these outcomes as $p = (0.25, 0.5, 0.25)$. Environmental stochasticity is represented by $\underline{r}(t) = (0.85r_{max}, 1.05r_{max}, 1.15r_{max})$ also with probabilities $\pi_i = (0.25, 0.5, 0.25)$. Assuming independence between partial harvest rates $h(t)$ and growth rates $r(t)$ the expectation of objective and dynamics follow as

$$J^*(\underline{x}(t), t)$$

$$= \max\left(\sum_i^3 F[x(t), d(t), t, h_i(t)]p_i + \sum_i^3\sum_j^3 J^*(x(t+1), t+1, h_i(t), r_j(t))p_i\pi_j)\right)$$

where

$$E\{x(t+1)\} = x(t) + \sum_{i=1}^3\sum_{j=1}^3\{r_j(t)N(t)[1 - N(t)/K] - h_i(t)N(t)\}p_i\pi_j\}.$$

Optimization then proceeds as before, again taking

$$J^*[50, t_{f-1}] = \frac{\text{Max}}{[d(t) \in D]}\left[\sum_{i=1}^3 50 \times h_i(t_{f-1}) + 0\right].$$

For the 6 possible harvest decisions $d(t) \in D = \{0, 0.1, 0.2, 0.3, 0.4, 0.5\}$, this results in expected values of $\{0, 4.8, 9.6, 14.4, 19.25, 24.06\}$, with the harvest decision that provides a maximum as $d(t_{f-1}) = 0.5$, but now with expected value of 24.06 instead of 25 as was the case under deterministic DP. Each of these different decisions will take the system to a different state at t_f, but that is irrelevant because of the absence of a terminal value in the objective functional. Stepping to t_{f-2}, we again examine the system at $x(t_{f-2}) = 50$. Here, the optimal state-specific decision maximizes

$$J^*(50, t_{f-2}) = \max\left(\sum_i^3 50 \times h_i(t_{f-2})]p_i + \sum_i^3\sum_j^3 J^*(x(t+1), t+1, h_i(t), r_j(t))p_i\pi_j)\right).$$

For the initial state $x(t_{f-2}) = 50$, the 6 possible harvest decisions expected states, respectively, at t_{f-1} of $x(t_{f-1}) = \{59.0, 54.18, 49.37, 44.56, 39.75, 34.93\}$ with values $J^*(x(t_{f-1}, t_{f-1}) = \{28.39, 25.98, 23.58, 21.65, 19.25, 16.84\}$. The immediate objective return at t_{f-2} is $50 \times d(t_{f-2}) = \{0, 4.8, 9.6, 14.4, 19.25, 24.06\}$. The optimal decision is that which maximizes the sum of these 2 indices is $\{28.39, 30.80, 33.20, 36.09, 38.50, 40.90\}$ that again is $d(t_{f-2}) = 0.5$ with value $J^* = 42$. As the algorithm steps backwards in time, the state-specific decisions become increasingly forward looking; in this example, after approximately 28 iterations the state-specific harvest decisions will be <0.2 for most states <60. Also notice that (at the equivalent stage) the cumulative objective return is lower for SDP than for DP. This is because stochasticity effectively acts as a constraint, whose removal (if possible) would inevitably result in better performance of the optimal decision.

Finally, finding the optimal decision – regardless of how it is arrived at – in stochastic systems should never be the last step. It is always a good idea to plug the optimal (usually state-specific) decision back into the dynamic model and simulate the populations change through time (in fact this is a good idea even if we are pretending the system changes deterministically).

Transition matrix formulation of SDP

Above, we explicitly defined dynamics in terms of state equations and random variable distributions. However, to solve the discrete Markovian SDP problem it is sufficient to describe dynamics in terms of **transition probabilities**

$$\pi(\underline{x}_{t+1|} \mid \underline{x}_t, \underline{d}_t) \tag{E.33}$$

defined as the probability of arriving in the state $\underline{x}_{t+1|}$ at time $\underline{t} + 1$, conditional on existence in the state $\underline{x}_t$ and implementation of decision $\underline{d}_t$ at time t. The recursive optimality equation is recast as

$$J^*[\underline{x}(t), t] = \frac{\text{Max}}{[\underline{d}(t) \in \underline{D}]} \left\{ \overline{F}(\underline{x}, \underline{d}, t) + \sum_{x_{t+1}} \pi(\underline{x}_{t+1} \mid \underline{x}_t, \underline{d}_t) J^*(\underline{x}(t+1), t+1) \right\} \tag{E.34}$$

where $\overline{F}(\underline{x}, \underline{d}, t)$ is the expected utility index for immediate objective return (so, averaged over random outcomes describing partial controllability, if applicable) and

$$\sum_{x_{t+1}} \pi(\underline{x}_{t+1} \mid \underline{x}_t, \underline{d}_t)$$

is understood to represent summation across the discrete state values at $t+1$, providing the expected optimal return over the remainder of the time horizon. To clarify, consider a univariate state x_t with 3 levels $x_t \in \{1,2,3\}$ and a single decision also with 3 levels $d_t \in \{1,2,3\}$. The state transition matrix conditional on a given decision (e.g., $d_t = 2$)

$$\pi(x_{t+1} \mid x_t, d_t = 2) =$$
$$\begin{bmatrix} \pi(x_{t+1} = 1 \mid x_t = 1, d_t = 2) & \pi(x_{t+1} = 2 \mid x_t = 1, d_t = 2) & \pi(x_{t+1} = 3 \mid x_t = 1, d_t = 2) \\ \pi(x_{t+1} = 1 \mid x_t = 2, d_t = 2) & \pi(x_{t+1} = 2 \mid x_t = 2, d_t = 2) & \pi(x_{t+1} = 3 \mid x_t = 2, d_t = 2) \\ \pi(x_{t+1} = 1 \mid x_t = 3, d_t = 2) & \pi(x_{t+1} = 2 \mid x_t = 3, d_t = 2) & \pi(x_{t+1} = 3 \mid x_t = 3, d_t = 2) \end{bmatrix}$$

and the future expected value averaged over the states at $t+1$ is provided by the row-column product of the above matrix and the vector of state-specific utilities

$$J^*(x(t+1), t+1) = \left[J^*(x(t+1) = 1, t+1) \quad J^*(x(t+1) = 2, t+1) \right.$$
$$\left. J^*(x(t+3) = 1, t+1) \right].$$

Casting the SDP problem with transition matrices first makes clear the Markovian nature of the decision problem, and makes clear that the decision problem can be completely specified (for discrete-state problems) by specification of

- a stage-return value function;
- a terminal value function (often null); and
- a matrix of state transition probabilities.

The last of these, in turn, may either be computed by the programming algorithm or computed elsewhere (e.g., in a stochastic population model) and then provided to the algorithm.

In addition to these advantages, this representation facilitates description of the more complex problems of optimization under multiple structural models, and reduction of structural uncertainty via adaptive management. However, it also illustrates one of the fundamental limitations of SDP: the "curses of dimensionality" (Bellman 1957). Given the example of a univariate state with 3 levels and 3 possible decisions, we must compute 9 state transitions for each decision, or 27 transitions altogether. A slightly more realistic problem with 10 state variable levels and 6 decision alternatives would require computation of 600 transitions. Thus, problems with many state (e.g., involving multiple species or habitat states, or spatial stratification) and decision combinations (e.g., both harvest regulations and habitat management) may be impracticable to solve using SDP.

Example

We illustrate SDP with a harvest problem involving a single state (abundance at 3 levels, $x_t = 5$, 10, or 15) and decision (harvest rates, also at 3 levels $d_t = 0.1, 0.2$, or 0.3 (Williams et al. 2002: 636). We assume perfect controllability, so the stage return function simple involves the product of state level with harvest rate, $F(x_t, d_t, t) = x_t d_t$; for example, a harvest decision of 0.3 for a state value of 15 provides a value of $F(x_t = 20, d_t = 0.3, t) = 4.5$. The decision-specific state-transition matrices are provided as

$$\pi(x_{t+1} \mid x_t, d_t = 0.1) = \begin{bmatrix} 0.2 & 0.5 & 0.3 \\ 0.2 & 0.3 & 0.5 \\ 0.1 & 0.3 & 0.6 \end{bmatrix}, \pi(x_{t+1} \mid x_t, d_t = 0.2) = \begin{bmatrix} 0.5 & 0.3 & 0.2 \\ 0.2 & 0.5 & 0.3 \\ 0.2 & 0.4 & 0.4 \end{bmatrix},$$

$$\pi(x_{t+1} \mid x_t, d_t = 0.3) = \begin{bmatrix} 0.7 & 0.3 & 0. \\ 0.6 & 0.3 & 0.1 \\ 0.3 & 0.5 & 0.2 \end{bmatrix}$$

Because there is no terminal value function, at $t = t_f$ it is obviously optimal to harvest at the maximal rate 0.3, so $J^*(x(t_f), t_f)' = [1.5, 3, 4.5]$. At t_{f-1} the HJB relationship requires maximization of

$$J(x(t_{f-1}), t_{f-1}) = x_{t_{f-1}} d_{t_{f-1}} + \pi(x_{t+1} \mid x_t, d_t) J^*(x(t_f), t_f).$$

At t_{f-1} the values of $J(x(t_{f-1}), t_{f-1})$ conditional on each decision $d_{t_{f-1}}$ are

$$J(x(t_{t_{f-1}}), t_{t_{f-1}} \mid d(t_{t_{f-1}})) = 0.1) = \begin{bmatrix} (0.1)5 \\ (0.1)10 \\ (0.1)15 \end{bmatrix} + \begin{bmatrix} 0.2 & 0.5 & 0.3 \\ 0.2 & 0.3 & 0.5 \\ 0.1 & 0.3 & 0.6 \end{bmatrix} \begin{bmatrix} 1.5 \\ 3 \\ 4.5 \end{bmatrix} = \begin{bmatrix} 3.65 \\ 4.45 \\ 5.25 \end{bmatrix},$$

$$J(x(t_{t_{f-1}}), t_{t_{f-1}} \mid d(t_{t_{f-1}})) = 0.2) = \begin{bmatrix} (0.2)5 \\ (0.2)10 \\ (0.2)15 \end{bmatrix} + \begin{bmatrix} 0.5 & 0.3 & 0.2 \\ 0.2 & 0.5 & 0.3 \\ 0.2 & 0.4 & 0.4 \end{bmatrix} \begin{bmatrix} 1.5 \\ 3 \\ 4.5 \end{bmatrix} = \begin{bmatrix} 3.55 \\ 5.15 \\ 6.30 \end{bmatrix},$$

$$J(x(t_{t_{f-1}}), t_{t_{f-1}} \mid d(t_{t_{f-1}})) = 0.3) = \begin{bmatrix} (0.3)5 \\ (0.3)10 \\ (0.3)15 \end{bmatrix} + \begin{bmatrix} 0.7 & 0.3 & 0. \\ 0.6 & 0.3 & 0.1 \\ 0.3 & 0.5 & 0.2 \end{bmatrix} \begin{bmatrix} 1.5 \\ 3 \\ 4.5 \end{bmatrix} = \begin{bmatrix} 3.45 \\ 5.25 \\ 7.35 \end{bmatrix},$$

The harvest decisions $d_{t_{f-1}} = (0.1, 0.0.3, 0.3)$ maximize this function, conditional on each of the respective states at $x_{t_{f-1}} = (5, 10, 15)$, providing $J^*(x(t_{f-1}), t_{f-1})$

$$J^*(x(t_{f-1}), t_{f-1}) = \begin{bmatrix} 3.65 \\ 5.25 \\ 7.35 \end{bmatrix}.$$

This value vector then becomes the optimal future utility at the next time step, and the algorithm repeats. After a few (<5) iterations, the algorithm converges to a stationary, state-specific optimal decision strategy of $d(x_t)' = (0.1, 0.2, 0.3)$ (Williams et al. 2002: 637). We have written R code to implement this example, provided in Appendix G.

Decision making under structural uncertainty

The above representation of stochastic dynamic programming assumes that dynamics, while stochastic, are predictable under a single statistical model. If, as is frequently the case, there exists **structural uncertainty** about the relationship between decision inputs and state outputs, then that uncertainty must be captured in the optimization framework, as it will affect optimal decision making.

We modify the SDP problem as described above by now allowing state dynamics to differ among several alternative structural models, $i = 1, \ldots .M$. Thus,

$$\pi_i(\underline{x}_{t+1} \mid \underline{x}_t, \underline{d}_t), i = 1, \ldots, M \qquad \text{(E.35)}$$

specifies dynamics for each alternative model. Further, we characterize structural uncertainty by a probability measure P_i for each alternative model, which in general may vary over time, with the condition

$$\sum_i^M P_i(t) = 1$$

or instance, if we have $M = 3$ models in our set describing structural uncertainty and $\underline{P}' = (0, 1, 0)$ we have no uncertainty (or complete certainty that model $M = 2$ is correct); if

$$\underline{P}' = (1/3, 1/3, 1/3)$$

then we are equally uncertain as to the identity of the correct model, etc.

Dynamic optimization proceeds as before, using the HJB relationship, but with expectations now involving both the state transitions conditional under each model, as well as averaging over the probability distribution $\underline{P}$.

$$J^*[\underline{x}(t), \underline{P}(t), t]$$
$$= \underset{[\underline{d}(t) \in \underline{D}]}{\text{Max}} \left\{ \overline{F}(\underline{x}, \underline{d}, t) + \sum_i \sum_{x_{t+1}} P_i(t)\pi_i(\underline{x}_{t+1} \mid \underline{x}_t, \underline{d}_t) J^*(\underline{x}(t+1), \underline{P}(t+1), t+1) \right\}$$
$$\text{(E.36)}$$

Finally, because the probability measures for the alternative models relate to the conditional transition probabilities for state transition, they are amenable to updating under **Bayes' theorem**:

$$P_i(t+1) = \frac{P_i(t)\pi_i(\underline{x}_{t+1} \mid \underline{x}_t, \underline{d}_t)}{\sum_m P_m(t)\pi_m(\underline{x}_{t+1} \mid \underline{x}_t, \underline{d}_t)} \qquad \text{(E.37)}$$

Thus, once a new state $\underline{x}_{t+1}$ is observed following implementation of decision $\underline{d}_t$ at an observed state $\underline{x}_t$, the above relationship provides a means to evolve the model probabilities through time. Here, the model probability $P_i(t)$ is viewed as a **prior probability**, from the standpoint that it is based on information prior to observing $\underline{x}_{t+1}$; the model-specific transition matrices $\pi_i(\underline{x}_{t+1} \mid \underline{x}_t, \underline{d}_t)$ are viewed as **likelihoods** of the observation of $\underline{x}_{t+1}$ under model i; and $P_i(t+1)$ is viewed as a **posterior probability** of model i.

Reduction of structural uncertainty

As seen above, this development can form the basis for making optimal decisions in stochastic systems subject to structural uncertainty, but additionally Eq. (E.37) provides a means to reduce uncertainty through time, and Eq. (E.34) permits the incorporation of the updated values $\underline{P}(t)$ into dynamic optimization.

Passive adaptation

This feedback of information from monitoring the system state through time is known as **adaptive optimization** (Williams et al. 2002). There are 2 principal alternatives for implementing adaptive optimization, in increasing mathematical and computational complexity. The first, known as **passive adaptive adaptation**, derives P_i optimal decision strategies for a fixed value $\underline{P}(t) = \underline{P}$; that is, once the values for $\underline{P}$ are obtained they are treated as constants in Eq. (E.34) and an optimal strategy is derived based on the current value of $\underline{P}$. Thus, Eq. (E.34) can be simplified as

$$J^*[\underline{x}(t), t] = \frac{\text{Max}}{[\underline{d}(t) \in \underline{D}]} \left\{ \overline{F}(\underline{x}, \underline{d}, t) + \sum_i \sum_{x_{t+1}} P_i \pi_i(\underline{x}_{t+1} \mid \underline{x}_t, \underline{d}_t) J^*(\underline{x}(t+1), t+1) \right\}$$

$$(E.38)$$

Thus, because $\underline{P}$ remains fixed over time, we are still seeking state-specific strategies as with SDP, with averaging over the constant vector of model probabilities. The valued for $\underline{P}$ may still be updated via Eq. (E.35), and then used to develop a new optimal, state-specific strategy.

Example – Passive adaptation

We return to the earlier harvest example, with the complication that now decision making operates under 2 alternative models. Under the first model, the decision-specific state transition matrices $\pi_1(x_{t+1} \mid x_t, d_t)$ are as before

$$\pi_1(x_{t+1} \mid x_t, d_t = 0.1) = \begin{bmatrix} 0.2 & 0.5 & 0.3 \\ 0.2 & 0.3 & 0.5 \\ 0.1 & 0.3 & 0.6 \end{bmatrix}, \pi_1(x_{t+1} \mid x_t, d_t = 0.2) = \begin{bmatrix} 0.5 & 0.3 & 0.2 \\ 0.2 & 0.5 & 0.3 \\ 0.2 & 0.4 & 0.4 \end{bmatrix},$$

$$\pi_1(x_{t+1} \mid x_t, d_t = 0.3) = \begin{bmatrix} 0.7 & 0.3 & 0. \\ 0.6 & 0.3 & 0.1 \\ 0.3 & 0.5 & 0.2 \end{bmatrix}$$

We now include a second set of matrices $\pi_2(x_{t+1}|x_t,d_t)$ under an alternative model

$$\pi_2(x_{t+1}\,|\,x_t,\,d_t = 0.1) = \begin{bmatrix} 0.3 & 0.5 & 0.2 \\ 0.3 & 0.3 & 0.4 \\ 0.2 & 0.3 & 0.5 \end{bmatrix},\ \pi_1(x_{t+1}\,|\,x_t,\,d_t = 0.2) = \begin{bmatrix} 0.7 & 0.3 & 0 \\ 0.4 & 0.5 & 0.1 \\ 0.5 & 0.4 & 0.1 \end{bmatrix},$$

$$\pi_1(x_{t+1}\,|\,x_t,\,d_t = 0.3) = \begin{bmatrix} 0.9 & 0.1 & 0 \\ 0.7 & 0.3 & 0 \\ 0.4 & 0.6 & 0 \end{bmatrix}.$$

Suppose that we start with initial model probabilities that are equal ($P_1(0) = P_2(0) = 0.5$). We form the HJB equations as before but now averaging over the 2 models with equal weight. Thus,

$$\bar{\pi}(x_{t+1}\,|\,x_t,\,d_t)) = 0.5 \times \pi_1(x_{t+1}\,|\,x_t,\,d_t)) + 0.5 \times \pi_2(x_{t+1}\,|\,x_t,\,d_t))$$

$$\bar{\pi}(x_{t+1}\,|\,x_t,\,d_t = 0.1) = \begin{bmatrix} 0.25 & 0.5 & 0.25 \\ 0.25 & 0.3 & 0.45 \\ 0.15 & 0.3 & 0.55 \end{bmatrix},$$

$$\bar{\pi}(x_{t+1}\,|\,x_t,\,d_t = 0.2) = \begin{bmatrix} 0.6 & 0.3 & 0.1 \\ 0.3 & 0.5 & 0.2 \\ 0.35 & 0.4 & 0.25 \end{bmatrix},$$

$$\bar{\pi}(x_{t+1}\,|\,x_t,\,d_t = 0.3) = \begin{bmatrix} 0.8 & 0.2 & 0.0 \\ 0.65 & 0.3 & 0.05 \\ 0.35 & 0.55 & 0.1 \end{bmatrix}.$$

At t_{f-1} the values of $J\left(x(t_{f-1}),t_{f-1}\right)$ conditional on each decision $d_{t_{f-1}}$ are

$$J(x(t_{t_{f-1}}),t_{t_{f-1}}\,|\,d(t_{t_{f-1}}) = 0.1) = \begin{bmatrix} (0.1)5 \\ (0.1)10 \\ (0.1)15 \end{bmatrix} + \begin{bmatrix} 0.25 & 0.5 & 0.25 \\ 0.25 & 0.3 & 0.45 \\ 0.15 & 0.3 & 0.55 \end{bmatrix}\begin{bmatrix} 1.5 \\ 3 \\ 4.5 \end{bmatrix} = \begin{bmatrix} 3.5 \\ 4.3 \\ 5.1 \end{bmatrix},$$

$$J(x(t_{t_{f-1}}),t_{t_{f-1}}\,|\,d(t_{t_{f-1}}) = 0.2) = \begin{bmatrix} (0.2)5 \\ (0.2)10 \\ (0.2)15 \end{bmatrix} + \begin{bmatrix} 0.6 & 0.3 & 0.1 \\ 0.3 & 0.5 & 0.2 \\ 0.35 & 0.4 & 0.25 \end{bmatrix}\begin{bmatrix} 1.5 \\ 3 \\ 4.5 \end{bmatrix} = \begin{bmatrix} 3.25 \\ 4.85 \\ 5.85 \end{bmatrix},$$

$$J(x(t_{t_{f-1}}),t_{t_{f-1}}\,|\,d(t_{t_{f-1}}) = 0.3) = \begin{bmatrix} (0.3)5 \\ (0.3)10 \\ (0.3)15 \end{bmatrix} + \begin{bmatrix} 0.8 & 0.2 & 0.0 \\ 0.65 & 0.3 & 0.05 \\ 0.35 & 0.55 & 0.1 \end{bmatrix}\begin{bmatrix} 1.5 \\ 3 \\ 4.5 \end{bmatrix} = \begin{bmatrix} 3.3 \\ 5.1 \\ 7.125 \end{bmatrix}.$$

For which the optimal state-specific decisions are $d(t_{t_{f-1}})' = (0.1, 0.3, 0.3)$, which can be shown to be the stationary state-specific strategy.

We can repeat this process for a different set of initial model probabilities, possibly updated following monitoring. For example, suppose that we select the optimal decision as above under equal model weighting and implement the decision to harvest at $d_t = 0.3$ given $x_t = 10$. In the following year we observe $x_{t+1} = 5$ and update the model probabilities by Eq. (E.35) as

$$P_i(t+1) = \frac{P_i(t)\pi_i(x_{t+1} = 5 \mid x_t = 10, d_t = 0.3)}{\sum_m P_m(t)\pi_m(x_{t+1} = 5 \mid x_t = 10, d_t = 0.3)} = \frac{0.5(0.6)}{0.5(0.6) + 0.5(0.7)} = 0.462.$$

We then use the new values of P to develop an optimal state-specific strategy from that point forward given the same state transition matrices for the 2 models. In this example the new values of $\underline{P}$ lead to convergence the same state-specific strategy as before. These now become the new prior probabilities for Eq. (E.35).

Active adaptation

The second approach, known as **active adaptive adaptation**, instead explicitly assumes that $\underline{P}(t)$ will evolve through time according to Eq. (E.37) and treats $\underline{P}(t)$ as another system state, sometimes referred to as an **information state**. Active adaptive optimization results in a single state-specific strategy, providing optimal decisions for combination of the states of nature ($\underline{x}$) and information state ($\underline{P}$). A consequence of this is that the future value of reduction in structural uncertainty on the objective return is anticipated in *current* decision making. This interplay between direct resource gain and information gain through decision making is also referred to as **dual control** (Williams et al. 2002). The HJB expression in Eq. (E.34)

$$J^*[\underline{x}(t), \underline{P}(t), t]$$
$$= \underset{[d(t) \in \underline{D}]}{\text{Max}} \left\{ \overline{F}(\underline{x}, \underline{d}, t) + \sum_i \sum_{x_{t+1}} P_i(t)\pi_i(\underline{x}_{t+1} \mid \underline{x}_t, \underline{d}_t)J^*(\underline{x}(t+1), \underline{P}(t+1), t+1) \right\}$$

should make it clear that we are now seeking an optimal decision strategy that is specific to each combination of system state $\underline{x}(t)$ and information state $\underline{P}(t)$. Again, the dynamics for the former are specified by the model-specific transition matrices, and for the latter by Eq. (E.37). In practice the likelihoods in Eq. (E.35) will be evaluated at $E[\underline{x}(t+1)]$, where

$$E[\underline{x}(t+1)\mid\underline{x}(t),\underline{d}(t)]=\sum_{i}\sum_{x_{t+1}}P_i(t)\pi_i(\underline{x}_{t+1}\mid\underline{x}_t,\underline{d}_t)$$

after linear interpolation to provide likelihood values between the discrete states.

Example – Active adaptation

We illustrate active adaptation with the same system representation as the above passive example (abundance at 3 levels and harvest decisions at 3 levels and with the dynamics represented by 2 alternative models as captured in the model-specific state transition matrices $\pi_i(x_{t+1}\mid x_t,d_t)$, $i = 1,2$. Structural uncertainty at time t is represented by $P_1(t) = P(t)$, $P_2(t) = 1-P(t)$, with the "information state" discretized to 9 levels $P(t) \in \{0.1,\ 0.2,0.3,0.4,0.5,0.6,0.7,0.8,0.9\}$. As before, the model- and decision-specific transition matrices are

$$\pi_1(x_{t+1}\mid x_t, d_t = 0.1) = \begin{bmatrix} 0.2 & 0.5 & 0.3 \\ 0.2 & 0.3 & 0.5 \\ 0.1 & 0.3 & 0.6 \end{bmatrix},\ \pi_1(x_{t+1}\mid x_t, d_t = 0.2) = \begin{bmatrix} 0.5 & 0.3 & 0.2 \\ 0.2 & 0.5 & 0.3 \\ 0.2 & 0.4 & 0.4 \end{bmatrix},$$

$$\pi_1(x_{t+1}\mid x_t, d_t = 0.3) = \begin{bmatrix} 0.7 & 0.3 & 0. \\ 0.6 & 0.3 & 0.1 \\ 0.3 & 0.5 & 0.2 \end{bmatrix}$$

and

$$\pi_2(x_{t+1}\mid x_t, d_t = 0.1) = \begin{bmatrix} 0.3 & 0.5 & 0.2 \\ 0.3 & 0.3 & 0.4 \\ 0.2 & 0.3 & 0.5 \end{bmatrix},\ \pi_1(x_{t+1}\mid x_t, d_t = 0.2) = \begin{bmatrix} 0.7 & 0.3 & 0 \\ 0.4 & 0.5 & 0.1 \\ 0.5 & 0.4 & 0.1 \end{bmatrix},$$

$$\pi_1(x_{t+1}\mid x_t, d_t = 0.3) = \begin{bmatrix} 0.9 & 0.1 & 0 \\ 0.7 & 0.3 & 0 \\ 0.4 & 0.6 & 0 \end{bmatrix}.$$

To illustrate, suppose we are at t_{f-1} with an information state of $P(t_f-1) = 0.5$ (similar calculations would be made for each of the other discrete information states).

At t_{f-1} the values of $J(x(t_{f-1}),P(t_{f-1}),t_{f-1})$ conditional on each decision $d_{t_{f-1}}$ and the information state are

$$J\left(x(t_{f-1}),P(t_{f-1}),t_{f-1}\right)= x_{t_{f-1}}d_{t_{f-1}} + P(t_{f-1})\pi_1(x_{t+1}\mid x_t, d_t)J^*\left(x(t_f), P(t_f), t_f\right)$$
$$+ (1-P(t_{f-1}))\pi_2(x_{t+1}\mid x_t, d_t)J^*\left(x(t_f), P(t_f), t_f\right)$$

or

$$J\left(x(t_{f-1}), P(t_{f-1}),t_{f-1}\right)= x_{t_{f-1}}d_{t_{f-1}} + \overline{\pi}(x_{t+1}\mid x_t, d_t, t)J^*\left(x(t_f), P(t_f), t_f\right)$$

where

$$\overline{\pi}(x_{t+1}\mid x_t, d_t, t) = 0.5\pi_1(x_{t+1|}\mid x_t, d_t) + (1-0.5)\pi_2(x_{t+1|}\mid x_t, d_t)$$

$$\overline{\pi}(x_{t+1}\mid x_t, d_t = 0.1, P(t_{f-1})=0.5) = \begin{bmatrix} 0.25 & 0.5 & 0.25 \\ 0.25 & 0.3 & 0.45 \\ 0.15 & 0.3 & 0.55 \end{bmatrix},$$

$$\overline{\pi}(x_{t+1}\mid x_t, d_t = 0.2, P(t_{f-1})=0.5) = \begin{bmatrix} 0.6 & 0.3 & 0.1 \\ 0.3 & 0.5 & 0.2 \\ 0.35 & 0.4 & 0.25 \end{bmatrix},$$

$$\overline{\pi}(x_{t+1}\mid x_t, d_t = 0.3, P(t_{f-1})=0.5) = \begin{bmatrix} 0.8 & 0.2 & 0 \\ 0.65 & 0.3 & 0.05 \\ 0.35 & 0.55 & 0.1 \end{bmatrix}.$$

The information state is irrelevant at t_f (since no "future value" is possible, by definition), so

$$J^*\left(x(t_f), P(t_f), t_f\right) = \begin{bmatrix} 1.5 & 1.5 & 1.5 & 1.5 & 1.5 & 1.5 & 1.5 & 1.5 & 1.5 \\ 3 & 3 & 3 & 3 & 3 & 3 & 3 & 3 & 3 \\ 4.6 & 4.6 & 4.6 & 4.6 & 4.6 & 4.6 & 4.6 & 4.6 & 4.6 \end{bmatrix}$$

where the rows index the system state at t_f and the columns the information states. Thus,

$$J(x(t_{t_{f-1}}), t_{t_{f-1}} \mid P(t_{f-1}) = 0.5, d(t_{t_{f-1}}) = 0.1) = \begin{bmatrix} (0.1)5 \\ (0.1)10 \\ (0.1)15 \end{bmatrix} + \begin{bmatrix} 0.25 & 0.5 & 0.25 \\ 0.25 & 0.3 & 0.45 \\ 0.15 & 0.3 & 0.55 \end{bmatrix} \begin{bmatrix} 1.5 \\ 3 \\ 4.5 \end{bmatrix}$$

$$= \begin{bmatrix} 3.5 \\ 4.3 \\ 5.1 \end{bmatrix},$$

$$J(x(t_{t_{f-1}}), t_{t_{f-1}} \mid P(t_{f-1}) = 0.5, d(t_{t_{f-1}}) = 0.2) = \begin{bmatrix} (0.2)5 \\ (0.2)10 \\ (0.2)15 \end{bmatrix} + \begin{bmatrix} 0.6 & 0.3 & 0.1 \\ 0.3 & 0.5 & 0.2 \\ 0.35 & 0.4 & 0.25 \end{bmatrix} \begin{bmatrix} 1.5 \\ 3 \\ 4.5 \end{bmatrix}$$

$$= \begin{bmatrix} 3.25 \\ 4.85 \\ 5.85 \end{bmatrix},$$

$$J(x(t_{t_{f-1}}), t_{t_{f-1}} \mid P(t_{f-1}) = 0.5, d(t_{t_{f-1}}) = 0.3) = \begin{bmatrix} (0.3)5 \\ (0.3)10 \\ (0.3)15 \end{bmatrix} + \begin{bmatrix} 0.8 & 0.2 & 0 \\ 0.65 & 0.3 & 0.05 \\ 0.35 & 0.55 & 0.1 \end{bmatrix} \begin{bmatrix} 1.5 \\ 3 \\ 4.5 \end{bmatrix}$$

$$= \begin{bmatrix} 3.3 \\ 5.1 \\ 7.125 \end{bmatrix}.$$

The objective functional is maximized by the state-specific decision conditional on $P(t_{f-1}) = 0.5$ of $d = (0.1, 0.3, 0.3)$. Similar computations across the other information states results in a matrix of values result in the system- and information-state combination of optimal decisions and associated values

$$J^* (x(t_{f-1}), P(t_{f-1}), t_{f-1})$$
$$= \begin{bmatrix} 3.38 & 3.41 & 3.44 & 3.47 & 3.5 & 3.53 & 3.56 & 3.59 & 3.62 \\ 4.98 & 5.01 & 5.04 & 5.07 & 5.10 & 5.13 & 5.16 & 5.19 & 5.22 \\ 6.95 & 6.99 & 7.04 & 7.08 & 7.13 & 7.17 & 7.22 & 7.26 & 7.31 \end{bmatrix}.$$

At the next time step t_{f-2}, these values now become the state-specific future values, i.e., the value that will be accrued by evolution of the system to the next system- and information-state combination. The future information state at t_{f-2} will be determined by the outcome that occurs following implementation of the state-specific decision d_{f-2} and subsequent observation of x_{f-1}. In turn, the transition to a new information state is determined from the likelihood of outcomes at t_{f-1}, conditional on the system state x_{tf-2}, decision d_{tf-2} and information state $P(t_{f-2})$ and application of Bayes' theorem. For instance, the expected outcome at t_{f-1} is

$$E(x_{t_{f-1}} \mid x_{t_{f-2}} = 10, d_{t_{f-2}} = 0.2, P(t_{f-2}) = 0.5) = 0.3(5) + 0.5(10) + 0.2(15) = 9.5$$

This value is used to obtain likelihoods under each alternative mode by linear interpolation from the likelihoods at the nearest neighboring states; in this example $L_1(x_{t_{f-1}} = 9.5) = 0.47$, $L_2(x_{t_{f-1}} = 9.5) = 0.49$. The updated information state value is

$$P(t_{f-1} \mid x_{t_{f-2}} = 2, d_{t_{f-2}} = 0.2, P(t_{f-2}) = 0.5) = \frac{0.5(0.47)}{0.5(0.47) + 0.5(0.49)} = 0.49,$$

which in turn determines the elements of $J^*(x(t_{f-1}), P(t_{f-1}), t_{f-1})$ to be used in the HJB relationship. For example,

$$J(x(t_{tf-2}), t_{tf-2} \mid P(t_{f-1}) = 0.49, d(t_{tf-2}) = 0.1) = \begin{bmatrix} (0.1)5 \\ (0.1)10 \\ (0.1)15 \end{bmatrix} + \begin{bmatrix} 0.251 & 0.5 & 0.249 \\ 0.251 & 0.3 & 0.449 \\ 0.151 & 0.3 & 0.549 \end{bmatrix} \begin{bmatrix} 3.5 \\ 5.10 \\ 7.17 \end{bmatrix}$$

$$= \begin{bmatrix} 5.713 \\ 6.627 \\ 7.499 \end{bmatrix},$$

$$J(x(t_{tf-2}), t_{t-2} \mid P(t_{f-1}) = 0.5, d(t_{tf-2}) = 0.2) = \begin{bmatrix} (0.2)5 \\ (0.2)10 \\ (0.2)15 \end{bmatrix} + \begin{bmatrix} 0.602 & 0.3 & 0.098 \\ 0.302 & 0.5 & 0.198 \\ 0.353 & 0.4 & 0.247 \end{bmatrix} \begin{bmatrix} 3.47 \\ 5.10 \\ 7.13 \end{bmatrix}$$

$$= \begin{bmatrix} 5.317 \\ 7.009 \\ 8.025 \end{bmatrix},$$

$$J(x(t_{tf-2}), t_{t-2} \mid P(t_{f-1}) = 0.5, d(t_{tf-2}) = 0.3) = \begin{bmatrix} (0.3)5 \\ (0.3)10 \\ (0.3)15 \end{bmatrix} + \begin{bmatrix} 0.8.02 & 0.198 & 0 \\ 0.651 & 0.3 & 0.049 \\ 0.351 & 0.551 & 0.098 \end{bmatrix} \begin{bmatrix} 3.47 \\ 5.07 \\ 7.13 \end{bmatrix}.$$

$$= \begin{bmatrix} 5.29 \\ 7.129 \\ 9.210 \end{bmatrix}.$$

Application of similar calculations to the other information states at t_{f-2} leads to optimal state- and information-specific decisions $d = (0.1, 0.3, 0.3)$ for all information states except the last ($P = 0.9$), at which $d = (0.1, 0.2, 0.3)$. Continuation of the algorithm leads to a stationary (no decision changes for 4 sequential stages) after 10 iterations, with the optimal states-specific decision $d = (0.1, 0.2, 0.3)$ for information states $P = (0.5, 0.6, 0.7, 1.0)$ and $d = (0.1, 0.3, 0.3)$ for the remaining information states.

Adaptive SDP is obviously more computationally demanding than SDP, due to the need to keep track of the "information state" and its dynamics. Thus, ASDP is only practicable for problems in which the system state and structural uncertainty can be reduced to relatively low dimensionality.

Generalizations of Markov decision processes

The dynamic programming approach described above is the typical approach to solving the class of optimization problems known as **Markov decision processes (MDP)**, which are characterized by the fact that the state transitions are dependent only on the current system state and decisions. This, in conjunction with the discretization of time, as well as typically space, leads to the recursive HJB approach described above.

An important generalization of MDP that is very relevant to conservation decision making is the allowance for partial observability of system states. That is, up to this point we have assumed that the state of the system $\underline{x}(t)$ is known without error (i.e., perfectly observed). Obviously this is not generally true, and instead of observing $\underline{x}(t)$ we instead have an estimate $\hat{\underline{x}}(t)$, which in turn is based on field data $\underline{y}(t)$. Clearly, decisions must now be made based on the estimates $\hat{\underline{x}}(t)$ rather than the actual (and unknown) $\underline{x}(t)$, leading to additional uncertainty in optimization. Williams (1996) and Williams et al. (2002) introduced these additional elements into the HJB relationship, incorporating the relationship between $\hat{\underline{x}}(t)$ and $\underline{x}(t)$ via conditional distributions, which are in turn derived from the statistical model specifying the assumed relationship of the data to parameter estimation. The state-specific transition probabilities must now be modified to reflect that decision making and monitoring are now based on $\hat{\underline{x}}(t)$. These expand as

$$\pi_i(\hat{\underline{x}}_{t+1} \mid \hat{\underline{x}}_t, d_t) = f_2(\underline{x}_t \mid \hat{\underline{x}}_t)\pi_i(\underline{x}_{t+1} \mid \underline{x}_t, d_t)f_1(\hat{\underline{x}}_{t+1} \mid \underline{x}_{t+1}).$$

The HJB is modified Eq. (E.34) as

$$J^*[\hat{\underline{x}}(t), \underline{P}(t), t] = \frac{\text{Max}}{[\underline{d}(t) \in \underline{D}]}\left\{\sum_{x_t} f_2(\underline{x}_t \mid \hat{\underline{x}}_t)\overline{F}(\underline{x}, d, t) \right.$$
$$\left. + \sum_i \sum_{\hat{x}_{t+1}} P_i(t)\pi_i(\hat{\underline{x}}_{t+1} \mid \underline{x}_t, \underline{d}_t)J^*(\underline{x}(t+1), \underline{P}(t+1), t+1)\right\}.$$

This problem is apparently not solvable via direct application of dynamic programming, but instead fits into the more general framework of **partially observable Markov decision processes** (POMDP; Sondik 1978). POMDP problems are usually solved by a backwards iteration ("value iteration") procedure invoking the HJB relationship as usual, but using approximation methods to evaluate the value function at each iteration. See recent work by Chades et al. 2008, 2011, McDonald-Madden et al. 2011, and Williams 2011 for applications in natural resource management and discussions of general algorithms for the computation of POMDP problems (Appendix F).

Heuristic methods

In this development, we have emphasized classical approaches for optimization and modern approaches such as dynamic programming. We have done so because these methods are known to provide global solutions to optimization problems, provided that the underlying assumptions of the procedures are met. However, the above methods can be difficult to apply or even inappropriate in some situations. For example, classical programming and by extension linear and nonlinear programming allow highly dimensioned systems with many decisions and constraints, but generally do not permit realistic modeling of system dynamics or uncertainty. Stochastic dynamic programming allows for more realism in system dynamics and incorporation of uncertainty (including structural uncertainty) but is generally limited to solving relatively low-dimensioned problems.

Heuristic methods apparently remove some or all of these restrictions and thus have attraction. However, because these procedures tend to be exploratory rather than based on a set of mathematical rules (e.g., Kuhn–Tucker conditions, HJB), there can be no assurance that the resulting "best" strategy is optimal. Here, we briefly list and comment on 3 of the more important heuristic approaches; this is by no means an exhaustive coverage.

Simulation-gaming/simulation-optimization

As the name implied, simulation gaming and **simulation-optimization** use simulation modeling to explore and compare candidate decision strategies. Simulation modeling can be used in combination with standard optimization techniques such as nonlinear programming to select among candidate decisions at various points in a simulated system, where the system state at time t is some function of previous system states and decision and random variables. There is no theoretical limit to the complexity of system representation and dynamics or decisions, hence the attraction. Certainly, simulation-gaming and simulation-optimization can be useful for exploring candidate decision scenarios, and we fully support the use of simulation modeling to explore the consequences of decisions that have been derived by true optimization approaches. However, we very much caution against the treatment of decisions derived by exploration as "optimal", given the general absence of rules or criteria for supporting such claims.

See Box 8.7 for an example of simulation-optimization applied to a conservation problem.

Genetic algorithms and machine learning

Genetic algorithms (Goldberg 1989) are search methods that mimic natural evolutionary processes. In genetic algorithms, decision trajectories are encoded via a population of strings that are analogous to chromosomes. These are combined into candidate "individuals" representing solutions, which in turn have varying degrees of "fitness" (objective value). In addition to a mechanism for specifying available decisions (the "genome") and a fitness (objective) function, the algorithm typically employees analogs of reproduction, crossover, and mutation to explore the genome (decision space). Genetic algorithms, like simulation-gaming, can involve very large and complex systems and many decisions, and solution algorithms may converge with relative efficiency to apparently optimal solutions. Genetic algorithms can also be viewed as part of a larger class of "autonomous learning" algorithms that include machine learning, often employed in the design of robotic systems (Goldberg 1989).

As with simulation-gaming, genetic algorithms are inherently exploratory and cannot be guaranteed to provide optimal solutions.

Simulated annealing

Simulated annealing is an exploration process that is based on analogy to the process of annealing in metallurgy, which involves the controlled cooling of a material to optimize crystal formation (Press et al. 2007). Simulated annealing can provide reasonably good approximations to the global optimum in many circumstances, and may be appropriate when the goal is "satisficing" rather than optimizing (Chapter 8). As with genetic algorithms and simulation-optimization, simulated annealing is inherently exploratory and cannot be guaranteed to provide optimal solutions.

References

Bellman, R.E. (1957) *Dynamic Programming*. Princeton University Press. Princeton, N.J.

Broyden, C.G. (1970) The convergence of a class of double-rank minimization algorithms, *Journal of the Institute of Mathematics and Its Applications* **6**, 76–90.

Chades, I., E. McDonald-Madden, M.A. McCarthy, B. Wintle, M. Linkie, and H.P. Possingham. (2008) When to stop managing or surveying cryptic threatened species. *PNAS* **105**, 13936–13940.

Chades, I., T.G. Martin, S. Nicol, M.A. Burgman, H.P. Possingham, and Y.M. Buckley. (2011) General rules for managing and surveying networks of pests, diseases, and endangered species. *PNAS* **108**, 8323–8328.

Goldberg, D.E. (1989) *Genetic Algorithms in Search, Optimization, and Machine Learning*. Addison-Wesley. Reading, MA.

Lubow, B.C. (1995) SDP: Generalized software for solving stochastic dynamic optimization problems. *Wildl. Soc. Bull.* **23**, 738–742.

Lunn, D.J., A. Thomas, N. Best, and D. Spiegelhalter. (2000) WinBUGS – a Bayesian modelling framework: concepts, structure, and extensibility. *Statistics and Computing* **10**, 325–337.

McDonald-Madden, E., I. Chadès, M.A. McCarthy, M. Linkie, and H.P. Possingham. (2011) Allocating conservation resources between areas where persistence of a species is uncertain. *Ecological Applications* **21**, 844–858.

Press, W.H., S.A Teukolsky, W.T. Vetterling, and B.P. Flannery, (2007) Section 10.12. Simulated Annealing Methods. *Numerical Recipes: The Art of Scientific Computing* (3rd edn). New York: Cambridge University Press.

Puterman, M.L. (1994) *Markov Decision Processes: Discrete Stochastic Dynamic Programming*. Wiley, New York.

Sondik, E.J. (1978) The optimal control of partially observable Markov processes over the infinite horizon: discounted cost". *Operations Research* **26** (2), 282–304.

Williams, B.K. (1996) Adaptive optimization of renewable natural resources: solution algorithms and a computer program. *Ecological Modelling* **93**, 101–111.

Williams, B.K. (2011) Resolving structural uncertainty in natural resource management using POMDP approaches. *Ecological Modelling* **222**, 1092–1102.

Williams, B.K., J.D. Nichols, and M.J. Conroy. (2002) *Analysis and Management of Animal Populations*. Elsevier-Academic. San Diego, CA.

Appendix F

Guide to Software

Program	Applications[1]	Source	Freeware?	Comments
AD MODEL BUILDER	PR, SIM, STAT	http://admb-project.org/	Y	Specializes in nonlinear modeling
EXCEL	PR, OPT, SIM	http://office.microsoft.com	N	Handles modest size problems well, not practical for large data sets or dynamic optimization problems
MATLAB	PR, OPT, SIM, STAT	http://www.mathworks.com	N	Efforts underway to write models for DP and POMDP
NETICA*	BBN	http://www.norsys.com	Y	>15-node models requires license
PYMC**	BAYES	http://pypi.python.org/pypi/pymc	Y	Growing community of support. Relative ease of de-bugging. Requires more user involvement to install and update than BUGS
R	PR, SIM, STAT,	http://www.r-project.org/	Y	Extensive community of user support
SAS®	PR, SIM, STAT	http://www.sas.com/	N	
SDP	DP	http://warnercnr.colostate.edu/~bruce/software.htm	Y	Updates not currently supported
STELLA®†	PR, SIM	http://www.iseesystems.com	N	Emphasizes graphical modeling
WinBUGS/ OpenBUGS‡	BAYES	http://www.mrc-bsu.cam.ac.uk/bugs	Y	Well documented, good community of support. Can be difficult to de-bug

[1]BAYES = bayesian analysis, BBN = Bayes belief networks and influence diagrams, DP = dynamic programming, PR = General modeling and programming, OPT = Optimization, SIM = Simulation modeling, STAT = Statistical analysis and modeling.
*Netica from Norsys Software Corp. (www.norsys.com);
**Patil, A., D. Huard and C.J. Fonnesbeck. 2010;
†STELLA® [Software]. isee systems, inc. Available from http://www.iseesystems.com;
‡Lunn et al (2000) with permission.

Appendix G

Electronic Companion to Book

See http://www.wiley.com/go/conroy/naturalresourcemanagement

Decision Making in Natural Resource Management: A Structured, Adaptive Approach,
First Edition. Michael J. Conroy and James T. Peterson.
© 2013 John Wiley & Sons, Ltd. Published 2013 by John Wiley & Sons, Ltd.

Glossary

Accuracy – mean squared distance of an estimate from the true parameter value. Accuracy is increased by either *bias* or *variance* of the estimate.

Action – management or other activity by humans that may influence a resource state. See *decision*.

Acyclic – a graphical model without a cycle or feedback loop, see *Bayesian belief network* and *influence diagram*.

Active ARM, active adaptive optimization – seeking of an optimal sequence of decisions that anticipates future reduction of *structural uncertainty*.

Adaptive optimization – incorporation of the reduction of *structural uncertainty* as part of the optimal control problem.

Adaptive harvest management – seeking a long-term harvest goal by means of adaptive optimization.

Adaptive [resource] management – incorporation of the reduction of structural uncertainty as part of optimal resource management via sequential decision making.

Adaptive stochastic dynamic programming (ASDP) – extension of stochastic dynamic programming to include the adaptive reduction of structural uncertainty.

Akaike's information criterion (AIC) – information theoretical statistic for model selection and multi-model inference.

Decision Making in Natural Resource Management: A Structured, Adaptive Approach, First Edition. Michael J. Conroy and James T. Peterson.
© 2013 John Wiley & Sons, Ltd. Published 2013 by John Wiley & Sons, Ltd.

Aleatory, aleatoric uncertainty – irreducible (but possibly quantifiable) uncertainty, such as demographic or environmental uncertainty or random variation among subjects.

Attribute – measure of a consequence or other response. Multiple *objective* attributes can be combined in a single value via an *objective function*.

Bayesian belief network (BBN) – a probabilistic network model, consisting of uncertainty or chance nodes connected by directed arcs that represent causality; an *influence diagram* without decision and utility nodes.

Bayesian credible interval (BCI) – Bayesian probability statement about the interval containing a parameter value.

Bayesian information criterion (BIC) – model selection criterion under Bayesian inference, analogous to AIC used under the *frequentist paradigm*; large-sample approximation of the *Bayes factor*.

Bayes factor – Bayesian measure of model comparison based on the ratio of the posterior marginal likelihoods averaged over the parameters.

Bayesian paradigm – inferential philosophy based entirely on conditional probability relationships and Bayes' theorem. See *frequentist paradigm*.

Bayesian inference, statistics – statistical inference under a *Bayesian paradigm*.

Bayes' theorem, Bayes' rule – central theorem of probability describing the relationship between prior and posterior knowledge via conditional probability relationships.

Bernoulli distribution – discrete probability distribution for modeling binary outcomes.

Best linear unbiased estimator (BLUE) – best (minimum variance) estimator of the class of unbiased linear estimators.

Beta distribution – continuous probability distribution bounded by 0 and 1, appropriate for modeling outcome x when x is on the unit interval. Frequently used to represent uncertainty and variability in probabilities.

Bias – expected difference between parameter estimate and true parameter value.

Binomial distribution – statistical distribution for modeling x the number of successes in n independent Bernoulli trials, each with a common probability of success p.

Bootstrap – procedure for drawing repeated random samples with replacement from a data set; used as an alternative approach for variance and confidence interval estimations.

Calculus of variations – an analytical approach for solving the *optimal control problem* based on maximizing an integral objective function that incorporates time.

Certain, certainty – outcome not subject to uncertainty. See *deterministic*.

Certainty equivalent – the size or magnitude of the outcome that the decision maker would accept under absolute certainty.

Chi-squared distribution – distribution of quadratic forms of normally distributed random variables. Large-sample approximation for goodness of fit and other test statistics.

Champion – an individual who has the authority, resources, and temerity to advance an SDM process in an organization.

Chance event – outcome subject to uncertainty. See *stochastic*.

Chance node – see *uncertainty node*.

Classical programming – a procedure for providing optimal solutions for constrained optimization problems, in which the objective function is differentiable and the constraints can be written as equalities.

Closed-loop policy – solution to optimal control problem that involves feedback from future (as of yet unobserved) realized states in the calculation of the current optimal decision.

Coefficient – parameter specifying the relationship between a factor and a response, as in for example the coefficients of a *linear model*.

Collectively exhaustive [events, states] – list of events or states such that no other events or states are possible.

Competing objectives – *fundamental objectives* that all require satisfaction, but insufficient resources exist to satisfy all.

Complementary [events], complementarity – event or random variables for which absence of membership in one subset of possible event outcomes dictates membership in the remaining (complementary) subset.

Conditional distribution – statistical distribution of a random variable whose values are dependent on the value of another random variable.

Conditional independence, conditionally independent – two random variables are conditionally independent if, after accounting for a third random variable that each is dependent on, knowledge of one of the outcomes provides no additional information about the other.

Conditional probability – probability of an event is dependent on the outcome of another event.

Conditional probability table – a table of probabilities describing conditional probability relationships in an influence diagram or decision network.

Conflicting objectives – fulfillment of one or more *fundamental objective* is in conflict with the other(s).

Conjugacy – property in which a prior distribution combined with a likelihood provides a posterior distribution of the same form as the prior distribution.

Conjugate distributions, conjugate pair – combination of prior distribution and likelihood resulting in conjugacy.

Consequence node – node in Bayesian belief or decision network describing an outcome as the result of inputs, usually including one or more decisions.

Constrained optimization – maximization or minimization of an *objective function* after accounting for constraints on the functions of the decisions.

Constraints – limits on the range or values of decisions or functions of decisions (including resulting state variable or objective outcomes).

Continuous distribution – distribution of random variable that may assume real values over the *support*.

Control – decision or other factor under the influence of a decision maker.

Correlation, product-moment correlation – measure of linear dependence between two random variables scale between −1 and +1.

Covariance – measure of linear dependence between two random variables.

Cross-validation – partitioning of data into subsets, some of which are used to estimate model parameters and the remaining used to test model predictions.

Cumulative distribution function (cdf) – probability that a random variable is equal or below a specific value.

Data – sample outcome, ordinarily (e.g., under random sampling) considered to be a random subset of a population of attributes.

Decision, decision alternative – choice among alternative *actions*, ordinarily involving an irrevocable allocation of resources, and directed toward achieving an *objective*.

Decision making under uncertainty – selection among alternative actions to achieve an objective when *uncertainty* exists about the resultant outcome.

Decision making under risk – sometimes distinguished from decision marking under uncertainty, to refer to a situation where probabilities of uncertain outcomes are known. We prefer instead to represent all uncertainty by means of probability distributions, and distinguish instead between reducible (*epistemic*, e.g., *structural*) and irreducible (*aleatoric*, e.g., environmental) forms of uncertainty.

Decision network – graphical representation of the relationship between *decisions, means objectives*, and *fundamental objectives*.

Decision nodes – node in *influence diagram* representing selection of a decision alternative as input to the conditional distribution of *consequence nodes*.

Decision space – list or range of decisions considered or permitted.

Decision trees – branching diagram illustrating a sequence of decisions followed by stochastic outcomes and their values. Can be used in place of an *influence diagram*.

Decision variable – see *control*.

Demographic stochasticity – form of *aleatoric uncertainty* due to the stochastic nature of birth–death processes.

Dependent variable – see *response*.

Deterministic – not subject to stochastic or random influences.

Deviance information criterion (DIC) – Bayesian analog to AIC that takes into account the hierarchical structure of parameters, e.g., under random effects models.

Dirichlet distribution – multivariate analog of the *beta distribution*; conjugate prior distribution for the *multinomial likelihood*.

Discrete distribution – distribution of random variable that assumes integer values over the *support*.

Direct elicitation – a process by which experts are asked to provide the probabilities or response values that are used to parameterize the relation between two or more components.

Directed arcs – arrows connecting nodes in an *influence diagram* or *Bayesian belief network* representing causality. An arc pointing from node A into B means that changes in the value of node A affect the value of node B.

Distribution – statistical model of a random variable, specified by one or more distribution *parameters*.

Dual – alternate formulation of the *linear programming* problem, formed by switching the roles of the *Lagrangian multipliers* and *decision variables*. See *primal*.

Dual control – positive feedback relationships between management and learning. See *active ARM*.

Dynamic programming – an algorithm, based on *Bellman's optimality principle* that permits solution to the *optimal control problem* via backwards induction from a terminal solution.

Dynamics – variation of a state variable over time (and sometimes space).

Elasticity – proportional *sensitivity* in relationship to proportional values of the input.

Ecological hierarchy – classification according to levels of biological organization (individuals, populations, communities, ecosystems, biomes) or functional groups (e.g., predators, herbivores, primary producers, detritivores, etc.).

Elasticity – proportional change that occurs in model output, given a small, proportional change in some model parameter or other input (like a decision variable). See *sensitivity*.

Elicitation – a process where experts are asked to provide information that is used to parameterize the relation between two or more model components.

Empirical Bayes estimation – a statistical procedure that estimates the prior distribution from the data.

Environmental stochasticity – form of *aleatoric uncertainty* due to stochastic variation in environmental factors.

Epistemic uncertainty – uncertainty that at least in principle can be reduced, such as *structural uncertainty*.

Equilibrium – condition where the *state* of a system is in balance through time.

Estimate – function of sample *data* used to evaluate a *parameter*.

Estimator – formula or algorithm used to obtain a statistical *estimate*.

Expected value – average value of an attribute over the distribution of values.

Explanatory factor – see *predictor*.

Exponential distribution – statistical distribution typically used to model the distribution of time elapsed until an event such as failure or arrival.

Feasible solutions – in *constrained optimization*, solutions that satisfy the constraining functions.

Fisher's F – test statistic based on quadratic forms of normally distributed variables; used in analysis of variance.

Fit [of model] – comparison of model predictions to sample outcomes.

Frequency elicitation – a form of *indirect elicitation* where experts are asked to provide an expected outcome in terms of a frequency, e.g., how many years in the last 100 did an event occur.

Frequentist paradigm – inferential philosophy based on concept of repeated sampling or experimentation from a fixed population. See *Bayesian paradigm*.

Full Bayesian – statistical inference based entirely on conditional probability relationships and Bayes' theorem. See *Empirical Bayes*.

Function elicitation – a form of *indirect elicitation* where experts are asked to provide the form of the relation between two or more components in a graphical format.

Fundamental objective – objective that relates to the core (fundamental) values of a decision maker or stakeholder.

Fundamental objective hierarchy – a hierarchical diagram displaying the relationship among *fundamental objectives* and their *attributes*.

Gamma distribution – generalization of the *exponential* distribution by the addition of a shape parameter.

Generalized linear model – generalization of *linear model* to allow non-normal error distributions and transformation of the response via link functions.

Genetic algorithm – heuristic method of optimization based on analogy with a genetic population.

Geometric distribution – discrete distribution describing the number of independent *Bernoulli* trials that occur before the first success.

Gibbs sampling – Bayesian updating method based on sequential sampling from full conditional posterior distributions of the unknown parameters or other quantities.

Global optimum – unique optimum (maximum or minimum) among *feasible solutions*.

Gradient – vector of first partial derivatives with respect to independent variables.

Gradient descent method – optimization procedure based on gradients.

Group elicitation – a process where a group of experts (rather than individuals) are asked to provide information that is used to parameterize the relation between two or more model components.

Hamiltonian – augmented objective function involving constraints by means of co-state variables, for solving multivariate dynamic optimization problems.

Hessian matrix – matrix of first and second partial derivatives used in optimization. See *information matrix*.

Heuristic methods – computationally intense methods used for obtaining very flexible solutions to optimization problems, but generally with no guarantee of a *globally optimal* solution.

Hidden objectives – *fundamental objectives* that have not been revealed as part of the decision-making process.

Hierarchical model – model in which some parameters are cast as functions (including statistical functions) of other parameters.

Histogram – plotting method for describing the frequency distribution of data.

Hypergeometric distribution – discrete distribution that arises in sampling without replacement from a finite population with two or more categories.

Hyper-parameter – parameter of a distribution whose outcomes are in turn parameters of other distributions. See *hierarchical model*.

Hypothesis – a proposed explanation for an outcome or relationship. See *prediction*.

Independent variable – *see predictor.*

Indirect elicitation – a process where experts are asked to provide information on the relation in a form such as frequencies between two or more model components used to estimate the parameters.

Influence diagrams – graphical model version of a decision model that consists of decisions, uncertainty, and utility nodes connected by *directed arcs* that describe the causal relations among components. Can be used in place of a *decision tree*.

Information matrix – matrix of first and second partial derivatives used in maximum likelihood estimation to optimize (see *Hessian matrix*), and to provide ML estimates of the *variance–covariance matrix*.

Information state – Measures of structural uncertainty incorporated as state variables for dynamic optimization. Results in incorporating the value of anticipated learning via management into current decision making.

Informative prior – *prior distribution* whose parameters reflect information about the distribution of the parameter being modeled.

Imperfect information – information reducing structural uncertainty in an incomplete fashion, often due to *partial observability*.

Interpolation, interpolate – equation or algorithm for predicting values between discrete model predictions.

Irreducible uncertainty – see *aleatoric uncertainty*.

Jackknife – data resampling method based on sequentially passing through the data, eliminating one observation at a time. Alternative method for variance and confidence interval estimation.

Jacobian – matrix of partial derivatives of the constraint function in *classical programming*.

Joint probability – probability of co-occurrence of two events or random variables.

Kuhn–Tucker conditions – sufficiency conditions for optimality of *nonlinear programming* problems.

Lagrangian multipliers – method used in classical programming to imposed equality constraints.

Law of total probability – provides the probability of a conditional outcome as the sum or integral over the conditional distribution, weighted by the probability (density) of the random variable upon which the outcome is conditioned.

Least squares estimation – estimation that minimizes the sum of quadratic distances between the predictions of a model and the data.

Likelihood – expression of probability distribution (pmf or pdf) as a function of unknown parameters, conditional on fixed, observed data. Optimization of this function provides the *maximum likelihood estimate*.

Likelihood ratio test (LRT) – null hypothesis test of a constrained versus unconstrained likelihood via comparison of their ratio given the maximum likelihood estimates under each likelihood.

Linguistic uncertainty – uncertainty due to imprecision of language.

Link function – transformation of response to allow linear modeling.

Linked decisions – *decisions* that are functionally linked or otherwise dependent in time or space.

Linear model – model in which the response is a linear function of *parameters* (*coefficients*) and *predictors*.

Linear programming – a special case of nonlinear programming, in which both the objective function and the constraints can be written as linear equations.

Local optimum – maximum or minimum over a restricted rage of *feasible solutions*.

Management – actions designed to effect outcomes in order to achieve conservation objectives.

Markovian, Markov process – [property of] a stochastic process in which the future behavior of the system depends only on the current system state and controls, and is independent of its past trajectories.

Markov chain Monte Carlo (MCMC) – procedure for generating provides a dependent sequence of samples that converges on a *stationary distribution* that behaves like a sample of the *posterior distribution*.

Markov decision process (MDP) – decision process in which the available decisions, objective returns, and transition probabilities depend only on the current system state.

Maximum likelihood estimation/estimate (MLE) – estimate/estimation of parameter of statistical distribution obtained by maximizing the likelihood function.

Maximum sustainable yield (MSY) – maximum yield that can be taken from a population at equilibrium growing according to the logistic model of density-dependent growth.

Mean – measure of central tendency. For a distribution, first central *moment*.

Means objective – objective that is intended to contribute to the fulfillment of one or more *fundamental objectives*, but does not in itself represent a fundamental objective.

Mean squared error – averaged (over the data) quadratic distance of predicted values from observed values.

Median – middle value in rank order; 50th percentile.

Method of moments – method of estimation based on equating population to sample *moments* and solving for the distribution parameters.

Metropolis–Hastings algorithm – form of *rejection sampling* used in *Markov chain Monte Carlo* (MCMC) sampling.

Minimum variance estimator/estimate – member of a class of estimators/estimates providing the smallest possible estimator variance.

Model – abstraction of reality used in prediction, statistical models, and decision analysis.

Model-averaged estimate – estimate computed by weighted averaging using AIC model weights; used to represent the influence of *model/ structural uncertainty* on parameter estimates and their confidence intervals.

Model uncertainty – uncertainty in relationship among model components, expressed as alternative models. See *structural uncertainty*.

Moment – parameter summarizing features of a statistical distribution such as location or dispersion. The kth moment of a distribution for a random variable x is defined as $E[x^k]$. The *mean* is the first moment around zero and the *variance* is the second moment around the mean. See *sample moment*.

Monitoring – tracking of a natural resource system through time to determine is status and trajectory, inform decision making, evaluate the outcomes of management, and reduce structural uncertainty.

Monte Carlo simulation – repeated generation of pseudorandom predictions via statistical distributions.

Multi-attribute valuation – valuation of a *fundamental objective* is based on combining the values of separate objective *attributes*. See also *fundamental objective hierarchy*.

Multi-model inference – parameter and prediction inference based on accounting for *model* or *structural uncertainty*.

Multinomial distribution – discrete distribution in which x_k represents the number of outcomes in each of k categories from a sample of n.

Multivariate normal distribution – continuous distribution in which several random variables follow a joint distributions represented by their means and a variance–covariance structure.

Mutually exclusive [events, states] – events or states that cannot simultaneously occur.

Necessary condition/ necessity – condition or factor required for an effect to occur. Existence of a necessary condition does not guarantee a resulting effect. See *sufficient condition*.

Negative binomial distribution – discrete distribution that models the number of failures x before the rth success. The *geometric distribution* is a special case with $r=1$.

Newton's Method – iterative procedure for maximizing unconstrained, nonlinear, but differentiable objective functions.

Node – a graphical element of a *Bayesian belief network* or *influence diagram* that represents components of a decision or system.

Normal regression – linear regression model in which error structure is assumed to follow a normal distribution.

Non-independent errors – model in which errors are not statistically independent, requiring specification of a covariance structure.

Non-informative prior – *prior distribution* whose parameters reflect a minimal amount of information about the distribution of the parameter being modeled; also referred to as a *vague prior*.

Nonlinear least squares – estimation that minimizes the sum of quadratic distances between the predictions of a nonlinear model and the data.

Nonlinear programming, – generalizes classical programming to allow for inequality constraints.

Objective – quantifiable outcomes that reflect the values of decision makers/ stakeholders, and are influenced by management *decisions*. See *fundamental objectives* and *means objectives*.

Objective function – mathematical formula for combining attribute values or utilities of different objective components into a common output for optimization.

Objective network – diagram for displaying the relationship among *fundamental* and *means objectives*.

Observational uncertainty – *see partial observability, statistical uncertainty.*

Opportunity set – set of permissible decisions after invoking any constraints.

Optimal control problem (OCP) – general procedure for determining optimal sequence of decisions in a stochastic, dynamics system.

Optimal decision – the decision or combination of decisions that results in the best results as defined by the **objective function** together with any **constraints.**

Optimization – process of seeking of values of a management control or other input that result in a desired value (usually a maximum or a minimum) of a specified objective function.

Ordinary least squares – least squares under assumptions of variance homogeneity, contrasted with *weighted least squares*.

Outcome – see *event*.

Overconfidence [of experts] – underestimation of uncertainty by experts in *elicitation* process. see *elicitation*.

Overdispersion – count data in which the data exhibit variance in excess of that predicted under the statistical model.

Parameter – constant or coefficient of a model.

Parametric bootstrapping – Monte Carlo simulation of data from a specified distribution or function of distributions, usually after fixing the parameter values to maximum likelihood or other estimates.

Partial controllability – uncertainty induced in decision making due to inability to perfectly control or implement intended decisions.

Partial observability – uncertainty induced in decision making due to inability to completely observe system states because of statistical uncertainty.

Partially observable Markov decision process (POMDP) – Markov decision process in which decisions are based on incompletely observed system states.

Passive ARM, passive adaptive optimization – seeking of an optimal sequence of decisions that does not anticipate future reduction of structural uncertainty.

Pontryagin's maximum principle – extends the calculus of variations to allow for complex constraints by means of co-state variables.

Posterior distribution – distribution representing uncertainty in parameters or other unknown quantities after the incorporation of sample data.

Posterior probability – probability of an event given sample data.

Prediction – statement, usually quantitative, deduced from a *hypothesis*, often by means of a *model*.

Predictor – factor used to predict a response, possibly in combination with other factors in a *linear model*.

Primal – usual formulation of the *linear programming* problem, in which a linear objective function of decision *variables*, is optimized with linear constraints. See *dual*.

Principle of optimality (Bellman's principle of optimality) – states that an optimal policy has the property that, regardless of initial conditions and decision from time t_0 to time t, all remaining decisions from t to t_f must also be optimal. Allows solution of dynamic optimization problems by backwards induction.

Prior distribution – distribution representing uncertainty in parameters or other unknown quantities before the incorporation of sample data.

Prior probability – probability of an event prior to the incorporation of sample data.

Probability – the measure of one's belief in the occurrence of a chance or outcome.

Probability density function (pdf) – statistical function representing probability that a continuous random variable has an outcome in a specific range.

Probability mass function (pmf) – statistical function representing probability of value outcomes for discrete random variables.

Probability distribution – statistical model representing the frequency distribution of discrete or continuous random variables. See *cumulative distribution function, probability density function,* and *probability mass function.*

Probability elicitation – a form of *direct elicitation* where experts are asked to provide an expected outcome in terms of a probability, e.g., the probability that event x will occur.

Proportional scoring – method for calculating *utility* of an objective on a linear, worst-to-best scale.

Quantitative [measures] – attributes whose values can be assessed on a numerical scale.

Qualitative [measures] – attribute whose values are assessed on a categorical, nominal, or other non-numeric scale.

Quantile – value below which a stated percentage of the data fall.

Random event – event [outcome] influenced by stochastic factors.

Random variable – attribute that varies stochastically according to a statistical distribution.

Reducible uncertainty – see *epistemic* and *linguistic uncertainty.*

Rejection sampling – Monte Carlo procedure in which samples are drawn from a proposal or enveloping distribution and compared to an acceptance criterion in order to achieve samples from a **target distribution** of interest.

Response – function of predictors and coefficients or other model inputs.

Reversible jump MCMC – a generalized *Metropolis–Hastings* procedure based on sampling over the model space and parameter space conditioned on the model. Used for Bayesian *multi-model inference.*

Risk – a state of uncertainty for which there exists an outcome that is undesirable.

Risk attitude – the chosen response of decision makers to uncertainty that matters and is a subjective judgment that reflects the values and perceptions of decision makers.

Risk premium – the difference between the expected value of the decision under uncertainty and the *certainty equivalent.*

Risk profiling – process for evaluating the tradeoffs between risk and return to the decision maker.

Risk tolerance – is the ability of the decision maker to accept or absorb risk.

Root node – a node in an *influence diagram* or *Bayesian belief network* that has no other nodes point into it, i.e., the value on the node does not depend on other nodes.

Sample moment – sample statistic that summarizes features of a statistical distribution such as location or dispersion; estimate of a moment. The kth sample in a distribution for a random variable x is defined as $\hat{M}_k = \frac{1}{n}\sum_{i-1}^{n} x_i^k$ from a random sample $x_1, \ldots x_n$. The sample mean $\bar{x}$ estimates the first moment around zero and the sample variance s^2 estimates the second moment around the mean. See *moment*.

Satisficing decision – converting from the original *objective* utility scale (whose optimum is ordinarily associated with a unique decision value) via an indicator function that maps a range of decisions into a range of outcomes that are sufficiently close to the objective.

Scatterplot – plot of observations of two random variables, used for visual inspection of linear or other functional relationships.

Sensitivity – change that occurs in model output, given a small change in some model parameter or other input (like a decision variable). See *elasticity*.

Sensitivity analysis – marginal change in an output given marginal change in an input.

Shrinkage – the process by which an estimate is pulled toward a prior based on the reliability (precision) of the estimate, see *empirical Bayes estimation*.

Shrinkage estimator – an estimator that incorporates the effects of *shrinkage*.

Simplex – algorithm for solving *linear programming* problems.

Simulation – generation of numerical predictions or other output as a function of initials conditions and parameters. When inputs involve stochastic variables it is referred to as *Monte Carlo simulation*.

Simulated annealing – heuristic procedure based on analogy to a cooling alloy.

Simulation-optimization – heuristic procedure combining Monte Carlo simulation with standard optimization procedures.

Standard normal distribution – normal distribution with mean zero and standard deviation one.

State – measurable attribute of a system, such as abundance or habitat conditions, that may vary over time and space according to *dynamic* equations.

State-dependent decision making – decisions are made after the system state is observed and the *optimal* or *satisficing decision* depends on the system state.

Static – system state not changing over time. see *dynamic*.

Stationary process – a stochastic process whose joint distribution does not change through time.

Stationary distribution – limiting distribution of a Markov chain. In *Markov chain Monte Carlo* this justifies considering the sample as from the limiting *posterior distribution*.

Stationary policy – solution optimal control problem, in which the state-specific optimal decision remains unchanged over time.

Statistic – a function of the sample *data*.

Statistical uncertainty – the uncertainty associated with sample *data* generally associated with parameter *estimates* and estimates of system *state*.

Stochastic – attribute varies according to a statistical distribution.

Stochastic dynamic programming (SDP) – extends *dynamic programming* to allow stochastic uncertainty in the objective and dynamics.

Stranded objectives – *fundamental objectives* lacking connection to *means objectives* or *decisions*.

Structural uncertainty – uncertainty in the relationship between model components, frequently expressed as alternative hypotheses or models. See *model uncertainty*.

Structured decision making – process for the systematic deconstruction and analysis of decision problems under which *objectives* and *decision alternatives* are explicitly defined and connected and analyzed via a *model*.

Student's t – statistic for null hypotheses testing with normally distributed data.

Subjective probability – probability derived from an individual or group's judgment about the likelihood of occurrence of a specific outcome.

Sufficient condition/ sufficiency – condition that alone guarantees that an effect will occur. See *necessary condition*.

Support [of a distribution] – region of x where the value of the pmf or pdf $f(x)$ is positive; interval or set of x whose complement has probability zero.

Sustainable yield – yield that when removed from a population results in the population being maintained at a stable equilibrium.

System dynamics – mathematical representation of processes determining changes in system states through time or space.

Systematic hierarchy – hierarchy determined by the nested membership of some systematic classifications in others, e.g., multiple species are members of the same genus, and genera of the same family, etc.

Transition probability, transition probability matrix – probability of transition from discrete system state i to state j over a defined time interval.

Uncertain, uncertainty – indetermination of an attribute, system state, or other feature. See *aleatoric*, *epistemic*, and *linguistic* uncertainty.

Uncertainty node – node in a *belief* or *decision network* that represents an uncertain state or outcome.

Unconditional probability – probability not dependent on the outcome of another random variable or other input.

Uniform distribution – discrete or uniform distribution in which values have equal probability of outcome over the *support*.

Utility – measure of satisfaction in an outcome.

Utility function – function that relates changes in satisfaction (utility) to changes in an objective attribute.

Vague prior – see *non-informative prior*.

Value(s) elicitation – a form of *direct elicitation* where experts are asked to provide an expected value of an outcome, e.g., what population size do you expect if management action A is executed? *Values elicitation* is also used to refers to the elicitation of utility values (see risk profiling).

Value of information – value, as measured in *objective function* units, of reduction of *epistemic (structural) uncertainty* in decision making.

Value of imperfect information – value of information to decision making reduced because of statistical or other *aleatory uncertainty*.

Value of perfect information – average value to decision making gained from reducing *epistemic* (principally, *structural*) *uncertainty* in the absence of other (irreducible or *aleatory*) uncertainty.

Variance [population moment] – measures of dispersion; second moment about the mean

Variance [of an estimator] – measure of repeatability of an estimate.

Variance–covariance matrix – matrix representing variance structure of a multivariate response, with variances of each response element on the principal diagonal and *covariances* as off-diagonal elements.

Weighted least squares – generalization of *least squares* to allow for non-homogeneous variances.

Wishart distribution – generalization of the *chi-squared distribution* to multiple dimensions. Used as conjugate prior for the inverse *variance-covariance* matrix of a *multivariate normal* distribution.

Index

Decision Making in Natural Resource Management: A Structured, Adaptive Approach,
First Edition. Michael J. Conroy and James T. Peterson.
© 2013 John Wiley & Sons, Ltd. Published 2013 by John Wiley & Sons, Ltd.